Selected Titles in This Series

177 **R. L. Dobrushin, R. A. Minlos, M. A. Shubin, and A. M. Vershik, Editors,** Topics in Statistical and Theoretical Physics (F. A. Berezin Memorial Volume)

176 **E. V. Shikin, Editor,** Some Questions of Differential Geometry in the Large

175 **R. L. Dobrushin, R. A. Minlos, M. A. Shubin, and A. M. Vershik, Editors,** Contemporary Mathematical Physics (F. A. Berezin Memorial Volume)

174 **A. A. Bolibruch, A. S. Merkur′ev, and N. Yu. Netsvetaev, Editors,** Mathematics in St. Petersburg

173 **V. Kharlamov, A. Korchagin, G. Polotovskiĭ, and O. Viro, Editors,** Topology of Real Algebraic Varieties and Related Topics

172 **K. Nomizu, Editor,** Selected Papers on Number Theory and Algebraic Geometry

171 **L. A. Bunimovich, B. M. Gurevich, and Ya. B. Pesin, Editors,** Sinai's Moscow Seminar on Dynamical Systems

170 **S. P. Novikov, Editor,** Topics in Topology and Mathematical Physics

169 **S. G. Gindikin and E. B. Vinberg, Editors,** Lie Groups and Lie Algebras: E. B. Dynkin's Seminar

168 **V. V. Kozlov, Editor,** Dynamical Systems in Classical Mechanics

167 **V. V. Lychagin, Editor,** The Interplay between Differential Geometry and Differential Equations

166 **O. A. Ladyzhenskaya, Editor,** Proceedings of the St. Petersburg Mathematical Society, Volume III

165 **Yu. Ilyashenko and S. Yakovenko, Editors,** Concerning the Hilbert 16th Problem

164 **N. N. Uraltseva, Editor,** Nonlinear Evolution Equations

163 **L. A. Bokut′, M. Hazewinkel, and Yu. G. Reshetnyak, Editors,** Third Siberian School "Algebra and Analysis"

162 **S. G. Gindikin, Editor,** Applied Problems of Radon Transform

161 **K. Nomizu, Editor,** Selected Papers on Analysis, Probability, and Statistics

160 **K. Nomizu, Editor,** Selected Papers on Number Theory, Algebraic Geometry, and Differential Geometry

159 **O. A. Ladyzhenskaya, Editor,** Proceedings of the St. Petersburg Mathematical Society, Volume II

158 **A. K. Kelmans, Editor,** Selected Topics in Discrete Mathematics: Proceedings of the Moscow Discrete Mathematics Seminar 1972–1990

157 **M. Sh. Birman, Editor,** Wave Propagation. Scattering Theory

156 **V. N. Gerasimov, N. G. Nesterenko, and A. I. Valitskas,** Three Papers on Algebras and Their Representations

155 **O. A. Ladyzhenskaya and A. M. Vershik, Editors,** Proceedings of the St. Petersburg Mathematical Society, Volume I

154 **V. A. Artamonov et al.,** Selected Papers in K-Theory

153 **S. G. Gindikin, Editor,** Singularity Theory and Some Problems of Functional Analysis

152 **H. Draškovičová et al.,** Ordered Sets and Lattices II

151 **I. A. Aleksandrov, L. A. Bokut′, and Yu. G. Reshetnyak, Editors,** Second Siberian Winter School "Algebra and Analysis"

150 **S. G. Gindikin, Editor,** Spectral Theory of Operators

149 **V. S. Afraĭmovich et al.,** Thirteen Papers in Algebra, Functional Analysis, Topology, and Probability, Translated from the Russian

148 **A. D. Aleksandrov, O. V. Belegradek, L. A. Bokut′, and Yu. L. Ershov, Editors,** First Siberian Winter School "Algebra and Analysis"

147 **I. G. Bashmakova et al.,** Nine Papers from the International Congress of Mathematicians, 1986

(Continued in the back of this publication)

Topics in Statistical and Theoretical Physics

F. A. Berezin Memorial Volume

Felix Alexandrovich Berezin
(1931–1980)

American Mathematical Society

TRANSLATIONS

Series 2 • Volume 177

Advances in the Mathematical Sciences — 32

(*Formerly Advances in Soviet Mathematics*)

Topics in Statistical and Theoretical Physics

F. A. Berezin Memorial Volume

R. L. Dobrushin
R. A. Minlos
M. A. Shubin
A. M. Vershik
Editors

American Mathematical Society
Providence, Rhode Island

Translation edited by A. B. Sossinsky

1991 *Mathematics Subject Classification.* Primary 81-XX, 82-XX, 83-XX; Secondary 47-XX.

ABSTRACT. The content of the volume is connected in various ways to the scientific heritage of F. A. Berezin (1931–1980), an outstanding Moscow mathematician who tragically died in an accident. F. A. Berezin discovered new notions and ideas in mathematical physics, representation theory, analysis, geometry and other areas of mathematics. Collected here are papers by many of his colleagues and students representing a wide spectrum of topics in statistical and theoretical physics and allied areas of mathematics. In particular, several papers discuss various aspects of quantum field theory, and related questions of supersymmetry, geometry, and representation theory. Other papers are devoted to problems of quasiclassical approximation and mathematical models of statistical physics.

Library of Congress Card Number 91-640741
ISBN 0-8218-0425-1
ISSN 0065-9290

∞ The paper used in this book is acid-free and falls within the guidelines
established to ensure permanence and durability.

10 9 8 7 6 5 4 3 2 1   01 00 99 98 97 96

Contents

Foreword

This volume is the second one in this series (for the first volume see *Contemporary Mathematical Physics*, Amer. Math. Soc. Transl. Ser. 2, vol. 151, 1996) devoted to the memory of F. A. Berezin, a remarkable mathematician from the heroic group of the creators of modern mathematical physics. The aims and the general character of this volume are similar to those of the previous one. As pointed out in the introduction to the previous volume, when we speak of Berezin's scientific heritage, we should always have in mind his two main achievements: quantization theory and supermathematics. But his name also remains in representation theory, in statistical physics, in operator theory, and in the geometry of homogeneous spaces. His papers were characterized by the breadth of their scope and the conceptuality of his approach. Now we are all witnesses of the development of Berezin's quantization and Berezin's Grassmannian analysis (supermathematics).

The present volume mostly contains papers of physical character, supermathematical models in physics, statistical physics, scattering theory and so on.

The technical and organizational work involved in the preparation of this volume was carried out by N. D. Vvedenskaya, to whom the editors express their gratitude.

We hope that the publication of this volume will be our tribute to the memory of F. A. Berezin and help popularize and develop his work.

R. Dobrushin,[1] R. Minlos, M. Shubin, A. Vershik

[1]Roland L'vovich Dobrushin died November 12, 1995. He was a friend of F. A. Berezin and one of the organizers of this collection.

Amer. Math. Soc. Transl.
(2) Vol. **177**, 1996

Berezin Quantization and Unitary Representations of Lie Groups

D. Bar-Moshe and M. S. Marinov

ABSTRACT. In 1974, F. A. Berezin proposed a quantum theory for dynamical systems having a Kähler manifold as their phase space. The system states were represented by holomorphic functions on the manifold. For any homogeneous Kähler manifold, the Lie algebra of its group of motions may be represented either by holomorphic differential operators ("quantum theory"), or by functions on the manifold with Poisson brackets generated by the Kähler structure ("classical theory"). The Kähler potentials and the corresponding Lie algebras are now constructed explicitly for all unitary representations of any compact simple Lie group. The quantum dynamics can be represented in terms of a phase-space path integral, and the action principle appears in the semiclassical approximation.

§1. Introduction

1.1. Historical background. In a series of papers [**1–4**], written in the beginning of the seventies, F. A. Berezin developed a new approach to the description of quantum dynamical systems on non-Euclidean phase-space manifolds. The approach originated essentially from two basic ingredients: functional formalism and complex analysis.

From the very beginning, quantum mechanics was expressed in terms of the Hilbert space of state vectors (represented, as a rule, by wave functions in the coordinate representation) and observables given by operators acting in that space. The fundamental observables, like energy, momentum, or angular momentum, were represented by differential operators, so that differential equations played the dominant role in the early quantum theory. Fock was probably the first [**5, 6**] who understood, still in the early days of quantum theory, that functional methods may be extremely powerful, especially in applications of the method of second quantization to quantum field theory. Schwinger's quantum action principle and Feynman's path integrals were introduced in quantum electrodynamics and led to its triumph in the late forties. Berezin was exposed to that brilliant development when he was a

1991 *Mathematics Subject Classification.* Primary 81S10; Secondary 17B81, 53C55, 58F06.

student in the fifties and took part in the Landau seminars at the Institute of Physical Problems in Moscow. At that time, there was evidently a missing element in the functional representation of quantum electrodynamics: a proper description of fermion fields. Schwinger employed *second-kind variables* [**7**] in his action principle in order to describe fermion fields, electrons in particular, but the formalism of anticommuting numbers had not yet been developed systematically. Berezin invented the integral in anticommuting (Grassmann) variables, and the unified functional approach to quantal systems with Bose and Fermi degrees of freedom was presented in a complete form in his thesis. His famous book [**8**] was based upon the thesis, and it has been the mathematical foundation for the future development of methods of supersymmetry in physics [**9**]. Later, spinning string models prompted an elaboration of the functional approach to particle spin dynamics and the Poisson brackets on super-phase spaces were defined [**10, 11**].

The second quantization in its functional form is naturally related to complex analysis. Fock [**6**], and later Bargmann [**14**], used the *holomorphic* representation of functionals in quantum theory. (For surveys, see [**12, 13**].) The investigation of properties of normal and antinormal symbols of the evolution operator enabled Berezin to derive sophisticated theorems about the spectra of the Hamiltonian [**1, 15**] for the Euclidean phase space. Those ideas were further developed later, e.g., in [**16**]. On the other hand, Berezin discovered that the formalism can be extended to *non-Euclidean* dynamical systems, provided the corresponding phase-space manifold has a complex Kählerian structure.

1.2. Quantization in Euclidean phase spaces. Berezin's arguments were fairly straightforward. It was known that in the Euclidean case, the states of any quantal system can be represented by holomorphic functions $\psi(z)$, where $z \equiv \{z_1, \dots, z_m\} \in \mathbb{C}^m$ and m is the number of degrees of freedom. The structure of the Hilbert space was introduced in the space of state vectors by means of the inner product

$$(\psi_2, \psi_1) = \int_{\mathbb{C}^m} \overline{\psi_2(z)}\, \psi_1(z)\, e^{-(z\cdot\bar{z})} d\mu(z, \bar{z})\,. \tag{1}$$

Here $(z \cdot \bar{z}) \equiv \sum_{\alpha=1}^{m} z_\alpha \bar{z}_\alpha$ is the Euclidean inner product, and the integral is taken with respect to the standard Liouville measure

$$d\mu(z, \bar{z}) \equiv \prod_{\alpha=1}^{m} \frac{dz_\alpha \wedge d\bar{z}_\alpha}{-2\pi i} = \prod_{\alpha=1}^{m} \frac{dq_\alpha\, dp_\alpha}{2\pi\hbar}, \qquad z = \frac{\mathbf{q} + i\mathbf{p}}{\sqrt{2\hbar}}, \tag{2}$$

where $\mathbf{q}$ and $\mathbf{p}$ are the usual coordinate and momentum vectors. The inner product is invariant under *translations* in the phase space provided the wave functions have appropriate transformation properties,

$$z \to z' = \varepsilon z + a \to \psi(z) \to \psi'(z) = e^{\varepsilon(z\cdot\bar{a}) + (a\cdot\bar{a})/2} \psi(z'), \tag{3}$$

where a is any complex m-vector and $|\varepsilon| = 1$. After three consecutive translations by a, b and $-a-b$, the phase space remains invariant, while each wave function acquires a constant phase shift

$$\psi(z) \to \psi'(z) = e^{i\varphi} \psi(z), \qquad \varphi = [(b \cdot \bar{a}) - (a \cdot \bar{b})]/2i\,. \tag{4}$$

Evidently, $\varphi/2$ is just the Euclidean area of the triangle built upon the vectors a and b. In general, any continuous translation in the phase space given by a closed line in $\mathbb{C}^m$ results in a phase shift, which is the Euclidean area of the minimal surface stretched upon the line. Thus the standard symplectic structure appears from equation (1), and one gets a geometric interpretation for canonical quantization: the space of state vectors is a line bundle based upon the phase space. The phase space translations and the phase shifts together constitute a Lie group $\mathcal{W}_m$ (Schwinger's *special canonical group* [**17**]). The corresponding Lie algebra (the Heisenberg–Weyl algebra) is given by the canonical commutation relations for its basis elements, which are the coordinate and momentum operators (or the creation and annihilation operators) and the unit operator that generates the phase shifts (the *gauge group* $U(1) \subset \mathcal{W}_m$).

The Berezin quantization stems from the extension of the holomorphic quantization to phase spaces which are Kähler manifolds having nonflat geometries. The inner product is defined using the integral with respect to an appropriate invariant measure, and the Hermitian inner product in the exponent is substituted for the Kähler potential (see equation (5) below).

1.3. Contents of this paper. The next section presents a short review of Berezin's quantization principle. Special attention is paid to the properties of cocycle functions, which define the structure of the line bundle representing the space of state vectors. Two examples are also given: the two-dimensional sphere and the Lobachevsky plane. Together with the usual phase plane given in the previous subsection, they represent all possible quantizations for one degree of freedom. The quantization is given by a nonnegative integer l, which specifies the *quantized* curvature of the phase space. In the compact case, i.e., for the sphere, the total number of states of the system is finite, $N = l+1$. Section 3 shows a correspondence between the quantization on a homogeneous Kähler manifold and the representation of the Lie algebra for its group of motions. The Lie algebra is represented in two alternative ways:

i) by holomorphic differential operators,
ii) by their symbols, i.e., functions on the phase space and the appropriate Poisson brackets.

This gives the actual meaning to the word "quantization". In section 4 the Kähler potentials are constructed explicitly for all compact simple Lie groups. Complex coordinates correspond to positive roots of the Lie algebra, and fundamental Kähler potentials are expressed in terms of generalized determinants (described in Appendix A). An approach to dynamics is described in section 5. An analog of the path integral is derived, which leads to an action functional in the semiclassical approximation, and the classical Hamilton equations of motion result from the corresponding variational principle.

This paper is mainly addressed to physicists, and its mathematical arguments are more intuitive than rigorous.

§2. Quantization on homogeneous Kähler manifolds

2.1. The Hilbert space of state vectors. We consider the m-dimensional Kähler manifold $\mathcal{M}$ with local complex coordinates z_α ($\alpha = 1, \dots, m$), where a group $\mathcal{G}$ acts holomorphically, i.e., $z \to gz$, $\forall g \in \mathcal{G}$ (notations mainly follow [**18**]).

State vectors of the quantal system are defined as holomorphic sections of the holomorphic line bundle $\mathcal{L}$ over $\mathcal{M}$. The state vectors are represented by locally holomorphic *wave functions* $\psi(z)$. The Hilbert space structure is assigned to $\mathcal{L}$ by means of the following $\mathcal{G}$-invariant inner product,

$$(\psi_2, \psi_1) = \int_{\mathcal{M}} \overline{\psi_2(z)}\, \psi_1(z) \exp[-K(z, \overline{z})]\, d\mu(z, \overline{z})\,. \tag{5}$$

Here $K(z, \overline{z})$ is the Kähler potential associated with the line bundle $\mathcal{L}$ and defined in any open coordinate neighborhood of the manifold $\mathcal{M}$. The meaning of the integral includes, of course, the sum over the neighborhoods covering $\mathcal{M}$. The invariance is imposed by the transformation law for the wave functions, which is consistent with that for the Kähler potential, namely for any $g \in \mathcal{G}$

$$K(z, \overline{z}) \to K(gz, \overline{gz}) = K(z, \overline{z}) + \Phi(z; g) + \overline{\Phi(z; g)}, \tag{6}$$

$$\psi(z) \to \psi(gz) = \exp[\Phi(z; g)](\hat{U}(g^{-1})\psi)(z), \tag{7}$$

where $\hat{U}(g)$ is a unitary operator representing the group element g in the Hilbert space $\mathcal{L}$, $\hat{U}(g_1)\hat{U}(g_2) = \hat{U}(g_1 g_2)$, and the *cocycle function* $\Phi(z; g)$ is (locally) holomorphic.

The (invariant) integration measure is expressed, as usual, in terms of the mth power of the corresponding Kähler $(1,1)$-form ω,

$$d\mu(z, \overline{z}) \equiv C \underbrace{\frac{\omega}{2\pi i} \wedge \cdots \wedge \frac{\omega}{2\pi i}}_{m \text{ times}}, \tag{8}$$

where C is a normalization constant (see equation (12) below) and the factor $2\pi i$ is introduced for future convenience. (We assume, of course, that the form ω is *nondegenerate*, so the integrals do not vanish identically.) The $(1,1)$-form ω has the following local representation

$$\omega \equiv \omega_{\alpha\bar{\beta}}(z, \overline{z})\, dz^\alpha \wedge d\overline{z}^\beta, \qquad \omega_{\alpha\bar{\beta}} = \partial_\alpha \partial_{\bar{\beta}} K, \tag{9}$$

where $\partial_\alpha \equiv \partial/\partial z^\alpha$, $\partial_{\bar{\beta}} \equiv \partial/\partial \overline{z}^\beta$. It follows from (6) that the form ω is closed, $\partial\omega = 0$, $\bar{\partial}\omega = 0$, and invariant under the group transformations.

Let us make two more assumptions.

(A) The group $\mathcal{G}$ acts transitively in $\mathcal{M}$, i.e., for any two points there is a group element transforming one of them into the other.

(B) Excluding from $\mathcal{M}$ a manifold $\mathcal{X}$ of a lower dimension, one gets a domain with simple topology, $\mathcal{M} \setminus X \equiv \mathbb{C}^m$.

It follows that any holomorphic function invariant under $\mathcal{G}$ is a constant. This property will be very important below. Various manifolds $\mathcal{X}$ that must be excluded in order to reduce $\mathcal{M}$ to $\mathbb{C}^m$ are transformed into each other by the group transformations. Because of (B), the inner product in (5) can be regarded as the integral over $\mathbb{C}^m$ and is independent of the choice of $\mathcal{X}$.

2.2. Symbols of linear operators. For the manifolds that we are concerned with here, the Kähler potential can be regarded as a boundary value of a function

$K(\zeta, \overline{z})$, holomorphic in the first variable and antiholomorphic in the second one. Berezin introduced a "super-complete set" of state vectors

$$\Psi_v(z) \equiv \exp[K(z, \overline{v})], \tag{10}$$

This set is an important class of *generalized coherent states* (an exposition and an abundant bibliography can be found in [**19, 20**]).

One can prove that, under a proper normalization of the integration measure, any element of the Hilbert space is reproduced by the following integral over the manifold,

$$\psi(\zeta) \equiv \int_{\mathcal{M}} \psi(z) \exp[K(\zeta, \overline{z}) - K(z, \overline{z})]\, d\mu(z, \overline{z}), \qquad \forall \psi \in \mathcal{L}. \tag{11}$$

The proof stems from the fact that, because of (6)–(7), the ratio of the integral on the right-hand side to $\psi(\zeta)$ is invariant under the group transformations. This number is independent of ψ, and setting a proper value of C in (8), one can make it equal to 1. In order to prove that, one can calculate (ψ, ψ), applying (10) first to a coherent state and then to ψ, getting an identity. Moreover, for the manifolds we are dealing with, $K(0, \overline{z}) \equiv 0$, and the following integral exists

$$\int_{\mathcal{M}} \exp[-K(z, \overline{z})]\, d\mu(z, \overline{z}) = 1. \tag{12}$$

In other words, all constant sections of $\mathcal{L}$ belong to the Hilbert space, and one can assume that $\psi_0(z) \equiv 1$ has the unit norm. (In typical physical problems, the constant ψ corresponds to "vacuum", i.e., the system ground state for a proper Hamiltonian.) Thus the constant C in (8) is expressed simply in terms of an integral over the manifold.

The reproducing kernel is specified by the Kähler potential in (11), and has an expansion in terms of *any* orthonormal basis $\{\phi_\nu(z)\}$ in $\mathcal{L}$,

$$\exp[K(\zeta, \overline{z})] = \sum_\nu \phi_\nu(\zeta) \overline{\phi_\nu(z)}. \tag{13}$$

This equality also gives an expansion of the coherent state (10) in the basis of the orthonormal states.

Linear operators in the Hilbert space, in particular those describing observables of the quantal system, are represented by their symbols in the following way: $\widehat{A} \to A(z, \overline{\zeta})$ means that

$$(\widehat{A}\psi)(\zeta) = \int_{\mathcal{M}} A(\zeta, \overline{z}) \psi(z) \exp[K(\zeta, \overline{z}) - K(z, \overline{z})]\, d\mu(z, \overline{z}). \tag{14}$$

The symbol representation has the following nice properties.

1. The symbol of the unit operator is just 1, $\hat{I} \to I(\zeta, \overline{z}) \equiv 1$.
2. The trace of any operator is given by the integral of its symbol,

$$\mathrm{tr}\,(\widehat{A}) = \int_{\mathcal{M}} A(z, \overline{z})\, d\mu(z, \overline{z}). \tag{15}$$

For compact manifolds, the trace of the unit operator $\hat{I}$ exists, the volume is finite and equals the total number of states N,

$$\operatorname{tr}(\hat{I}) \equiv N = \int_{\mathcal{M}} d\mu(z,\bar{z}) \,. \tag{16}$$

3. Hermitian conjugation in the Hilbert space is represented by complex conjugation of the symbol and the transposition of its arguments,

$$\widehat{A}^{\dagger} \to A^{*}(\zeta,\bar{z}) = \overline{A(z,\bar{\zeta})} \,. \tag{17}$$

4. The symbol for the product of operators is given by an integral of the product of their symbols (the $*$-product)

$$\begin{aligned}\widehat{A}\widehat{B} \to (A*B)(\zeta,\bar{\eta}) \equiv \int_{\mathcal{M}} & A(\zeta,\bar{z})\, B(z,\bar{\eta}) \exp[K(\zeta,\bar{z}) - K(z,\bar{z}) \\ & + K(z,\bar{\eta}) - K(\zeta,\bar{\eta})]\, d\mu(z,\bar{z}) \,.\end{aligned} \tag{18}$$

In particular, one has an analog of the Gaussian integral,

$$\exp[K(\zeta,\bar{\zeta})] = \int_{\mathcal{M}} \exp[K(\zeta,\bar{z}) - K(z,\bar{z}) + K(z,\bar{\zeta})]\, d\mu(z,\bar{z}) \,. \tag{19}$$

The inner products of the coherent states (10) are obtained from (19).

In general, any system state is given by a (positive semidefinite) density operator $\hat{\rho}$, which can also be represented with its symbol $\rho(z,\bar{z})$. For any operator $\widehat{A}$, its expectation value in the given state is

$$\begin{aligned}\langle A\rangle_{\rho} \equiv \operatorname{tr}(\widehat{A}\hat{\rho}) = \iint_{\mathcal{M}} & A(\zeta,\bar{z})\, \rho(z,\bar{\zeta}) \exp[K(\zeta,\bar{z}) - K(z,\bar{z}) \\ & + K(z,\bar{\zeta}) - K(\zeta,\bar{\zeta})]\, d\mu(z,\bar{z})\, d\mu(\zeta,\bar{\zeta}) \,.\end{aligned} \tag{20}$$

Thus the associative algebra of observables for the quantal system is constructed completely in terms of operator symbols.

2.3. Cocycle functions. The cocycle functions, defined in equation (6), have the following properties,

$$\Phi(z,e) = 0, \qquad e \text{ is the unit in } \mathcal{G}, \tag{21}$$

$$\Phi(gz;g^{-1}) = -\Phi(z;g), \qquad \forall g \in \mathcal{G}, \tag{22}$$

$$\Phi(z;g_2g_1) = \Phi(g_2z;g_1) + \Phi(z;g_2), \qquad \forall g_1,g_2 \in \mathcal{G} \,. \tag{23}$$

Writing the latter condition in infinitesimal form, one gets differential equations for the cocycle functions. In order to write them down we need some notation.

Let $\mathbf{g}$ be the Lie algebra of the group $\mathcal{G}$. Introducing a basis τ_a in $\mathbf{g}$, with $a = 1,\ldots,n \equiv \dim \mathbf{g}$, one obtains Cartesian coordinates for the group elements $g = \exp(-\xi^{a}\tau_a)$, and the corresponding (left) Lie derivatives D_a on the group manifold. The action of the group $\mathcal{G}$ on the manifold $\mathcal{M}$ determines the holomorphic Killing fields

$$\nabla_a = \kappa_a^{\alpha}(z)\,\partial_{\alpha}, \qquad \kappa_a^{\alpha}(z) \equiv D_a(gz)^{\alpha}|_{g=e} \,. \tag{24}$$

The conjugate Killing field and the differential operator $\overline{\nabla}_a$ are defined similarly. The Lie derivative and the Killing derivative satisfy the commutation relations of the Lie algebra $\mathbf{g}$,

$$[\tau_a, \tau_b] = f_{ab}^c \tau_c, \qquad [D_a, D_b] = f_{ab}^c D_c, \qquad [\nabla_a, \nabla_b] = f_{ab}^c \nabla_c . \tag{25}$$

A holomorphic vector field is associated with the cocycle functions,

$$\varphi_a(z) = D_a \Phi(z; g)|_{g=e} . \tag{26}$$

Now the differential equations for the cocycle functions are written as follows:

$$D_a \Phi = \nabla_a \Phi + \varphi_a . \tag{27}$$

The equations are consistent, provided the functions $\varphi_a(z)$ satisfy the linear differential equations

$$\nabla_a \varphi_b - \nabla_b \varphi_a = f_{ab}^c \varphi_c . \tag{28}$$

In equation (6), the cocycle functions are defined up to an arbitrary imaginary term, but this term determines the phase shift of the wave functions in (7). Since the wave functions are assumed to be single-valued on $\mathcal{M}$, the additional term must be a multiple of $2\pi i$, and one gets a boundary condition on Φ. Namely, let us consider a *compact* one-parameter subgroup of $\mathcal{G}$, $u(t) \in U(1) \subset \mathcal{G}$, where $0 \leqslant t < 2\pi$. For *any* such a subgroup, one must have

$$\Phi(z; u(2\pi)) = 2\pi i l, \tag{29}$$

where l is an integer, depending, in principle, on the equivalence class of the subgroups $U(1)$. This is a *quantization condition* for the Kähler manifolds, which is necessary for the consistency of the quantum theory based on the line bundle structure presented above. The quantization condition is of exactly the same nature as the Dirac quantization for magnetic charge [**21**], as explained by Wu and Yang [**22**]. If this condition holds, the corresponding Kähler potential is called *integral.* One can see that the integral of ω over any two-dimensional cycle spanning a line homotopic to the group trajectory is given by (29).

2.4. Sphere and pseudosphere. The simplest examples presented by Berezin [**4**] for $m = 1$ are the two-dimensional sphere S^2 and the pseudosphere H^2 (the Lobachevsky plane). The following table is a summary for these two manifolds,

(30)

$\mathcal{M}$	S^2	H^2
$\mathcal{G}$	$SU(2)$	$SU(1,1)$
$K(z,\bar{z})$	$l \log(1 + z\bar{z})$	$-(l+1)\log(1 - z\bar{z})$
$d\mu(z,\bar{z})$	$(l+1)(1+z\bar{z})^{-2} dz \wedge d\bar{z}/2\pi i$	$l(1-z\bar{z})^{-2} dz \wedge d\bar{z}/2\pi i$
gz	$(\alpha z - \beta)/(\bar{\beta} z + \bar{\alpha})$ $\alpha\bar{\alpha} + \beta\bar{\beta} = 1$	$(\alpha z + \beta)/(\bar{\beta} z + \bar{\alpha})$ $\alpha\bar{\alpha} - \beta\bar{\beta} = 1$
$\Phi(z;g)$	$-l \log(\bar{\beta} z + \bar{\alpha})$	$(l+1)\log(\bar{\beta} z + \bar{\alpha})$

In both cases, l is a positive integer, α and β are complex group parameters. An orthonormal basis in $\mathcal{L}$ is given by the functions $\phi_\nu = c_\nu z^\nu$, where ν is a nonnegative integer and c_ν is a normalization constant. For the compact case, S^2, the manifold has a finite volume equal to the total number of states, $V = N = l+1$. (The norm of ϕ_ν given by (5), does not exist if $\nu > l$.) For the pseudosphere, the Hilbert space is infinite-dimensional, and the domain in $\mathbb{C}$ is noncompact, $z\bar{z} < 1$. The monomials given above are always eigenfunctions of the Hermitian operator $\widehat{H} = zd/dz$, which has an integer spectrum, like the one-dimensional harmonic oscillator in the Euclidean case. Thus we have obtained all the unitary representations of $SU(2)$ (corresponding to the angular momentum values $l/2$), and the representations of the discrete series for $SU(1,1)$.

The usual quantum mechanics in one degree of freedom is the boundary case between S^2 and H^2. Actually, if the motion in the system is confined to a restricted domain $|z| \ll 1$, it is possible to introduce the usual phase space and write $z = (q+ip)/\sqrt{2}R$, so that for $R \to \infty$ one gets the Hilbert space of equation (1) for $m=1$, $\hbar = R^2/l$ and $l \gg 1$. Berezin noted that $\hbar^{-1}$ must have a discrete spectrum for S^2; he remarked that this condition "seems extravagant" [**4**]. In fact, the "quantization" of the Kähler structures is typical; it results from equation (29) if one has a nontrivial representation of a compact subgroup in the group of motions. Here the subgroup is $U(1)$: $\beta = 0$, $\alpha = e^{it}$.

Quantization on the sphere was proposed by Souriau [**23**]. However, a decade before, Klauder [**24**] was probably the first who considered quantization on the sphere (actually, on the direct product of infinitely many spheres) for the spinor representation (the particular case $l = 1$), introduced the coherent states, and used this method in quantum field theory. Berezin [**3**] considered classical symmetric complex domains, and the Lobachevsky plane was the simplest particular case.

§3. The Lie algebra and the Poisson brackets

The group action in $\mathcal{L}$ is represented by the unitary operator

$$\widehat{U}(g)\psi(z) = \exp[-\Phi(z;g^{-1})]\,\psi(g^{-1}z), \tag{31}$$

which leads to a representation of the Lie algebra $\mathbf{g}$ in terms of the holomorphic first-order differential operators,

$$\tau_a \to \widehat{T}_a = \nabla_a - \varphi_a(z), \qquad [\widehat{T}_a, \widehat{T}_b] = f_{ab}^c \widehat{T}_c \tag{32}$$

(cf. equations (25) and (28)).

The symbol representation for the operators is obtained from (11),

$$(\widehat{U}(g)\psi)(\zeta) = \int_{\mathcal{M}} U_g(\zeta,\bar{z})\psi(z)\exp[K(\zeta,\bar{z}) - K(z,\bar{z})]\,d\mu(z,\bar{z}), \tag{33}$$

$$\begin{aligned} U_g(\zeta,\bar{z}) &= \exp[K(g^{-1}\zeta,\bar{z}) - K(\zeta,\bar{z}) - \Phi(\zeta;g^{-1})] \\ &\equiv \exp[K(\zeta,\overline{gz}) - K(\zeta,\bar{z}) + \overline{\Phi(z;g)}], \end{aligned} \tag{34}$$

$$\begin{aligned} T_a(\zeta,\bar{z}) &= \nabla_a K(\zeta,\bar{z}) - \varphi_a(\zeta) \equiv -\overline{T_a(z,\bar{\zeta})} \\ &= -\overline{\nabla}_a K(\zeta,\bar{z}) + \overline{\varphi_a(z)}\,. \end{aligned} \tag{35}$$

(The second equality in (34) is true for the real basis, where $\widehat{T}_a = -\widehat{T}_a^\dagger$.) Thus the symbols of the Lie algebra are expressed in terms of the Killing fields and the symplectic one-form generated by the Kähler structure,

$$T_a(\zeta,\overline{z}) = \kappa(\zeta)_a^\alpha \Lambda_\alpha(\zeta,\overline{z}) - \varphi_a(\zeta), \qquad dK(\zeta,\overline{z}) = \Lambda_\alpha d\zeta^\alpha + \Lambda_{\bar\beta}\, d\overline{z}^\beta. \tag{36}$$

Assuming that the $(1,1)$-form ω is nondegenerate, one can define the Poisson brackets in $\mathcal{M}$ by means of the field ϖ dual to ω. Namely, for any two symbols $A(z,\overline{z})$ and $B(z,\overline{z})$, one has

$$\{A,B\}_{\text{P.B.}} \equiv \varpi^{\alpha\bar\beta}(\partial_\alpha A \partial_{\bar\beta} B - \partial_\alpha B \partial_{\bar\beta} A) = -\{B,A\}_{\text{P.B.}} \tag{37}$$

$$\omega_{\alpha\bar\beta}\varpi^{\alpha\bar\gamma} = \delta_{\bar\beta}^{\bar\gamma}, \qquad \omega_{\alpha\bar\beta}\varpi^{\gamma\bar\beta} = \delta_\alpha^\gamma.$$

The Jacobi identity results from the fact that ω defined in (9) is a closed form. Using (35), one can show that the Poisson brackets yield a representation for the Lie algebra $\mathbf{g}$,

$$\{T_a, T_b\}_{\text{P.B.}} = \nabla_a T_b - \nabla_b T_a = f_{ab}^c T_c\,. \tag{38}$$

The comparison between equations (32) and (38) shows the meaning of the "quantization" performed. The Lie multiplication in $\mathbf{g}$ given by equations (25) and (32) play the same role as the *canonical commutation relations* in the standard quantum theory [**25**]: there is a one-to-one correspondence between the commutators and the Poisson brackets for the *canonical variables* T_a. Of course, the correspondence is maintained necessarily only for the canonical variables, as it holds for coordinates and momenta in the usual theory.

The coherent states (10) are solutions to the differential equations

$$\widehat{T}_a \Psi_v(z) = T_a(z,\overline{v})\,\Psi_v(z)\,. \tag{39}$$

As in ordinary quantum mechanics, they minimize the uncertainty relations (cf. [**20**]).

§4. Kähler structures for compact Lie groups

4.1. Borel coordinates. The Kähler potentials have been constructed explicitly for all compact simple Lie groups [**26–29**]. The complex parameters are introduced following Borel's method [**30**].

Let $\mathcal{G}$ be a compact simple Lie group, and $\mathcal{T}$ be its maximal Abelian subgroup (the maximal torus); the coset space $\mathcal{F} = \mathcal{G}/\mathcal{T}$ is a flag manifold. The local complex parametrization in $\mathcal{F}$ is introduced by means of the canonical diffeomorphism $\mathcal{G}/\mathcal{T} \cong \mathcal{G}^c/\mathcal{P}$. Here $\mathcal{G}^c$ is the complex extension of $\mathcal{G}$; the parabolic subgroup $\mathcal{P}$ is a semidirect product of $\mathcal{T}$ and the Borel subgroup $\mathcal{B} \subset \mathcal{G}^c$. We shall use the canonical basis $\{\tau_a\} = \{h_j, e_{\pm\alpha}\}$ in the Lie algebra $\mathbf{g}$, where $j = 1,\ldots,r \equiv \operatorname{rank}(\mathbf{g})$ and $\{\alpha\} \in \Delta_{\mathbf{g}}^+$ are the positive roots of $\mathbf{g}$ (see, e.g., [**31**]). The number of positive roots is $n_+ = [\dim(\mathbf{g}) - \operatorname{rank}(\mathbf{g})]/2$. The Lie products of the basis elements, important in the sequel, are

$$[h_j, h_k] = 0, \quad [h_j, e_\alpha] = (\alpha\cdot\mathbf{w}_j)\,e_\alpha, \quad [e_\alpha, e_\beta] = \chi(\alpha,\beta)\,e_{\alpha+\beta}\,. \tag{40}$$

Here $\chi(\alpha,\beta)$ is a function on the root lattice, which vanishes if $\alpha+\beta \notin \Delta_{\mathbf{g}}^+$, and $\mathbf{w}_j$ are the fundamental weight vectors. For any unitary irreducible group representation, its dominant weight $\mathbf{l}$ is given by the sum of the fundamental weights with nonnegative integer coefficients,

$$\mathbf{l} = \sum_{j=1}^{r} l^j \mathbf{w}_j \, . \tag{41}$$

Using the canonical basis, the Lie algebra $\mathbf{g}^c$ splits into three subalgebras, $\mathbf{g}^c = \mathbf{g}^- \oplus \mathbf{t}^c \oplus \mathbf{g}^+$, corresponding to three subsets of the basis elements, $\{e_{-\alpha}\}$, $\{h_j\}$, $\{e_\alpha\}$. Respectively, the Lie algebra $\mathbf{g}^+$ generates a nilpotent subgroup $\mathcal{G}^+ \subset \mathcal{G}^c$, and the Lie algebra $\mathbf{p} = \mathbf{g}^- \oplus \mathbf{t}^c$ generates $\mathcal{P}$. The complex parameters introduced in $\mathcal{F}$ correspond to the positive roots of $\mathbf{g}$, and (complex) $\dim(\mathcal{F}) = n_+$,

$$f(z) = \exp\left(\sum_{\alpha \in \Delta_{\mathbf{g}}^+} z^\alpha e_\alpha \right) \in \mathcal{G}^+, \qquad z^\alpha \in \mathbb{C} \, . \tag{42}$$

Similarly, one can write out the elements of the parabolic subgroup. Any element of the complex group has a unique Mackey decomposition,

$$\begin{gathered} g = fp, \quad f \in \mathcal{G}^+, \, p \in \mathcal{P}, \\ p = \exp\left(\sum_{j=1}^{r} x^j h_j + \sum_{\alpha \in \Delta_{\mathbf{g}}^+} y^\alpha e_{-\alpha} \right). \end{gathered} \tag{43}$$

(The decomposition is valid for all g except for a subset of lower dimension). Since $f(z)$ is an element of a nilpotent group, its matrix representations are polynomials of z^α. The local form (42) for f is valid in a neighborhood of the point $z^\alpha = 0$, i.e., the origin of the coordinate system in $\mathcal{F}$. Transition to other domains of $\mathcal{F}$ covering the manifold completely can be performed using the group transformations. In order to obtain the decomposition (43) for any given g, one must first calculate $f(z)^{-1}g = p$ using the fact that $f(z)^{-1} \in \mathcal{G}^+$ is a polynomial of z_α in any group representation. Setting $p = \exp(\eta)$, $\eta \in \mathbf{g}^c$, one requires that η have no e_α-components. Thus one gets n_+ algebraic equations for n_+ variables z_α. As soon as the equations are solved and the z's are found as functions of g, one gets $f(z)$ and $p = f^{-1}g$.

The group $\mathcal{G}^c$ acts on $\mathcal{F}$ by left multiplications. Actually, for any g and z, there is a unique decomposition,

$$gf(z) = f(gz)p(z;g), \qquad p(z;g) \in \mathcal{P}, \tag{44}$$

where gz is a *rational* function of z. For any element g that does not drive the point with coordinates z^α outside the coordinate neighborhood (containing the origin) where (42) is valid, gz and $p(z;g)$ can be obtained from (44) by means of algebraic operations. Performing two consecutive transformations, as in (23), one gets

$$p(z; g_2 g_1) = p(g_2 z; g_1) p(z, g_2), \qquad \forall g_1, g_2 \in \mathcal{G}^c. \tag{45}$$

The decomposition in (44) shows the way to the desired Kähler structure.

4.2. Kähler potentials. In a work by Bando, Kuramoto, Maskawa, and Uehara [**26**], the Kähler potentials were expressed in terms of the fundamental unitary representation of $\mathcal{G}$. The representation $g \to \hat{g}$ (in this section, the caret stands for the matrix) is also a representation of $\mathcal{G}^c$, but the elements of $\mathcal{G}^+$ and $\mathcal{P}$ are not unitary if they do not belong to $\mathcal{G}$. A *partial* solution of equations (6), (7) is given in terms of the generalized determinant (see Appendix A),

$$K^j(\zeta, \bar{z}) = \log \det{}'(\hat{f}(\zeta)\,\theta_j \hat{f}(z)^\dagger), \tag{46}$$

$$\Phi^j(z; g) = -\log \det{}'(\hat{g}\hat{f}(z)\,\theta_j \hat{f}(gz)^{-1}). \tag{47}$$

Here, θ_j is a projection matrix that satisfies the following conditions:

$$\theta_j = \theta_j^\dagger, \quad \theta_j^2 = \theta_j, \quad \det{}'\theta_j = 1, \qquad \theta_j \hat{p}\theta_j = \hat{p}\theta_j, \quad \forall p \in \mathcal{P}. \tag{48}$$

There are r independent matrices of this kind ($j = 1, \dots, r$) having different ranks. Using (44) and (80), one can prove for each $\theta_j \equiv \theta$ and $\tilde{\theta} \equiv p^{-1}\theta(p^{-1})^\dagger$ that as soon as $\hat{g}^\dagger = \hat{g}^{-1}$,

$$\begin{aligned} \frac{\det{}'[\hat{f}(g\zeta)\,\theta\hat{f}(gz)^\dagger]}{\det{}'\theta} &= \frac{\det{}'[\hat{g}\hat{f}(\zeta)\,\tilde{\theta}\hat{f}(z)^\dagger\hat{g}^\dagger]}{\det{}'\tilde{\theta}}\,\frac{\det{}'\tilde{\theta}}{\det{}'(\hat{p}\tilde{\theta}\hat{p}^\dagger)} \\ &= \frac{\det{}'[\hat{f}(\zeta)\,\theta\hat{f}(z)^\dagger]\det{}'\theta}{\det{}'[\theta\hat{p}(\zeta; g)]\det{}'[\theta\hat{p}(z; g)^\dagger]}. \end{aligned} \tag{49}$$

The denominator is the product of a holomorphic function of ζ and an antiholomorphic function of z. One can see (cf. equation (50) below) that $\operatorname{rank}(\theta_j)$ is equal to the number of those weights $\mathbf{w}$ in the fundamental representation that satisfy the condition $(\mathbf{w} \cdot \mathbf{w}_j) = (\mathbf{w}_1 \cdot \mathbf{w}_j)$. (Here $\mathbf{w}_1$ is the dominant weight of the fundamental representation, i.e., the lowest fundamental weight.) In Appendix B, the ranks of θ_j are given for all the classical groups and for the exceptional group G_2.

A matrix having the properties described above can be constructed for any subgroup $U(1) \subset \mathcal{G}$, and the basis matrices θ_j correspond to the components of the maximal torus $\mathcal{T}$. In order to see that, let us take (noncompact) subgroup $q_j(t) = \exp(-ith_j) \in \mathcal{G}^c$ and consider the transformation $p \to p_t = q_j(t)pq_j(t)^{-1}$. (Note that $\hat{h}_j$ is anti-selfadjoint for unitary representations.) It is easy to see that $p_t \in \mathcal{P}$ and its parameters are $y_t^\alpha = y^\alpha \exp[-t(\alpha \cdot \mathbf{w}_j)]$. Since $(\alpha \cdot \mathbf{w}_j) \geqslant 0$ for all positive roots α, only those parameters y^β for which $(\beta \cdot \mathbf{w}_j) = 0$ survive in the limit $t \to \infty$. In other words, p_∞ is reduced to the subgroup $\mathcal{P}_j$ generated by $e_{-\beta}$ for the roots β satisfying the above condition. This observation suggests the following construction of the projection matrices in terms of elements of the Cartan subalgebra. For any unitary representation, where $\hat{h}_j$ is diagonal, θ_j is also diagonal and has 1 at the sites where $-i\hat{h}_j$ has its maximum eigenvalue, which is $v_j \equiv (\mathbf{w}_j \cdot \mathbf{w}_1)$. All the other elements of θ_j are 0. Thus the fundamental Kähler potentials can be also written in the following invariant form

$$K^j(\zeta, \bar{z}) = \lim_{\lambda \to 0} \lim_{t \to \infty} \log \frac{\det[\hat{f}(\zeta)\,\hat{q}_j(t)\,\hat{f}(z)^\dagger - \lambda \exp(v_j t)]}{\det[\hat{q}_j(t) - \lambda \exp(v_j t)]}. \tag{50}$$

Of course, the order of the limits cannot be changed.

Evidently, the sum of two Kähler potentials also has the desired transformation property (6), so the fundamental potentials generate a lattice. Ultimately, the *general* solution is given for any group representation with a dominant weight **l**,

$$K^{(\mathbf{l})}(\zeta,\overline{z}) = \sum_{j=1}^{r} l_j K^j(\zeta,\overline{z}), \tag{51}$$

$$\Phi^{(\mathbf{l})}(z;g) = \sum_{j=1}^{r} l_j \Phi^j(z;g)\,. \tag{52}$$

The fundamental potentials (46) are logarithms of polynomials in ζ and z, so the function $\exp[K^{(\mathbf{l})}]$, which appears in the integrals in §2, is a polynomial, and its degree is determined by **l**.

If any unitary representation $\mathrm{R}_{\mathbf{l}}$ is used in (46) instead of the fundamental one R_1, the result can be reduced to a sum of the fundamental potentials K^j, with integer coefficients, as soon as $\mathrm{R}_{\mathbf{l}}$ can be extracted from a direct product of R_1's.

Having the expression for the Kähler potential, one immediately obtains symbols for elements of the group $\mathcal{G}$ and its Lie algebra **g**,

$$U_g^{(\mathbf{l})}(\zeta,\overline{z}) = \prod_{j=1}^{r}\left[\frac{\det'(\hat{g}\hat{f}(\zeta)\theta_j\hat{f}(z)^\dagger)}{\det'(\hat{f}(\zeta)\theta_j\hat{f}(z)^\dagger)}\right]^{l_j}, \tag{53}$$

$$T_a^{(\mathbf{l})}(\zeta,\overline{z}) = \mathrm{tr}\,(\Theta^{(\mathbf{l})}(\zeta,\overline{z})\widehat{\tau}_a)\,. \tag{54}$$

Here the trace is taken in the fundamental representation, and the matrix Θ, sometimes called the *momentum map*, was described in [**29**]. It is the sum of the fundamental components,

$$\Theta^{(\mathbf{l})}(\zeta,\overline{z}) = \sum_{j=1}^{r} l_j \Theta_j(\zeta,\overline{z})\,. \tag{55}$$

The fundamental momentum maps $\Theta_j(z,\overline{z})$ may be regarded as the projections θ_j transported from the origin of $\mathcal{M}$ to an arbitrary point. In other words, there exists an element $v(\zeta,\overline{z}) \in \mathcal{G}$ such that

$$\Theta_j(\zeta,\overline{z}) = \hat{v}\theta_j\hat{v}^{-1}, \qquad \hat{v}^{-1} = \hat{v}^\dagger. \tag{56}$$

It was shown [**29**] that $\Theta_j(\zeta,\overline{z})\exp[K^j(\zeta,\overline{z})]$ is a polynomial in ζ and z. The Killing fields and the Lie algebra cocycles φ_a, which are present in the general expression for T_a in (36), were also given in [**29**].

It is easy to see that the K^j satisfy the following differential equations:

$$\nabla_\beta(\zeta)K^j(\zeta,\overline{z}) - \overline{\nabla_\beta(z)}K^j(\zeta,\overline{z}) = 0, \qquad \forall\beta:\ (\beta\cdot\mathbf{w}_j) = 0\,. \tag{57}$$

Therefore the $(1,1)$-form ω, defined by (9), is degenerate if there is a nonempty set of roots σ for which $K^{(\mathbf{l})}$ satisfies (57),

$$(\mathbf{l}\cdot\sigma) = 0, \qquad \sigma \in \Delta_{\mathbf{s}}^+ \subset \Delta_{\mathbf{g}}^+\,. \tag{58}$$

Actually, the set $\Delta_{\mathbf{s}}^+$ is generated by the primitive roots γ_j for which $l_j = 0$. Now the Kähler structure is introduced in the subset of $\mathcal{F}$ given by the constraints $z_\sigma = 0$. The flag manifold $\mathcal{F}$ can be regarded as a fiber bundle, $\mathcal{M}$ being its base, where the unitary group representation R_l generates a nondegenerate Kähler structure. The local coordinates on $\mathcal{M}$ are z_α, where $\alpha \in \Delta^+_{\mathbf{g}} \setminus \Delta^+_{\mathbf{s}}$. Respectively, the little group of $\mathcal{M}$ is larger than the maximal torus,

$$\mathcal{M} = G/H, \qquad H = S \otimes T', \tag{59}$$

where $\mathcal{S}$ is the semisimple Lie group having $\mathbf{s}$ as its Lie algebra, and $\mathcal{T}' \subset T$ is the torus generated by those basis elements h_j for which $l_j \neq 0$. This construction has a clear interpretation in terms of Dynkin graphs [**28**]. Given a group representation R_l, one must eliminate the nodes for which $l_j \neq 0$ from the Dynkin graph. The remaining nodes determine a semi-simple Lie algebra $\mathbf{s} \subset \mathbf{h} \subset \mathbf{g}$. The parabolic subgroup $\mathcal{P}$ is extended similarly; its Lie algebra is $\mathbf{p} = \mathbf{g}^- + \mathbf{t}'^{\mathrm{c}} + \mathbf{s}^{\mathrm{c}}$.

4.3. Unitary groups. For $\mathcal{G} = SU(r+1)$, the fundamental representation is $(r+1)$-dimensional. The primitive roots and the fundamental weights are given in terms of an orthonormal basis $\{\varepsilon_i\}$ in Euclidean space $\mathbb{R}^{r+1}$ (see, e.g., [**31**])

$$\gamma_j = \varepsilon_j - \varepsilon_{j+1}, \qquad \mathbf{w}_j = \sum_{i=1}^{j} \varepsilon_i - \frac{j}{r+1} \sum_{i=1}^{r+1} \varepsilon_i \,. \tag{60}$$

Now $v_j = 1 - j/(r+1)$, and $\operatorname{rank}(\theta_j) = j$. The local coordinates corresponding to the positive roots $\alpha = \varepsilon_j - \varepsilon_k$ are elements of a triangular matrix $\hat{z}$ with entires z_{jk}, $1 \leqslant j < k \leqslant (r+1)$ (other entries of the matrix are zero), and the complex dimension of $\mathcal{F}$ is $r(r+1)/2$. The matrix $\hat{f}(z)$ is triangular; its diagonal elements are 1, and polynomials in z^α stand above the diagonal. (The general form of the polynomials can be written out easily.) The manifold $\mathcal{F}$ has an additional symmetry under the reflection of the root space for which $j \to r-j$, and $\theta_{r-j} \to v(I-\theta_j)v^{-1}$, where v is the matrix reversing the order of components in the representation space; it corresponds to the automorphism of the root system specific for unitary groups.

If the number of nonzero components l_j is $k < r$, the phase space is a section of the flag manifold, $\mathcal{M} \subset F$. Now the gauge group is larger than $\mathcal{T}_r$, namely

$$\mathcal{H} = T_k \otimes \prod SU(r_i + 1), \quad \text{where } \sum r_i = r - k \,.$$

The corresponding Dynkin graph is obtained by eliminating k links from the chain describing $SU(r+1)$. If only one component, say $l_q = l$, is nonzero, then one obtains the Grassmann manifold

$$\mathcal{M} = \mathrm{Gr}(p,q) \equiv U(p+q)/U(p) \otimes U(q) \qquad (1 \leqslant q \leqslant p < r+1 \equiv p+q),$$

which is a rank-one section of $\mathcal{F}$. The complex dimension of the manifold is pq, and the local coordinates are elements of a $p \times q$ matrix $\hat{z}$, so that the elements of $\hat{f}$ are $f_{jn} = \delta_{jn} + z_{j,n-p}$, where $1 \leqslant j \leqslant p$, and $p+1 \leqslant n \leqslant p+q$. The resulting Kähler potential is

$$K(z,\overline{z}) = l \log\det(I_q + \hat{z}^\dagger \hat{z}) \,. \tag{61}$$

For $q = 1$, $\hat{z}$ is a complex vector, $\mathcal{M} \equiv \mathrm{CP}^r$ is the complex projective space, and one gets the familiar Fubini–Study metric [**18**]. The metric is "quantized", since l is an integer.

Extending the arguments presented in 2.4, one can claim that if l is large it evaluates the area of two-dimensional cross sections of the phase space $\mathcal{M}$ in the

$\hbar$ units, i.e. $\hbar \sim l^{-1}$. If the rank of $\mathcal{M}$ is $k > 1$, one has several nonzero numbers l_j. Introducing a local system of coordinates and momenta, one gets an apparent anisotropy in $\hbar$, which cannot be eliminated because of the boundary conditions.

§5. Dynamics

Until this section, we have been discussing *quantum kinematics* (as defined in Schwinger's book [**17**]). In order to introduce *dynamics*, one needs a Hamiltonian $\widehat{H}$, to be given as a function of the canonical variables $\widehat{T}_a$, just as Hamiltonians of the standard theory are expressed in terms of coordinates and momenta. Thus $\widehat{H}$ belongs to the universal enveloping algebra of $\mathbf{g}$. The problem is to get the evolution operator $\widehat{G}(t) = \exp(-it\widehat{H})$, where t stands for *time*, and to find solutions to the Heisenberg and/or Schrödinger equations of motion,

$$(62) \qquad d\widehat{T}_a/dt = -i[\widehat{T}_a, \widehat{H}], \qquad \widehat{T}_a(t) = \widehat{G}(t)^{-1}\widehat{T}_a(0)\,\widehat{G}(t),$$

$$(63) \qquad d\hat{\rho}/dt = i[\hat{\rho}, \widehat{H}], \qquad \hat{\rho}(t) = \widehat{G}(t)\,\hat{\rho}(0)\,\widehat{G}(t)^{-1},$$

where $\hat{\rho}$ is the system density operator. In the functional approach, one must calculate the partition function $Z(t)$ and the generating functional F_η, which shows the system response to external (time-dependent) *source terms* in the Hamiltonian,

$$(64) \qquad i\partial\widehat{G}_\eta/\partial t = [\widehat{H} + i\eta^a(t)\widehat{T}_a]\,\widehat{G}_\eta, \qquad \widehat{G}_\eta(t_0, t_0) = \hat{I},$$

$$(65) \qquad F_\eta = \lim_{t_\pm \to \pm\infty} \operatorname{Tr}\widehat{G}_\eta(t_+, t_-), \qquad Z(t) = \operatorname{Tr}\widehat{G}_0(t_0 + t, t_0)\,.$$

If the symbols are used, the equations of motion are linear integro-differential equations, and the trace is an integral in $\mathcal{M}$. In constructing possible Hamiltonians as functions of $\widehat{T}_a$, one should have in mind that for compact spaces the group representations are finite-dimensional, and there are some identities for any given representation. In particular, each Casimir operator $\widehat{C}_n$ is the unit operator times a number that depends on $\mathbf{l}$.

The solution is obtained immediately for Hamiltonians that are elements of $\mathbf{g}$ combined with the Casimir operators, i.e.,

$$\widehat{H} = \sum \mu_n \widehat{C}_n + i\xi^a\widehat{T}_a,$$

where μ_n and ξ^a are real provided that $\widehat{H}$ is selfadjoint. In this case, the evolution operator is just an element of $\mathcal{G}$, its symbol is given by equation (53), and $Z(t)$ is the representation character multiplied by $\exp(-it\sum \mu_n\lambda_n)$, where λ_n are eigenvalues of $\widehat{C}_n$.

Let us consider, for instance, the Hamiltonian $\widehat{H} = \mu\widehat{C}_2 + i\xi^a\widehat{T}_a$ for the sphere $S^2 \equiv SU(2)/U(1)$. Its symbol is

$$(66) \quad H(z,\bar{z}) = \mu l\,(l+2)/4 + l\,(1+z\bar{z})^{-1}[\xi^1(z+\bar{z}) - i\xi^2(z-\bar{z}) + \xi^3(1-z\bar{z})]\,.$$

Using a change of coordinates (rotation), one can get $\xi^1 = 0 = \xi^2$, and the spectrum is obtained immediately, $\varepsilon_\nu = \text{const} + \xi\nu$, where $\xi = |\xi^3|$ and $\nu = 0, 1, \ldots, l$. The constant μ may be chosen so as to have $\varepsilon_0 = 0$. The partition function is

$$(67) \qquad Z(t) = \exp(-il\xi t)\,\frac{\sin(l+1)\,\xi t}{\sin \xi t}\,.$$

The classical trajectories are the parallels of latitude on the sphere, namely, $z(t) = z(0)e^{-i\xi t}$. In the limit discussed in 2.4, where $l \to \infty$ and $z = (q+ip)/\sqrt{2l\hbar}$, we get the standard Hamiltonian of the one-dimensional oscillator, retaining terms of the order of $z\overline{z}$, i.e., $\hbar H(p,q) = \xi(p^2+q^2) + \text{const}$.

The semiclassical approximation usually works in situations where large parameters are present in exponential integrands, and the integrals are evaluated by means of the steepest descent method. This is also true for dynamics on homogeneous Kähler manifolds. The large parameter appears if some components of $\mathbf{l}$ are large. Incidentally, the total number of states, which is proportional to the volume of $\mathcal{M}$, is also large in this case. The large parameter must also be present in the Hamiltonian H; otherwise, the classical dynamics would be trivial.

In order to get the semiclassical approximation, let us construct an analog of the path integral [**33**]. The standard method is to use the identity

$$\hat{U}_t \equiv e^{-i\widehat{H}t} = (e^{-i\widehat{H}t/\mathcal{N}})^{\mathcal{N}}, \qquad \forall \mathcal{N}, \tag{68}$$

and to write an approximate expression for the symbol of the evolution operator at small times, $\tau = t/\mathcal{N}$, where $\mathcal{N} \to \infty$,

$$e^{-i\tau\widehat{H}} \to e^{-i\tau H(z,\overline{z})} + O(\tau^2). \tag{69}$$

In principle, this is not correct for our manifolds, because each symbol must be a rational function, i.e., a polynomial divided by $\exp(K)$, and this form does not survive the exponentiation. In the limit of large $\mathbf{l}$, however, the polynomials are of a very high degree, and this condition is not too restrictive. Multiplying $\mathcal{N}$ operators (69) and calculating the trace (cf. equation (20)), one gets the partition function

$$\begin{aligned} Z(t) = \lim_{\mathcal{N}\to\infty} \int_{\mathcal{M}} \cdots \int \prod_{n=1}^{\mathcal{N}} d\mu(z_n,\overline{z}_n) \\ \times \exp\left\{ \sum_{n=1}^{\mathcal{N}} [K(z_n,\overline{z}_{n+1}) - K(z_n,\overline{z}_n) - i\tau H(z_n,\overline{z}_{n+1})] \right\}. \end{aligned} \tag{70}$$

The boundary condition is $\overline{z}_{\mathcal{N}+1} \equiv \overline{z}_1$, so the sequence of points on $\mathcal{M}$ can be regarded as closed. The exponent has an extremum under the following conditions: for $n = 1, \ldots, \mathcal{N}$,

$$\begin{aligned} \Lambda_\alpha(z_n,\overline{z}_{n+1}) - \Lambda_\alpha(z_n,\overline{z}_n) - i\tau\partial_\alpha H(z_n,\overline{z}_{n+1}) = 0, \\ \overline{\Lambda}_\alpha(z_{n-1},\overline{z}_n) - \overline{\Lambda}_\alpha(z_n,\overline{z}_n) - i\tau\partial_{\bar{\alpha}} H(z_{n-1},\overline{z}_n) = 0, \end{aligned} \tag{71}$$

where Λ and $\overline{\Lambda}$ are the partial derivatives of K, as given in (36). Under reasonable conditions on the Hamiltonian, in the limit $\tau \to 0$, the region $z_{n+1} \to z_n$ contributes predominantly to the integral, so that

$$\Lambda_\alpha(z_n,\overline{z}_{n+1}) - \Lambda_\alpha(z_n,\overline{z}_n) \approx \omega_{\alpha\bar{\beta}}(z_n,\overline{z}_n)(\overline{z}_{n+1} - \overline{z}_n)^{\overline{\beta}}, \tag{72}$$

and equations (71) become the usual Hamiltonian equations of motion, which can be written in the familiar form with the Poisson brackets defined in (37),

$$\frac{dz^\alpha}{dt} = \{H, z^\alpha\}_{\text{P.B.}}, \qquad \frac{d\overline{z}^\alpha}{dt} = \{H, \overline{z}^\alpha\}_{\text{P.B.}}. \tag{73}$$

These equations of motion can be also derived from the variational principle applied to the action resulting from the sum in (70) if the difference of the two Kähler potentials is replaced by the differential,

$$\mathcal{A}_t = \int_0^t \left\{ \frac{i}{2} [\Lambda^\alpha(z,\bar{z})\, dz^\alpha - \overline{\Lambda}^\alpha(z,\bar{z})\, d\bar{z}^\alpha] - H(z,\bar{z})\, d\tau \right\}. \tag{74}$$

(We added the total derivative $dK/2$ to the integrand to make the action manifestly real. Note that K is not defined globally on $\mathcal{M}$, but this addition is possible because of the cocycle condition.) The action functional is calculated on closed trajectories, $z(0) = z(t)$.

The arguments presented above support the meaning of the construction as a method of quantization: the quantum theory has its classical limit described by the action functional with the Kähler symplectic form. The classical theory is local and does not require integer coefficients at the fundamental Kähler potentials.

§6. Concluding remarks

Berezin's method results in the following construction. For any unitary representation $\mathrm{R_l}$ of a compact simple group $\mathcal{G}$, one has a compact homogeneous Kähler manifold $\mathcal{M} \equiv G/H$ and a Hilbert space $\mathcal{L}$ of (locally) holomorphic functions which can be regarded as a line bundle over $\mathcal{M}$. According to the Borel–Weil–Bott theorem, the Hilbert space representation is unitary. The Lie algebra $\mathbf{g}$ is realized by means of linear differential operators in $\mathcal{L}$, or by means of Poisson brackets for functions on $\mathcal{M}$. This is the essence of *quantization*. The Kähler potentials have been constructed explicitly for all compact Lie groups.

Until now, the Berezin quantization had no real *physical applications*, except for some "model building". An extension to a new theory of quantized fields will probably be the next step.

We conclude with a few remarks on subjects which have been beyond the scope of this paper.

1. *Noncompact groups.* The method can be extended easily to noncompact (locally compact) groups which are subgroups of $\mathcal{G}^c$. Let η be a nondegenerate matrix in the fundamental representation space of $\mathcal{G}$. A subgroup $\widetilde{\mathcal{G}} \in \mathcal{G}^c$ can be specified by the pseudo-unitarity condition $\hat{g}\eta\hat{g}^\dagger = \eta$. The group $\widetilde{\mathcal{G}}$ is noncompact if η is not positive definite. The arguments of §4 are valid if instead of equation (46) the fundamental potentials are defined by

$$K^j(\zeta,\bar{z}) = \log \det{}'(\hat{f}(\zeta)\,\theta_j \hat{f}(z)^\dagger \eta) \tag{75}$$

(it is assumed that $\det{}'(\theta_j\eta) = 1$). The pseudosphere in 2.4 is obtained in this way from $SU(1,1) \subset SL(2,\mathbb{C})$ with $\theta = \mathrm{diag}\,(1,0)$ and $\eta = \mathrm{diag}\,(1,-1)$. More information on noncompact homogeneous Kähler manifolds can be found elsewhere [**33**]. The manifold is noncompact, as it is the image of a bounded domain in $\mathbb{C}^m$, where $\det{}'$ in (75) is positive. The boundary of the domain is an analog of the absolute for the Lobachevsky plane. The Kähler manifolds appear for discrete series of unitary (infinite-dimensional) representations of noncompact groups. In general, homogeneous manifolds of Lie groups cannot be supplied with a Kähler structure, their geometry is too complicated (see, e.g., [**34**]). In particular, it is impossible for $\widetilde{\mathcal{G}}/\widetilde{\mathcal{H}}$, if $\widetilde{\mathcal{H}}$ is noncompact.

2. *Infinite-dimensional limit.* The main building blocks of this method admit an extension to the limit of infinite-dimensional groups. For example, infinite-dimensional Hilbert–Schmidt Grassmannians have been considered [**33**]. It was found that the classical Hamiltonians generate a Lie subalgebra of the total Poisson bracket Lie algebra, isomorphic to the central extension of $\mathbf{g}$ which includes the (quantum) anomaly. Geometric quantization in the $N \to \infty$ limit has been also considered in a recent paper [**35**].

3. *Supermanifolds.* For the compact case, representations of the Lie groups can also be constructed in supermanifolds, i.e., in the Grassmann algebra with anticommuting generators (the exterior algebra [**17**]). The example of $SO(3)$ was considered and applied to the description of spinning particles [**11**]. Different groups appear for supermanifolds with different numbers of Fermi-type degrees of freedom. Two approaches, based on Kähler manifolds and supermanifolds, are equivalent but quite different technically. In field theory the equivalence is known as "bosonization of fermion models".

4. *Integrals on Kähler manifolds.* Starting from known transformation properties, one can calculate a class of integrals on $\mathcal{M}$, as in (19). A particular result is Weyl's formula for the representation's dimension and a similar formula for volumes of manifolds. The Weyl group can be also realized in $\mathcal{M}$. (This result will be published elsewhere.)

5. *Integrable systems.* Investigation of Lie subalgebras of the universal enveloping algebra of $\mathbf{g}$ may lead to a new insight in the theory of integrable systems (cf. the example in §5).

6. *Additional references.* Only a part of a great number of works dealing with the Berezin approach to quantization has been mentioned here. Besides those mentioned, see, for instance, [**36–39**] and the references therein.

Appendix

A. Generalized determinant. For any selfadjoint positive semidefinite linear operator A, its generalized determinant, $\det' A$, is defined as the product of all its *nonzero* eigenvalues. Given a pair of semidefinite operators of the same rank, A and B, one has

$$\frac{\det' A}{\det' B} = \lim_{\lambda \to 0} \frac{\det(A - \lambda)}{\det(B - \lambda)}. \tag{76}$$

Applying the identity

$$\log \frac{\alpha}{\beta} = -\int_0^\infty (e^{-t\alpha} - e^{-t\beta}) \frac{dt}{t}, \qquad \forall \alpha, \beta > 0, \tag{77}$$

to the eigenvalues of A and B, and summing, one gets

$$\log \frac{\det' A}{\det' B} = -\int_0^\infty [\operatorname{Tr}(e^{-tA}) - \operatorname{Tr}(e^{-tB})] \frac{dt}{t}. \tag{78}$$

This equality enables one to define generalized determinants for elliptic operators having definite partition functions; it has been used in quantum field theory [**40**].

Evidently, for any nonsingular operator B, $\det'(BAB^{-1}) = \det' A$. In a special block representation of operators A and B,

$$A = \begin{pmatrix} a & 0 \\ 0 & 0 \end{pmatrix}, \qquad B = \begin{pmatrix} b & c \\ c' & b' \end{pmatrix}, \tag{79}$$

where a, b, and b' are square matrices (linear operators), c, c' are, in general, nonsquare, and $\det a \neq 0$, $\det b \neq 0$, one has

$$\frac{\det' BAB^\dagger}{\det' A} = \det(bb^\dagger + cc^\dagger)\,. \tag{80}$$

This equality was used in §4.

B. Projection matrices for classical groups and G_2. We shall use the notations of [**31**]. Roots and weights for simple Lie algebras of rank r are given in terms of the orthonormal basis $\{\epsilon_j\}$ in the root space, except for A_r and G_2, where the root space is described as a r-dimensional hyperplane in $(r+1)$-dimensional Euclidean space (the root space is normal to the vector $\varepsilon = \sum_{j=1}^{r+1} \epsilon_j$). The primitive roots γ_j are

$$\begin{array}{lll} A_r: & \epsilon_j - \epsilon_{j+1} & (1 \leqslant j \leqslant r), \\ B_r: & \epsilon_j - \epsilon_{j+1},\ \epsilon_r & (1 \leqslant j < r), \\ C_r: & \epsilon_j - \epsilon_{j+1},\ 2\epsilon_r & (1 \leqslant j < r), \\ D_r: & \epsilon_j - \epsilon_{j+1},\ \epsilon_{r-1} + \epsilon_r & (1 \leqslant j < r), \\ G_2: & \epsilon_1 - 2\epsilon_2 + \epsilon_3,\ \epsilon_2 - \epsilon_3\,. & \end{array} \tag{81}$$

The corresponding fundamental weights $\mathbf{w}_j$ (the lowest vector $\mathbf{w}_1$ is the last in the line) are

$$\begin{array}{lll} A_r: & \sum_{i=1}^{j} \epsilon_i - \frac{j}{r+1}\varepsilon & (1 \leqslant j \leqslant r), \\ B_r: & \sum_{i=1}^{j} \epsilon_i,\ \frac{1}{2}(\epsilon_1 + \cdots + \epsilon_r) & (1 \leqslant j \leqslant r-1), \\ C_r: & \sum_{i=1}^{j} \epsilon_i & (1 \leqslant j \leqslant r), \\ D_r: & \sum_{i=1}^{j} \epsilon_i,\ \frac{1}{2}(\epsilon_1 + \cdots + \epsilon_r),\ \frac{1}{2}(\epsilon_1 + \cdots - \epsilon_r) & (1 \leqslant j \leqslant r-2), \\ G_2: & 2\epsilon_1 - \epsilon_2 - \epsilon_3,\ \epsilon_1 - \epsilon_3\,. & \end{array} \tag{82}$$

The weights of the fundamental representation, having $\mathbf{w}_1$ as its dominant weight, and its dimension d_f are given by

$$
\begin{array}{lll}
 & A_r: \quad \epsilon_j - \dfrac{1}{r+1}\varepsilon, & d_f = r+1, \\
 & B_r: \quad \dfrac{1}{2}(\pm\epsilon_1 \pm \epsilon_2 \cdots \pm \epsilon_r), & d_f = 2^r, \\
(83) & C_r: \quad (\pm\epsilon_1 \pm \epsilon_2 \cdots \pm \epsilon_r), & d_f = 2^r; \\
 & D_r: \quad \dfrac{1}{2}(\pm\epsilon_1 \pm \epsilon_2 \cdots \pm \epsilon_r), & d_f = 2^{r-1}, \\
 & G_2: \quad \pm(\epsilon_1 - \epsilon_3),\ \pm(\epsilon_1 - \epsilon_2),\ \pm(\epsilon_2 - \epsilon_3),\ \mathbf{0}, & d_f = 7.
\end{array}
$$

(For D_r the number of minuses in each weight vector is odd.)

Ultimately, the ranks of the projection matrices θ_j corresponding to the fundamental weights $\mathbf{w}_j$ are given in the following Dynkin graphs. The rank of θ_1, corresponding to $\mathbf{w}_1$, is 1.

$$
\begin{array}{lrcccccc}
 & & r & & r-1 & 2 & & 1 \\
 & SU(r+1) \sim A_r: & \bigcirc & - & \bigcirc - \cdots - & \bigcirc & - & \bigcirc \\
 & & 2^{r-1} & & 2^{r-2} & 2 & & 1 \\
 & SO(2r+1) \sim B_r: & \bigcirc & - & \bigcirc - \cdots - & \bigcirc & \Rightarrow & \bigcirc \\
 & & 2^{r-1} & & 2^{r-2} & 2 & & 1 \\
(84) & Sp(2r) \sim C_r: & \bigcirc & - & \bigcirc - \cdots - & \bigcirc & \Leftarrow & \bigcirc \\
 & & 2^{r-2} & & 2^{r-3} & 2 & & 1 \\
 & SO(2r) \sim D_r: & \bigcirc & - & \bigcirc - \cdots - & \bigcirc & - & \bigcirc \\
 & & & & & | & & \\
 & & & & r & \bigcirc & & \\
 & & 2 & & 1 & & & \\
 & G_2: & \bigcirc & \Rrightarrow & \bigcirc & & &
\end{array}
$$

Similarly, one can calculate the ranks of θ_j for the other 4 exceptional groups. It is remarkable that all the projection matrices can be constructed recursively.

Acknowledgements. This work was completed when the second author was visiting at the University of Texas at Austin. It is a pleasure to thank Cécile DeWitt–Morette and other colleagues at the Center for Relativity and the Department of Physics for their kind hospitality. The support for the research from the NSF grant PHY9120042, G.I.F. and the Technion V.P.R. Fund is gratefully acknowledged.

References

1. F. A. Berezin, *Covariant and contravariant symbols of operators*, Izv. Akad. Nauk SSSR Ser. Mat. **36** (1972), no. 5, 1134–1167; English transl., Math. USSR-Izv. **6** (1972), 1117–1151.
2. ———, *Quantization*, Izv. Akad. Nauk SSSR Ser. Mat. **38** (1974), no. 5, 1116–1175; English transl., Math. USSR-Izv. **8** (1974), 1109–1165.
3. ———, *Quantization in complex symmetric spaces*, Izv. Akad. Nauk SSSR Ser. Mat. **39** (1975), no. 2, 363–402; English transl., Math. USSR-Izv. **9** (1975), 341–379.

4. ———, *General concept of quantization*, Comm. Math. Phys. **40** (1975), 153–174.
5. V. A. Fock, *Verallgemeinerung und Lösung der Diracschen statistischen Gleichung*, Z. Phys. **49** (1928), 339–357.
6. ———, *Konfigurationsraum und zweite Quantelung*, Z. Phys. **75** (1932), 622–647.
7. J. Schwinger, *A note on the quantum dynamical principle*, Philos. Mag. **44** (1953), 1171–1179.
8. F. A. Berezin, *The method of second quantization*, "Nauka", Moscow, 1965; English transl., Academic Press, New York, 1966.
9. B. DeWitt, *Supermanifolds*, 2nd edition, Cambridge Univ. Press, 1992.
10. F. A. Berezin and M. S. Marinov, *Classical spin and Grassmann algebra*, Pis′ma Zh. Èksper. Teoret. Fiz. **21** (1975), 678–680; English transl., JETP Lett. **21** (1975), 320–321.
11. ———, *Particle spin dynamics as the Grassmann variant of classical mechanics*, Ann. Phys. **104** (1977), 336–362.
12. J. V. Novozhilov and A. V. Tulub, *Die Methode der Funktionale in der Quantenfeldtheorie*, Fortschr. Phys. **5** (1958), 50–107.
13. L. D. Faddeev, *Introduction to functional methods*, Methods in Field Theory (R. Balian and J. Zinn-Justin, eds.), North Holland, Amsterdam, 1976, pp. 1–40.
14. V. Bargmann, *Remarks on a Hilbert space of analytic functions*, Proc. Nat. Acad. Sci. USA **48** (1962), 199–204.
15. F. A. Berezin, *Convex operator functions*, Mat. Sb. **88** (1972), 268–276; English transl., Math. USSR-Sb. **17** (1972), 269–277.
16. B. Simon, *The classical limit of quantum partition functions*, Comm. Math. Phys. **71** (1980), 247–276.
17. J. Schwinger, *Quantum kinematics and dynamics*, W. A. Benjamin, New York, 1970.
18. S. Kobayashi and K. Nomizu, *Foundations of differential geometry*, vol. 2, Interscience, New York, 1969.
19. J. R. Klauder and B.-S. Skagerstam, *Coherent states: Applications in physics and mathematical physics*, World Sci. Publ., Singapore, 1985.
20. A. M. Perelomov, *Generalized coherent states and their applications*, Springer-Verlag, New York, 1986.
21. P. A. M. Dirac, *Quantized singularities in the electromagnetic fields*, Proc. Roy. Soc. London Ser. A **133** (1931), 60–72.
22. T. T. Wu and C. N. Yang, *Concept of nonintegrable phase factors and global formulation of gauge fields*, Phys. Rev. D **12** (1975), 3845–3857.
23. J.-M. Souriau, *Structure des systèmes dynamiques*, Dunod, Paris, 1970.
24. J. R. Klauder, *The action option and a Feynman quantization of spinor fields in terms of ordinary c-numbers*, Ann. Phys. **11** (1960), 123–168.
25. P. A. M. Dirac, *The principles of quantum mechanics*, Clarendon Press, Oxford, 1958.
26. M. Bando, T. Kuramoto, T. Maskawa, and S. Uehara, *Structure of nonlinear realization in supersymmetric theories*, Phys. Lett. B **138** (1984), 94–98.
27. K. Itoh, T. Kugo, and H. Kunitomo, *Supersymmetric nonlinear realization for arbitrary Kählerian coset space G/H*, Nuclear Phys. B **263** (1986), 295–308.
28. M. Bordemann, M. Forger, and H. Römer, *Homogeneous Kähler manifolds: Paving the way towards new supersymmetric sigma models*, Comm. Math. Phys. **102** (1986), 605–647.
29. D. Bar-Moshe and M. S. Marinov, *Realization of compact Lie algebras in Kähler manifolds*, J. Phys. A **27** (1994), 6287–6298.
30. A. Borel, *Kählerian coset spaces of semisimple Lie groups*, Proc. Nat. Acad. Sci. USA **40** (1954), 1147–1151.
31. N. Bourbaki, *Groupes et algèbres de Lie*, Masson, Paris, 1981.
32. M. S. Marinov, *Path integrals on homogeneous manifolds*, J. Math. Phys. **36** (1995), 2458–2469.
33. D. Bar-Moshe, *Geometric quantization on Kählerian systems*, Ph.D. Thesis, Technion, Haifa, 1993.
34. V. V. Gorbatsevich and A. L. Onishchik, *Lie transformation groups*, Lie Groups and Lie Algebras. I (A. L. Onishchik, ed.), Springer-Verlag, Berlin–Heidelberg–New York, 1993, pp. 95–229.

35. M. Bordemann, E. Meinrenken, and M. Schlichenmaier, *Toeplitz quantization of Kähler manifolds and $gl(N)$ $N \to \infty$ limits*, Freiburg THEP 93/22.
36. C. Moreno, *Geodesic symmetries and invariant star products on Kähler symmetric spaces*, Lett. Math. Phys. **13** (1987), 245–257.
37. G. M. Tuynman, *Quantization: Towards a comparison between methods*, J. Math. Phys. **28** (1987), 2829–2840.
38. A. Odzijewicz, *On reproducing kernels and quantization of states*, Comm. Math. Phys. **114** (1988), 577–597.
39. M. Cahen, S. Gutt, and J. H. Rawnsley, *Quantization of Kähler manifolds,*, J. Geom. Phys. **7** (1990), 45–62; Trans. Amer. Math. Soc. **337** (1993), 73–98; Lett. Math. Phys. **30** (1994), 291–305.
40. A. S. Schwarz, *Quantum field theory and topology*, Springer-Verlag, Berlin–Heidelberg–New York, 1993.

D. Bar-Moshe: Department of Physics, Technion-Israel Institute of Technology Haifa 32000, Israel

M. Marinov: Department of Physics, Technion-Israel Institute of Technology Haifa 32000, Israel and Center for Relativity, University of Texas at Austin, Austin, Texas, 78712-1081, USA

Amer. Math. Soc. Transl.
(2) Vol. **177**, 1996

Generalized Field-Antifield Formalism

I. A. Batalin and I. V. Tyutin

ABSTRACT. A generalized version of the field-antifield BV formalism is proposed. The quantum master equation and hypergauge-fixing procedure are reformulated so as to be invariant with respect to arbitrary reparametrizations of the field-antifield variables. The multilevel formulation is constructed inductively by including the hypergauge Lagrangian multipliers into the extended field-antifield phase space. The nth level functional integral is shown to be gauge independent and reduced to the $(n-1)$th level one.

§1. Introduction

It is an honor for us to have the opportunity to contribute an article to the memorial volume devoted to Felix Alexandrovich Berezin. For many years both authors had the priceless privilege of being in human and scientific contact with Felix Alexandrovich, and we regard this as a great piece of luck in our lives.

Felix Alexandrovich was an outstanding scientist who originated new fundamental concepts in pure mathematics as well as in theoretical physics. One can mention, for instance, the well-known concept of supermathematics or the general quantization concept.

In the present paper we describe the field-antifield quantization formalism, which is substantially based on the ideas of supermathematics. As usual in a supertheory, we deal here with Bosonic and Fermionic quantities that must be treated in an uniform way. However, the specific feature of the field-antifield formalism is that the presence of Fermions is unavoidable, so that a purely Bosonic counterpart cannot exist in principle.

The field-antifield formalism was originally constructed in order to solve the general problem of covariant (Lagrangian) quantization of gauge theories. This problem has a long history, beginning with the famous works of Feynman [**1**], Faddeev and Popov [**2**] and DeWitt [**3**].

A unique closed approach to the covariant quantization problem was proposed in the work of Batalin and Vilkovisky [**4**]. These authors introduced the field-antifield phase space concept as well as the antibracket operation, which is the

1991 *Mathematics Subject Classification.* Primary 81T70.

antisymplectic counterpart of the well-known Poisson bracket. Moreover, a nilpotent second-order differential operator that differentiates the antibracket according to the Leibniz rule was discovered. Due to the mentioned property, henceforth we shall refer to this remarkable operator as the "antisymplectic differential".

The authors of [**4**] have formulated a general quantization principle to be applied directly to the Lagrangian formalism. The principle requires that the exponential of $i/\hbar$ times the quantum action be annihilated by the antisymplectic differential. Thus the quantum master equation acquires a great importance. The corresponding classical master equation requires that the classical master action commute with itself (in the antibracket sense) to give zero.

In its own turn, the classical master equation defines a universal gauge hypertheory whose hypergauge generators are always nilpotent at the classical hyperextremals. The classical master action that possesses the minimal possible hypergauge degeneracy is called "proper". This minimal degeneracy is eliminated exactly by imposing the standard BV hypergauge conditions requiring that all the antifields be equal to the corresponding field derivatives of a Fermionic function. Due to the quantum master equation, the functional integral does not depend formally on the hypergauge Fermionic function variations. If one uses the proper master action together with the standard BV hypergauge, then the functional integral is certainly nondegenerate and thus can be calculated using the loop expansion technique.

The general strategy of the BV approach is to involve a given gauge theory into the universal hypertheory determined by the proper master action. Moreover, the necessary spectrum of the field-antifield variables is determined in a way that ensures that the master action be proper. This is the mechanism by means of which all the ghost generations appear naturally.

The above-mentioned strategy has been applied successfully to gauge theories with irreducible open algebras [**4**] and to theories with linearly dependent gauge generators [**5**] as well. Moreover, the recent developments [**6–11**] in secondary-quantized string field theory are substantially based on the BV approach.

Many authors have contributed to the development and application of the field-antifield formalism. For detailed references, see the survey of Henneaux [**12**]. The important contributions of Zinn–Justin [**13**], Kallosh [**14**], de Wit and van Holten [**15**] were instrumental in revealing the general status of the classical master equation. Witten [**16**] has given a deep geometric interpretation to the quantum master equation. An $Sp(2)$-covariant version of the BV formalism has been proposed by Batalin, Lavrov, and Tyutin [**17–19**]. Henneaux [**20**] has extended Witten's interpretation to cover the $Sp(2)$-covariant formalism. A relationship between the Hamiltonian BFV and Lagrangian BV formalisms has been revealed by Grigoryan, Grigoryan and Tyutin [**21**]. These authors have used a functional-integral counterpart of the operator method originally proposed by Batalin and Fradkin [**22**]. Schwarz et al. [**23–25**] studied the general geometric and algebraic aspects of the BV formalism, as well as the semiclassical approximation in it. Independently of the gauge field quantization problem, an invariant geometric description of the symplectic and antisymplectic structures on Kählerian superspaces has been given by Khudaverdyan and Nersessyan [**26, 27**]. Volkov et al. [**28, 29**] studied the antibracket reformulation and quantization of supersymmetric mechanics.

In the present work, we describe further steps undertaken in developing the field-antifield formalism, according to the recent papers [**30–33**] of the present authors. The main motivation is to naturally extend the field-antifield phase space as far as possible from the viewpoint of the symmetry properties and physical content of the theory.

The first problem is to reformulate the quantum master equation, as well as the hypergauge-fixing procedure, so as to make them explicitly invariant with respect to arbitrary reparametrizations of the field-antifield phase variables.

The second problem is to assign the antifields to the hypergauge Lagrangian multipliers and thus to initiate the hierarchical proliferation process that introduces the hypergauges of higher levels.

The meaning of the higher level hypergauges can be explained in more detail as follows. By definition, the formalism is called "of the first level" when the hypergauge conditions are imposed directly on the original field-antifield variables. The corresponding Lagrangian multipliers do not possess their own antifields at the first level. In order to move up to the second level formalism, we assign antifields to the first level Lagrangian multipliers and then include these new anticanonical pairs into the extended phase space. To retain the correct number of physical degrees of freedom, we need new, second-level, hypergauge conditions to be imposed on the new antifields. Thus we introduce new, second-level, Lagrangian multipliers which do not possess, at this level, their own antifields, which will appear at the third level, and so on.

At each level one should provide for the overall gauge independence of the corresponding functional integral. To solve this problem, we write out the nth level quantum master equation, which guarantees the formalism to be invariant with respect to the generalized BRST-type transformations.

The nth level quantum master equation turns out to be also a generating mechanism for the $(n-1)$th level hypergauge algebra. This mechanism in a remarkable way combines the characteristic features of the Hamiltonian and Lagrangian gauge algebra generating equations, while the anticanonical pairs of the $(n-1)$th level Lagrangian multipliers play the role of the Hamiltonian ghost variables.

Finally, we show that the nth level functional integral is gauge-independent and reduce it to the $(n-1)$th level one.

In the Appendix, we study local quantum deformations of the antisymplectic differential and show that they are trivial.

As usual, we denote by $\varepsilon(A)$ the Grassmann parity of the expression A, and $\mathrm{Ber}(K)$ stands for the Berezinian (superdeterminant) of a supermatrix K. Other notation will be clear from the context.

§2. Antisymplectic differential and antibrackets

Suppose

$$\Gamma^A, \quad A = 1, \dots, 2N, \qquad \varepsilon(\Gamma^A) \equiv \varepsilon_A, \tag{2.1}$$

is a set of field-antifield variables:

$$\{\Gamma^A\} = \{\varphi^a, \varphi^*_a \mid a = 1, \dots, N,\ \varepsilon(\varphi^*_a) = \varepsilon(\varphi^a) + 1\}. \tag{2.2}$$

We regard the variables (2.2) as the local coordinates of the corresponding field-antifield phase space $\mathcal{M}$.

Let Δ be a most general Fermionic second-order differential operator without the derivativeless term,

$$\Delta \equiv (1/2)(-1)^{\varepsilon_A}(\partial_A + F_A)E^{AB}\partial_B, \qquad \varepsilon(\Delta) = 1, \tag{2.3}$$

where $E^{AB}(\Gamma)$, $\varepsilon(E^{AB}) = \varepsilon_A + \varepsilon_B + 1$, is a metric assumed to be adjoint-antisymmetric,[1]

$$E^{AB} = -E^{BA}(-1)^{(\varepsilon_A+1)(\varepsilon_B+1)}, \tag{2.4}$$

and nondegenerate, while $F_A(\Gamma)$, $\varepsilon(F_A) = \varepsilon_A$, is a "connection" field.

Let us require that the operator Δ be nilpotent,

$$\Delta^2 = 0. \tag{2.5}$$

The nilpotency condition itself immediately yields

$$\Delta(-1)^{\varepsilon_C}(\partial_C + F_C)E^{CD} = 0, \tag{2.6}$$

$$E^{DB}E^{CA}[\partial_A F_B - \partial_B F_A(-1)^{\varepsilon_A\varepsilon_B}](-1)^{\varepsilon_B(\varepsilon_C+1)} = 0, \tag{2.7}$$

$$(-1)^{(\varepsilon_A+1)(\varepsilon_C+1)}E^{AD}\partial_D E^{BC} + \text{cycle}(A, B, C) = 0. \tag{2.8}$$

Since the metric E^{AB} is nondegenerate, equation (2.7) gives locally

$$F_A = \partial_A \ln M(\Gamma), \tag{2.9}$$

so that equation (2.6) determines, in fact, the function $M(\Gamma)$. By substituting (2.9) into (2.6) and (2.3), we obtain

$$\Delta(-1)^{\varepsilon_C}M^{-1}\partial_C M E^{CD} = 0, \tag{2.10}$$

$$\Delta = (1/2)(-1)^{\varepsilon_A}M^{-1}\partial_A M E^{AB}\partial_B. \tag{2.11}$$

It is natural to regard the operator Δ as being scalar with respect to the general transformations of the field-antifield variables $\Gamma^A \to \Gamma'^A$:

$$\Delta' F'(\Gamma') = \Delta F(\Gamma), \qquad F'(\Gamma') = F(\Gamma). \tag{2.12}$$

Then the matrix E^{AB} behaves like a contravariant tensor of the second rank, while the function M possesses the transformation property of a measure density:

$$E'^{AB}(\Gamma') = \Gamma'^A\overleftarrow{\partial_C}E^{CD}(\Gamma)\overrightarrow{\partial_D}\Gamma'^B, \tag{2.13}$$

$$M'(\Gamma')\,\text{Ber}\,(\partial\Gamma') = M(\Gamma). \tag{2.14}$$

The invariant measure $d\mu(\Gamma)$ on the phase space $\mathcal{M}$,

$$d\mu(\Gamma) \equiv M(\Gamma)\,d\Gamma, \tag{2.15}$$

is naturally associated with the density $M(\Gamma)$.

[1] Note that the adjoint-antisymmetry property (2.4) is not to be imposed imperatively. When it is nonzero, the adjoint-symmetric part of E^{AB} can be absorbed into the redefinition of the "connection" field F_A.

In turn, the cyclic equation (2.8) is just the antisymplecticity property of the metric E^{AB}. This property allows one to introduce, in a natural way, the invariant antibracket operation

$$(A, B) \equiv A\overleftarrow{\partial_C} E^{CD} \overrightarrow{\partial_D} B \tag{2.16}$$

with the following algebraic properties:

$$\varepsilon((A, B)) = \varepsilon(A) + \varepsilon(B) + 1, \tag{2.17}$$

$$(A, B) = -(B, A)(-1)^{(\varepsilon(A)+1)(\varepsilon(B)+1)}, \tag{2.18}$$

$$(A, BC) = (A, B)C + B(A, C)(-1)^{(\varepsilon(A)+1)\varepsilon(B)}, \tag{2.19}$$

$$(AB, C) = (A, B)C + B(A, C)(-1)^{\varepsilon(A)\varepsilon(B)}, \tag{2.20}$$

$$((A, B), C)(-1)^{(\varepsilon(A)+1)(\varepsilon(C)+1)} + \text{cycle}(A, B, C) = 0. \tag{2.21}$$

Besides, the formula

$$\Delta(A, B) = (\Delta A, B) + (A, \Delta B)(-1)^{\varepsilon(A)+1} \tag{2.22}$$

shows that the operator Δ is a differentiation in the algebra formed by functions with the antibracket as the multiplication operation. This property allows one to call Δ an "antisymplectic differential".

On the other hand, applying the operator Δ to the ordinary product AB, we obtain:

$$\Delta(AB) = (\Delta A)B + (A, B)(-1)^{\varepsilon(A)} + A(\Delta B)(-1)^{\varepsilon(A)}. \tag{2.23}$$

Let us note that one can regard (2.23) as *defining* the antibracket (A, B). Then the corresponding properties (2.18), (2.21), and (2.22) can be derived without making use of the explicit form (2.16), by proceeding from the nilpotency of Δ and assuming that Δ is a differential operator of the second order.

§3. General functional integral of the first level

The concept of the antisymplectic differential plays the key role in constructing the general form for the Lagrangian functional integral. In fact, the basic idea of the BV approach is to involve an initially given gauge theory into the universal hypertheory whose hypergauge generators are always nilpotent at the classical hyperextremals. To define the above-mentioned hypertheory in a most natural and effective way, one should require that the exponential of $i/\hbar$ times quantum action be annihilated by the antisymplectic differential Δ. Thus we arrive at the quantum master equation.

To make the functional integral nondegenerate, one needs hypergauge-fixing. If the hypergauge conditions are imposed directly on the basic field-antifield variables (2.1), then, by definition, the functional integral is said to be "of the first level".

However, it will be shown in the next section that the hypergauge Lagrangian multipliers of the first level can be supplied with their own antifields which turn out to be new hypergauge variables requiring new, second-level, hypergauge conditions for themselves. In that case, by definition, the functional integral is said to be "of the second level", and so on.

So, we propose the following basic formula for the general Lagrangian functional integral of the first level:

$$Z = \int \exp\left\{\frac{i}{\hbar}\left[W(\Gamma;\hbar) + G_a(\Gamma;\hbar)\pi^a\right] - H(\Gamma;\hbar)\right\} d\mu(\Gamma)\, d\pi, \tag{3.1}$$

where the quantum action W satisfies the master equation

$$\Delta \exp\{(i/\hbar)W\} = 0, \tag{3.2}$$

or equivalently

$$(W,W)/2 = i\hbar\Delta W, \tag{3.3}$$

the hypergauge functions

$$G_a(\Gamma;\hbar), \quad a = 1,\ldots,N, \qquad \varepsilon(G_a) = \varepsilon_a, \tag{3.4}$$

should eliminate a gauge degeneracy of the action W by singling out the Lagrangian surface $G_a = 0$ in the field-antifield phase space. This condition implies that the antibracket involution relations must be satisfied:

$$(G_a, G_b) = G_c U^c_{ab}; \tag{3.5}$$

the function H satisfies the equations

$$(H, G_a) = \Delta G_a - U^b_{ba}(-1)^{\varepsilon_b} - G_b V^b_a, \tag{3.6}$$

$$\Delta H - (H,H)/2 + V^a_a = G_a \widetilde{G}^a. \tag{3.7}$$

It is crucial that the total set of equations (3.2), (3.5)–(3.7) implies that the integrand of (3.1) is invariant under the generalized BRST-type transformations

$$\delta\Gamma^A = (\Gamma^A, -W + G_a\pi^a + i\hbar H)\,\mu, \tag{3.8}$$

$$\delta\pi^a = (-U^a_{bc}\pi^c\pi^b(-1)^{\varepsilon_b} + 2i\hbar V^a_b\pi^b + 2(i\hbar)^2\widetilde{G}^a)\,\mu, \tag{3.9}$$

where $\mu = \text{const}$, $\varepsilon(\mu) = 1$.

Choosing the Fermionic parameter μ to be an arbitrary function

$$\mu = (i/2\hbar)\,\delta X(\Gamma), \tag{3.10}$$

and performing the additional variations

$$\delta\Gamma^A = (\Gamma^A, \delta X)/2, \qquad \delta\pi^a = \delta\Lambda^a_b\pi^b, \tag{3.11}$$

where $\delta\Lambda^a_b(\Gamma)$ are arbitrary functions too, one generates the following effective changes in the integrand of (3.1):

$$\delta G_a = (G_a, \delta X) + G_b\delta\Lambda^b_a, \tag{3.12}$$

$$\delta H = -\Delta\delta X + (H, \delta X) - \delta\Lambda^a_a(-1)^{\varepsilon_a}, \tag{3.13}$$

where the first and second term in right-hand side of (3.12) describes, respectively, the most general change of the hypergauge surface and its basis.

On the other hand, the transformations (3.12), (3.13) retain the form of equations (3.5)–(3.7) by inducing the following variations of structure coefficients:

$$\begin{aligned}\delta U^a_{bc} &= (U^a_{bc}, \delta X) + [(\delta\Lambda^a_b, G_c) + U^a_{bd}\delta\Lambda^d_c] \\ &\quad - [(\delta\Lambda^a_c, G_b) + U^a_{cd}\delta\Lambda^d_b](-1)^{(\varepsilon_b+1)(\varepsilon_c+1)} - \delta\Lambda^a_d U^d_{bc},\end{aligned} \tag{3.14}$$

$$\begin{aligned}\delta V^a_b &= (V^a_b, \delta X) + V^a_d\delta\Lambda^d_b - \delta\Lambda^a_d V^d_b \\ &\quad + [\Delta\delta\Lambda^a_b - (H, \delta\Lambda^a_b)](-1)^{\varepsilon_a},\end{aligned} \tag{3.15}$$

$$\delta\widetilde{G}^a = (\widetilde{G}^a, \delta X) - \delta\Lambda^a_b\widetilde{G}^b. \tag{3.16}$$

It will be shown in §6 that the functional integral (3.1) does not depend on the arbitrariness of the solution to equations (3.5)–(3.7). Thus we conclude that (3.1) does not depend on the choice of the functions G_a.

Let us note the following interesting circumstance. One can rewrite the variation (3.13) in the form

$$\delta H = -D\delta X - \delta\Lambda^a_a(-1)^{\varepsilon_a} \tag{3.17}$$

where D is the analog of the covariant derivative,

$$DG \equiv \Delta G - (H, G). \tag{3.18}$$

This operator squared is a "multiplication" operator in the algebra formed by functions with the antibracket as the product:

$$D^2G = (F, G), \tag{3.19}$$

$$F \equiv -\Delta H + (H, H)/2. \tag{3.20}$$

It is natural to regard H and F as a connection and its stress function. Then the flat connections are naturally defined as satisfying the quantum master equation.

Let us transform H by the law that is characteristic for a connection,

$$\delta H = -D\delta X, \tag{3.21}$$

and let G be transformed uniformly:

$$\delta G = (G, \delta X). \tag{3.22}$$

Then DG and F are transformed uniformly as well:

$$\delta F = (F, \delta X), \qquad \delta(DG) = (DG, \delta X). \tag{3.23}$$

Now let us determine conditions that provide for the hypergauge functions G_a to eliminate a degeneracy of the classical action. Let us look for a solution to equation (3.3) in the form of $\hbar$-power series expansion

$$W = S + i\hbar W_1 + \dots. \tag{3.24}$$

We have

$$(S, S) = 0, \tag{3.25}$$

$$(S, W_1) = \Delta S, \tag{3.26}$$

and so on.

The classical master equation (3.25) exactly determines the universal hypertheory with nilpotent hypergauge generators. To see this, let us differentiate equation (3.25) to find the Noether identities

$$R^B_A \partial_B S = 0, \tag{3.27}$$

where the hypergauge generators are

$$R^B_A \equiv [2(\overrightarrow{\partial}_A S \overleftarrow{\partial}_C) + S\overrightarrow{\partial}_A \overleftarrow{\partial}_C] E^{CB}. \tag{3.28}$$

Differentiating now the identities (3.28) we find the desired nilpotency property

$$R^B_A R^C_B |_{\partial S=0} = 0, \tag{3.29}$$

that gives

$$\operatorname{rank} \|R^B_A\| |_{\partial S=0} \leqslant N. \tag{3.30}$$

Next, let us write down the total classical action that enters the functional integral (3.1):

$$\text{Classical action} = S + G_a \pi^a. \tag{3.31}$$

The corresponding equations of motion are

$$\partial_A S + (\partial_A G_a)\pi^a = 0, \qquad G_a = 0, \tag{3.32}$$

that yield

$$R^B_A (\partial_B G_a)\pi^a = 0, \tag{3.33}$$

because of the identities (3.27).

A solution to equations (3.33) for the Lagrangian multipliers π^a is unique iff the following uniqueness condition is satisfied:

$$\operatorname{rank} \|R^B_A \partial_B G_a\| |_{\partial S=G=0} = N, \tag{3.34}$$

so that we have

$$\operatorname{rank} \|R^B_A\| |_{\partial S=G=0} \geqslant N, \tag{3.35}$$

$$\operatorname{rank} \|\partial_B G_a\| |_{\partial S=G=0} = N. \tag{3.36}$$

The conditions (3.30) and (3.35) are compatible with each other iff the equality

$$\operatorname{rank} \|R^B_A\| |_{\partial S=G=0} = N \tag{3.37}$$

holds. By definition, the master action S is proper, since condition (3.37) is satisfied.

Let us show that the proposed geometrically covariant formulation is equivalent to the standard BV formalism. First of all, let us note that if the metric E^{AB} is given, then the measure M is defined nonuniquely. Let the nilpotent operator Δ_1 be constructed by using another measure $M_1 \equiv TM$:

$$\Delta_1 = (1/2)(-1)^{\varepsilon_A} M_1^{-1} \partial_A M_1 E^{AB} \partial_B = \Delta + t\overleftarrow{\partial}_A E^{AB} \partial_B, \tag{3.38}$$

$$T \equiv \exp(2t). \tag{3.39}$$

It follows from the nilpotency property of Δ and Δ_1 that t satisfies the quantum master equation

$$\Delta \exp(t) = 0. \tag{3.40}$$

Let the functional integral Z_1 be constructed according to the formula

$$Z_1 = \int \exp\Big\{\frac{i}{\hbar}(W_1 + G_a\pi^a) - H_1\Big\}M_1\, d\Gamma\, d\pi, \tag{3.41}$$

$$\Delta_1 \exp\{(i/\hbar)W_1\} = 0, \tag{3.42}$$

where H_1 satisfies the equations

$$(H_1, G_a) = \Delta_1 G_a - U^b_{ba}(-1)^{\varepsilon_b} - G_b V^b_{1a}, \tag{3.43}$$

$$\Delta_1 H_1 - (H_1, H_1)/2 + V^a_{1a} = G_a \widetilde{G}^a_1. \tag{3.44}$$

Rewrite Z_1 in the form

$$Z_1 = \int \exp\Big\{\frac{i}{\hbar}(W' + G_a\pi^a) - H'\Big\}M\, d\Gamma\, d\pi, \tag{3.45}$$

$$W' = W_1 - i\hbar t, \qquad H' = H_1 - t. \tag{3.46}$$

Then one can check easily that W' satisfies the quantum master equation (3.2), and H' satisfies the set of equations (3.6), (3.7) with the functions V^a_{1b} and $\widetilde{G}^a_1$ replacing V^a_b and $\widetilde{G}^a$. It is natural to identify $W' = W$ and $H' = H$. Thus the arbitrariness in determining the operator Δ can be compensated by redefining the quantum action W and function H.

Let us transform the expression (3.1) for Z to the Darboux variables φ, φ^* in which the metric is $E^{AB} = \text{const}$. Let us assume that the hypergauge conditions $G_a = 0$ are solvable with respect to the variables φ^*. Then for Z we have:

$$Z = \int \exp\Big(\frac{i}{\hbar}\widetilde{W} - \sigma\Big)\delta(g_a)\, d\varphi\, d\varphi^*, \qquad \widetilde{W}(\varphi, \varphi^*) = W(\Gamma), \tag{3.47}$$

$$G_a(\Gamma) = g_b\lambda^b_a(\varphi) + O(g^2), \tag{3.48}$$

$$H(\Gamma) = h(\varphi) + O(g), \tag{3.49}$$

$$\sigma \equiv h + \ln \mathrm{Ber}\,(\lambda), \tag{3.50}$$

$$g_a \equiv \varphi^*_a - f_a(\varphi). \tag{3.51}$$

Due to the above arguments we can, without loss of generality, choose the value $M = 1$ to appear in the operator Δ in the Darboux variables. It follows from equations (3.5), (3.6) that

$$f_a(\varphi) = \partial\Psi(\varphi)/\partial\varphi^a, \tag{3.52}$$

$$\partial\sigma(\varphi)/\partial\varphi^a = 0, \qquad \sigma(\varphi) = \text{const}. \tag{3.53}$$

Thus we have proved that the generalized formalism is equivalent to the standard BV formulation. At the same time we have established that the functional integral (3.1) does not depend on the choice of hypergauge functions.

We conclude this section with the following remark. At the first level, it appears to be just possible to construct the extra measure factor $\exp(-H)$ in an explicitly covariant closed form, that is[2]

$$\exp(-H) = \{JM^{-1}\,\mathrm{Ber}\,[(F,G)]\}^{1/2}, \tag{3.54}$$

where G_a satisfy (3.5), F^a, $a = 1,\ldots,N$, $\varepsilon(F^a) = \varepsilon_a + 1$, are certain functions for which the replacement $\Gamma^A \to \bar{\Gamma}^A \equiv \{F^a; G_a\}$ is an invertible reparametrization whose Jacobian is denoted by J,

$$J \equiv \mathrm{Ber}\,(\partial\bar{\Gamma}). \tag{3.55}$$

It is an important property that the expression (3.54), being taken on the hypergauge surface, does not depend on F^a. One can also show that this expression represents the general solution to the equations (3.6), (3.7).

Substituting (3.54) into (3.1), we obtain the following representation for the first level functional integral:

$$Z = \int \exp\left(\frac{i}{\hbar}W\right)\delta(G)\{JM\,\mathrm{Ber}\,[(F,G)]\}^{1/2}d\Gamma. \tag{3.56}$$

Due to the presence of the square root $\{\quad\}^{1/2}$, the complete measure factor in the integrand of (3.56) cannot be, in general, parametrized by means of an integral over new fields with a local action.

§4. The nth level formalism

It is a remarkable circumstance that equations (3.5)–(3.7) determining the functions G_a and H can be obtained by reducing the so-called second level hypergauge theory formulated in a more extended field-antifield phase space. In turn, the equations determining the hypergauge functions of the second level theory can be obtained by reducing the third level hypergauge theory, and so on.

In this section we construct inductively the nth level functional integral for $n = 2, 3, \ldots$.

First, let us define recursively the nth level set of variables of the field-antifield phase space

$$\Gamma^{(n)A_{(n)}} \equiv \{\Gamma^{(n-1)A_{(n-1)}}; \pi^{(n-1)a}, \pi_a^{*(n-1)}\}, \tag{4.1}$$

where

$$\Gamma^{(1)A_{(1)}} \equiv \Gamma^A, \qquad \pi^{(1)a} \equiv \pi^a; \tag{4.2}$$

here $\pi^{(n-1)a}$ and $\pi_a^{*(n-1)}$ are the $(n-1)$th level Lagrangian multipliers and their conjugated antifields respectively, so that

$$\varepsilon(\pi^{(n)a}) = \varepsilon(\pi_a^{*(n)}) + 1 = \varepsilon_a + n - 1. \tag{4.3}$$

[2]O. M. Khudaverdyan, private communication.

In what follows all the antibrackets $(\ \cdot, \cdot\ )$ are understood to include the totally extended set (4.2) of field-antifield variables, and the only nonzero elementary antibrackets for the Lagrangian multipliers are

$$(\pi^{(n)a}, \pi^{*(n)}_b) = \delta^{(m)(n)}\delta^a_b. \tag{4.4}$$

Further, one constructs recursively the nilpotent operators $\Delta^{(n)}$:

$$\Delta^{(n)} \equiv \Delta^{(n-1)} + (-1)^{(\varepsilon_a+n)} \frac{\partial}{\partial\pi^{(n-1)a}} \frac{\partial}{\partial\pi^{*(n-1)}_a}, \tag{4.5}$$

$$\Delta^{(1)} \equiv \Delta. \tag{4.6}$$

Let us introduce the nth level Planck constant $\hbar^{(n)}$, $\varepsilon(\hbar^{(n)}) = 0$, $n \geqslant 2$, in addition to the usual one $\hbar$, and then define the new quantum number, called the *Planck parity* $\mathrm{Pl}^{(n)}$,

$$\mathrm{Pl}^{(n)}(\Gamma^{(n-1)}) = \mathrm{Pl}^{(n)}(\hbar) = 0, \tag{4.7}$$

$$\mathrm{Pl}^{(n)}(\hbar^{(n)}) = \mathrm{Pl}^{(n)}(\pi^{(n-1)}) = -\,\mathrm{Pl}^{(n)}(\pi^{*(n-1)}) = 1. \tag{4.8}$$

The nth level quantum action $W^{(n)}(\Gamma^{(n)};\hbar;\hbar^{(n)})$ is defined as satisfying the quantum master equation

$$\Delta^{(n)} \exp\left\{\frac{i}{\hbar^{(n)}} W^{(n)}(\Gamma^{(n)};\hbar;\hbar^{(n)})\right\} = 0. \tag{4.9}$$

The action $W^{(n)}$ possesses the quantum numbers

$$\varepsilon(W^{(n)}(\Gamma^{(n)};\hbar;\hbar^{(n)})) = 0, \qquad \mathrm{Pl}^{(n)}(W^{(n)}(\Gamma^{(n)};\hbar;\hbar^{(n)})) = 1, \tag{4.10}$$

and has the following series expansion in powers of $\hbar^{(n)}$, $\pi^{(n-1)}$, $\pi^{*(n-1)}$:

$$\begin{aligned} W^{(n)}(\Gamma^{(n)};\hbar;\hbar^{(n)}) &= \Omega^{(n)}(\Gamma^{(n)};\hbar) + i\hbar^{(n)}\Xi^{(n)}(\Gamma^{(n)};\hbar) \\ &\quad + (i\hbar^{(n)})^2\widetilde{\Omega}^{(n)}(\Gamma^{(n)};\hbar) + \dots, \end{aligned} \tag{4.11}$$

$$\begin{aligned} \Omega^{(n)}(\Gamma^{(n)};\hbar) &= G^{(n-1)}_a(\Gamma^{(n-1)};\hbar)\,\pi^{(n-1)a} \\ &\quad + (1/2)\,\pi^{*(n-1)}_c U^{(n-1)c}_{ab}(\Gamma^{(n-1)};\hbar) \\ &\qquad \times \pi^{(n-1)b}\pi^{(n-1)a}(-1)^{(\varepsilon_a+n)} + \dots, \end{aligned} \tag{4.12}$$

$$\begin{aligned} \Xi^{(n)}(\Gamma^{(n)};\hbar) &= H^{(n-1)}(\Gamma^{(n-1)};\hbar) \\ &\quad + \pi^{*(n-1)}_a V^{(n-1)a}_b \\ &\qquad \times (\Gamma^{(n-1)};\hbar)\,\pi^{(n-1)b} + \dots, \end{aligned} \tag{4.13}$$

$$\widetilde{\Omega}^{(n)}(\Gamma^{(n)};\hbar) = \pi^{*(n-1)}_a \widetilde{G}^{(n-1)a}(\Gamma^{(n-1)};\hbar) + \dots. \tag{4.14}$$

The nth level functional integral is defined as follows:

$$\begin{aligned} Z^{(n)} = \int \exp\Big\{\frac{i}{\hbar}\,[W^{(n)}(\Gamma^{(n)};\hbar;\hbar) \\ + G^{(n)}_a(\Gamma^{(n)};\hbar)\,\pi^{(n)a}] - H^{(n)}(\Gamma^{(n)};\hbar)\Big\} d\mu^{(n)}, \end{aligned} \tag{4.15}$$

where the measure $d\mu^{(n)}$ is defined recursively:

$$d\mu^{(n)} = d\mu^{(n-1)} d\pi^{*(n-1)} d\pi^{(n)}, \qquad n \geqslant 2, \tag{4.16}$$

$$d\mu^{(1)} \equiv d\mu, \tag{4.17}$$

the final functions $G_a^{(n)}(\Gamma^{(n)};\hbar)$ and $H^{(n)}(\Gamma^{(n)};\hbar)$ satisfy the system of equations

$$(G_a^{(n)}, G_b^{(n)}) = G_c^{(n)} U_{ab}^{(n)c}, \tag{4.18}$$

$$(H^{(n)}, G_a^{(n)}) = \Delta^{(n)} G_a^{(n)} + U_{ba}^{(n)b}(-1)^{(\varepsilon_b+n)} - G_b^{(n)} V_a^{(n)b}, \tag{4.19}$$

$$\Delta^{(n)} H^{(n)} - (H^{(n)}, H^{(n)})/2 + V_a^{(n)a} = G_a^{(n)} \widetilde{G}^{(n)a}. \tag{4.20}$$

Besides, a natural boundary condition to be imposed on $H^{(n)}$ is considered in §6.

The following remark is relevant here. Being the nth level theory under consideration, the final functions $G_a^{(n)}$ and $H^{(n)}$ are subordinated to the relations (4.18)–(4.20), while the preceding functions $G_a^{(k)}$, $H^{(k)}$, $\widetilde{G}^{(k)a}$, $U_{bc}^{(k)a}$, $V_b^{(k)a}$, $1 \leqslant k \leqslant n-1$, are to be found by solving the equations for $W^{(k)}$, $2 \leqslant k \leqslant n-1$. Furthermore, all the functions $H^{(k)}$, $1 \leqslant k \leqslant n-1$, are restricted by the boundary conditions analogous to the one imposed on the function $H^{(n)}$.

It will be shown below that the functional integral $Z^{(n)}$ does not depend on the choice of $G_a^{(n)}$, and in a special hypergauge coincides with the $(n-1)$th level functional integral $Z^{(n-1)}$, where the functions $G_a^{(n-1)}$ fix the final hypergauge in this integral.

§5. Gauge invariance of the nth level formalism

In this section we show that the functional integral $Z^{(n)}$ is $G^{(n)}$-independent and equivalent to $Z^{(n-1)}$.

The integrand of (4.15) is invariant under the transformations

$$\delta\Gamma^{(n)A_{(n)}} = (\Gamma^{(n)A_{(n)}}, -W^{(n)} + G_a^{(n)}\pi^{(n)a} + i\hbar H^{(n)})\mu^{(n)}, \tag{5.1}$$

$$\begin{aligned}\delta\pi^{(n)a} = {} & \big[U_{bc}^{(n)a}\pi^{(n)c}\pi^{(n)b}(-1)^{(\varepsilon_b+n)} \\ & + 2i\hbar V_b^{(n)a}\pi^{(n)b} + 2(i\hbar)^2\widetilde{G}^{(n)a}\big]\mu^{(n)},\end{aligned} \tag{5.2}$$

where $\mu^{(n)} = \text{const}$, $\varepsilon(\mu^{(n)}) = 1$.

Choosing an arbitrary function for the Fermionic parameter $\mu^{(n)}$

$$\mu^{(n)} = (i/2\hbar)\,\delta X^{(n)}, \tag{5.3}$$

and performing the additional variations

$$\delta\Gamma^{(n)A_{(n)}} = (\Gamma^{(n)A_{(n)}}, \delta X^{(n)})/2, \qquad \delta\pi^{(n)a} = \delta\Lambda_b^{(n)a}\pi^{(n)b}, \tag{5.4}$$

where $\delta\Lambda_b^{(n)a}(\Gamma)$ are arbitrary functions too, we induce the following variations in the integrand of (4.15):

$$\delta G_a^{(n)} = (G_a^{(n)}, \delta X^{(n)}) + G_b^{(n)}\delta\Lambda_a^{(n)b}, \tag{5.5}$$

$$\delta H^{(n)} = -\Delta^{(n)}\delta X^{(n)} + (H^{(n)}, \delta X^{(n)}) + \delta\Lambda_a^{(n)a}(-1)^{(\varepsilon_a+n)}. \tag{5.6}$$

The equations (4.18)–(4.20) for $G_a^{(n)}$ and $H^{(n)}$ retain their form under the variations (5.5), (5.6) by inducing the following transformations for the structure coefficients:

$$\begin{aligned}\delta U_{bc}^{(n)a} &= (U_{bc}^{(n)a}, \delta X^{(n)}) + [(\delta\Lambda_b^{(n)a}, G_c^{(n)}) + U_{bd}^{(n)a}\delta\Lambda_c^{(n)d}] \\ &\quad - [(\delta\Lambda_c^{(n)a}, G_b^{(n)}) + U_{cd}^{(n)a}\delta\Lambda_b^{(n)d}](-1)^{(\varepsilon_b+n)(\varepsilon_c+n)} \\ &\quad - \delta\Lambda_d^{(n)a}U_{bc}^{(n)d}, \end{aligned} \tag{5.7}$$

$$\begin{aligned}\delta V_b^{(n)a} &= (V_b^{(n)a}, \delta X^{(n)}) + V_d^{(n)a}\delta\Lambda_b^{(n)d} - \delta\Lambda_d^{(n)a}V_b^{(n)d} \\ &\quad - [\Delta^{(n)}\delta\Lambda_b^{(n)a} - (H^{(n)}, \delta\Lambda_b^{(n)a})](-1)^{(\varepsilon_a+n)}, \end{aligned} \tag{5.8}$$

$$\delta\widetilde{G}^{(n)a} = (\widetilde{G}^{(n)a}, \delta X^{(n)}) - \delta\Lambda_b^{(n)a}\widetilde{G}^{(n)b}. \tag{5.9}$$

Besides, it will be shown in the next section that $Z^{(n)}$ does not depend on the arbitrariness of the solution to equations (4.18)–(4.20) for $G_a^{(n)}$ and $H^{(n)}$. Thus we conclude that $Z^{(n)}$ does not depend on the choice of $G_a^{(n)}$.

Let us suppose that the functions $G_a^{(n)}$ are solvable with respect to the antifields $\pi_a^{*(n)}$:

$$\operatorname{Ber}\left(\frac{\partial_l G_a^{(n)}}{\partial \pi_b^{*(n-1)}}\right)\bigg|_{G_a^{(n)}=0} \neq 0. \tag{5.10}$$

Now choose the simplest hypergauge

$$G_a^{(n)} = \pi_a^{*(n)} \tag{5.11}$$

of class (5.10). It will be shown in the next section that the representation

$$H^{(n)} = -(i/\hbar)\,W^{(n-1)}(\Gamma^{(n-1)};\hbar;\hbar) + O(\pi^*) \tag{5.12}$$

holds, provided the hypergauge (5.11) and the appropriate boundary condition for $H^{(n)}$ chosen. Thus $Z^{(n)}$ reduces to the form

$$Z^{(n)} = \int \exp\left\{\frac{i}{\hbar}\,[W^{(n-1)} + G_a^{(n-1)}\pi^{(n-1)a}] - H^{(n-1)}\right\}d\mu^{(n-1)}. \tag{5.13}$$

To identify this representation with the $(n-1)$th level functional integral $Z^{(n-1)}$, it is sufficient to show that the functions $G_a^{(n-1)}$, $H^{(n-1)}$ satisfy relations of the form (4.18)–(4.20). Substituting the expansion (4.11) for $W^{(n)}$ into the quantum master equation (4.9), we find the following equations for the functions $\Omega^{(n)}$, $\Xi^{(n)}$, $\widetilde{\Omega}^{(n)}$, $n \geqslant 2$:

$$(\Omega^{(n)}, \Omega^{(n)}) = 0, \tag{5.14}$$

$$(\Omega^{(n)}, \Xi^{(n)}) = \Delta^{(n)}\Omega^{(n)}, \tag{5.15}$$

$$(\Omega^{(n)}, \widetilde{\Omega}^{(n)}) = \Delta^{(n)}\Xi^{(n)} - (\Xi^{(n)}, \Xi^{(n)})/2. \tag{5.16}$$

For the lowest orders in $\pi^{(n-1)}$, $\pi^{*(n-1)}$ these equations give

$$(G_a^{(n-1)}, G_b^{(n-1)}) = G_c^{(n-1)} U_{ab}^{(n-1)c}, \tag{5.17}$$

$$(H^{(n-1)}, G_a^{(n-1)}) = \Delta^{(n-1)} G_a^{(n-1)} + U_{ba}^{(n-1)b}(-1)^{(\varepsilon_b+n-1)} - G_b^{(n-1)} V_a^{(n-1)b}, \tag{5.18}$$

$$\Delta^{(n-1)} H^{(n-1)} - (H^{(n-1)}, H^{(n-1)})/2 + V_a^{(n-1)a} = G_a^{(n-1)} \widetilde{G}^{(n-1)a}. \tag{5.19}$$

Thus the quantum master equation automatically yields the relations required to use $G_a^{(n-1)}$ and $H^{(n-1)}$, as well as the lower level function $G_a^{(k)}$ and $H^{(k)}$, as the hypergauge-fixing functions.

Thus we obtain

$$Z^{(n)} = Z^{(n-1)}, \qquad n \geqslant 2, \tag{5.20}$$

$$Z^{(n)} = Z^{(1)} \equiv Z, \tag{5.21}$$

where we identify

$$W^{(1)}(\Gamma^{(1)}; \hbar; \hbar) \equiv W(\Gamma; \hbar), \quad G_a^{(1)} \equiv G_a, \quad H^{(1)} \equiv H. \tag{5.22}$$

§6. Structure of the hypergauge conditions

In this section we consider in detail the structure of the hypergauge functions that follows from relations (4.18)–(4.20), including the transformation properties, boundary conditions, and equivalence between the simplest hypergauge $G_a^{(n)} = \pi_a^{*(n-1)}$ and arbitrary ones.

In what follows we study nth level hypertheory, $n = 2, 3, \ldots$. The label n will be omitted. For instance, W, G, H, π, π^*, Γ, Δ denote $W^{(n-1)}$, $G^{(n)}$, $H^{(n)}$, $\pi^{(n-1)}$, $\pi^{*(n-1)}$, $\Gamma^{(n-1)}$, $\Delta^{(n)}$, respectively.

Since the integrand of Z contains the final hypergauges G_a inside the δ-function only, we are only interested in their properties on the hypergauge surface

$$G_a = 0. \tag{6.1}$$

Suppose these equations possess the solution

$$g_a \equiv \pi_a^* - f_a(\Gamma, \pi) = 0. \tag{6.2}$$

Let us expand G_a and H in g-power series:

$$G_a(\Gamma, \pi, \pi^*) = g_b \lambda_a^b(\Gamma, \pi) + g_c g_b \lambda_a^{bc}(\Gamma, \pi) + \ldots, \tag{6.3}$$

$$H(\Gamma, \pi, \pi^*) = h(\Gamma, \pi) + g_a h^a(\Gamma, \pi) + \ldots. \tag{6.4}$$

It can be checked easily that equations (4.18)–(4.20) are solvable exactly, since each of them is satisfied in the lowest order in g_a. The corresponding equations are

$$(g_a, g_b) = 0, \tag{6.5}$$

$$(\sigma, g_a) = \Delta g_a, \tag{6.6}$$

$$\Delta\sigma - (\sigma, \sigma)/2 = 0, \tag{6.7}$$

$$\sigma \equiv h + \ln \mathrm{Ber}(\lambda). \tag{6.8}$$

These equations can be solved explicitly. It follows from (6.5) that the general solution for g_a is of the form

$$g_a = \exp(\operatorname{ad}\Psi)\,\pi^*_a, \tag{6.9}$$

$$\Psi \equiv \Psi(\Gamma,\pi), \qquad \operatorname{ad}\Psi G \equiv (\Psi, G). \tag{6.10}$$

Without loss of generality one can suppose that

$$\Psi(\Gamma, 0) = 0, \tag{6.11}$$

where $\Psi(\Gamma,\pi)$ is an arbitrary Fermionic function in other respects. The equation (6.6) gives

$$\sigma = E(\operatorname{ad}\Psi)\,\Delta\Psi + \exp(\operatorname{ad}\Psi)\,\sigma_0, \tag{6.12}$$

where $E(x) \equiv x^{-1}(\exp(x) - 1)$, and $\sigma_0 \equiv \sigma_0(\Gamma)$ is an arbitrary Bosonic function. Finally, it follows from (6.7) that σ_0 satisfies the quantum master equation

$$\Delta\sigma_0 - (\sigma_0, \sigma_0)/2 = 0. \tag{6.13}$$

It is natural to identify σ_0 and $-iW/\hbar$. This can be formulated by imposing the boundary condition

$$\{H(\Gamma,\pi,\pi^*) + \ln \operatorname{Ber}[(\pi, G(\Gamma,\pi,\pi^*))]\}|_{G=0,\pi=0} = -(i/\hbar)\,W(\Gamma;\hbar;\hbar) \tag{6.14}$$

on the required solution to equations (4.18)–(4.20). This boundary condition is invariant with respect to the gauge transformations (5.5), (5.6), provided the functions δX satisfy the condition

$$\delta X(\Gamma,\pi,\pi^*)|_{G=0,\pi=0} = 0. \tag{6.15}$$

This condition still retains the possibility of varying Ψ arbitrarily; this arbitrariness follows, in its turn, from the formula for the gauge variation $\delta\Psi$,

$$\delta\Psi = -E^{-1}(\operatorname{ad}\Psi)\,\delta X_0, \tag{6.16}$$

$$\delta X_0 \equiv \delta X|_{G_a=0}. \tag{6.17}$$

The expression (6.16) shows that the correspondence between $\delta\Psi$ and δX is one-to-one. This means that the function Ψ can be gauged-out to take the value zero by performing a finite gauge transformation.

It follows from the form of the gauge variations δG_a and δH that the corresponding variation $\delta\sigma$ is of the form

$$\delta\sigma = (\sigma, \delta X_0) - \Delta\delta X_0. \tag{6.18}$$

This expression shows that the boundary condition (6.14) is invariant with respect to the gauge transformations.

Let us perform a finite gauge transformation that results in the functions G_a taking the form

$$G_a = \pi^*_a. \tag{6.19}$$

Then for the function H we have

$$H(\Gamma,\pi,\pi^*) = -(i/\hbar)\,W(\Gamma) + O(\pi^*). \tag{6.20}$$

Thus we have established the existence of the gauge transformation that was used in §5 to carry out the reduction $Z^{(n)} \to Z^{(n-1)}$.

It also follows from the expression (6.12) that the function $h(\Gamma, \pi)$ is determined uniquely, provided G_a is given and the boundary condition is fixed. All of the arbitrariness of the solution to equations (4.18)–(4.20) disappears on the hypergauge surface $G_a = 0$, so that the functional integral does not depend on this arbitrariness.

Appendix. Local deformation triviality theorem

In this appendix we describe local quantum deformations of the antisymplectic differential operator. In fact, we study locally the Fermionic differential operators of the form

(A.1) $$\Delta = \sum_{n=2}^{\infty} \hbar^{n-2} \Delta_n, \qquad \varepsilon(\Delta) = 1,$$

where

(A.2) $$\begin{aligned}\Delta_2 &= (1/2)(-1)^{\varepsilon_A} M^{-1} \partial_A M E^{AB} \partial_B \\ &= (1/2)(-1)^{\varepsilon_A} E^{AB} \partial_B \partial_A + (1/2)(-1)^{\varepsilon_A} M^{-1} (\partial_A M E^{AB}) \partial_B,\end{aligned}$$

(A.3) $$\Delta_n = \sum_{m=0}^{n} \Delta_{n|m}^{A_m \cdots A_1}(\Gamma) \partial_{A_1} \cdots \partial_{A_m}, \qquad n \geqslant 3.$$

We require that the operator (A.1) satisfy the nilpotency equation

(A.4) $$\Delta^2 = 0$$

formally in all orders in $\hbar$.

In order zero, equation (A.4) is certainly satisfied because the operator (A.2) coincides with the one in (2.11) and hence is nilpotent:

(A.5) $$\Delta_2^2 = 0.$$

It follows from the cyclic relations (2.21) that locally one can always choose the so-called Darboux coordinates, for which the matrix E^{AB} is constant:

(A.6) $$E^{AB}(\Gamma) = \bar{E}^{AB} = \text{const}.$$

It is assumed in what follows that the phase variables Γ^A are chosen so as to coincide with the Darboux ones, and hence the operator Δ_2 has the form

(A.7) $$\Delta_2 = (1/2)(-1)^{\varepsilon_A} M^{-1} \partial_A \bar{E}^{AB} M \partial_B.$$

In turn, the nilpotency equation (A.5) implies

(A.8) $$\bar{\Delta}_2 \sqrt{M} = 0,$$

where

(A.9) $$\bar{\Delta}_2 = (1/2)(-1)^{\varepsilon_A} \bar{E}^{AB} \partial_B \partial_A, \qquad \bar{\Delta}_2^2 = 0.$$

The following general remark is relevant here. Let us suppose that we have constructed a nilpotent operator Δ. Then the operators

(A.10) $$\Delta' = \exp(-U)\Delta\exp(U) + \delta, \qquad \delta = \text{const},$$

are also nilpotent, where the Bosonic operator U and Fermionic constant δ are both arbitrary.

Specifically, choosing $U = -\ln\sqrt{M}$, $\delta = 0$ in the transformation law (A.10) applied to the operator (A.1), we ensure that the corresponding operator Δ_2' coincides with $\bar{\Delta}_2$. Thus we conclude that it is sufficient for our purposes to solve the nilpotency equation (A.4) with the operator Δ_2 in the form

(A.11) $$\Delta_2 = \bar{\Delta}_2 = (1/2)(-1)^{\varepsilon_A}\bar{E}^{AB}\partial_B\partial_A, \qquad \bar{E}^{AB} = \text{const}.$$

Let us consider equation (A.4) in the first order in $\hbar$. We have

(A.12) $$[\bar{\Delta}_2, \Delta_3] = 0.$$

In this equation and later we use the standard supercommutator, that is

(A.13) $$[F, G] \equiv FG - GF(-1)^{\varepsilon(F)\varepsilon(G)}.$$

The following lemma will be proved at the end of this appendix.

LEMMA. *Let X_n be a local differential operator of order n that satisfies the equation*

(A.14) $$[\bar{\Delta}_2, X_n] = 0.$$

Then the following representation holds locally for the operator X_n:

(A.15) $$X_n = [\bar{\Delta}_2, Y_{n-1}] + y, \qquad y = \text{const},$$

(A.16) $$\varepsilon(Y_{n-1}) = \varepsilon(X_n) + 1, \qquad \varepsilon(y) = \varepsilon(X_n),$$

where Y_{n-1} is a local differential operator of order $n-1$.

As applied to equation (A.12), the lemma stated above means that

(A.17) $$\Delta_3 = [\bar{\Delta}_2, U_2] + \delta_3, \qquad \delta_3 = \text{const},$$

(A.18) $$\varepsilon(U_2) = 0, \qquad \varepsilon(\delta_3) = 1,$$

where U_2 is a local second-order differential operator.

Let us consider the operator

(A.19) $$\Delta^{(2)} = \exp(\hbar U_2)\Delta\exp(-\hbar U_2) - \hbar\delta_3$$

that must satisfy the nilpotency equation (A.4) and have a power series expansion in $\hbar$ of the form

(A.20) $$\Delta^{(2)} = \bar{\Delta}_2 + \hbar^2\Delta_4^{(2)} + O(\hbar^3),$$

where $\Delta_4^{(2)}$ is a local fourth-order differential operator.

Due to the nilpotency property of the operator $\Delta^{(2)}$, the following equation holds for the operator $\Delta_4^{(2)}$:

(A.21) $$[\bar{\Delta}_2, \Delta_4^{(2)}] = 0,$$

so that the lemma stated above yields:

$$\Delta_4^{(2)} = [\bar{\Delta}_2, U_3] + \delta_4, \qquad \delta_4 = \text{const}, \tag{A.22}$$

$$\varepsilon(U_3) = 0, \qquad \varepsilon(\delta_4) = 1, \tag{A.23}$$

where U_3 is a local third-order differential operator.

Let us introduce the operator

$$\Delta^{(3)} = \exp(\hbar^2 U_3)\Delta^{(2)}\exp(-\hbar^2 U_3) - \hbar^2\delta_4, \tag{A.24}$$

which satisfies the nilpotency equation and has a power series expansion in $\hbar$ of the form

$$\Delta^{(3)} = \bar{\Delta}_2 + \hbar^3\Delta_5^{(3)} + O(\hbar^4), \tag{A.25}$$

so that

$$[\bar{\Delta}_2, \Delta_5^{(3)}] = 0. \tag{A.26}$$

Then we apply the lemma again, and so on.

Finally, using induction, we conclude that the general solution to the nilpotency equation in the Darboux variables has the form

$$\Delta = \exp(-U)\bar{\Delta}_2\exp(U) + \delta, \qquad \delta = \text{const}, \tag{A.27}$$

$$\varepsilon(U) = 0, \qquad \varepsilon(\delta) = 1, \tag{A.28}$$

where U is an arbitrary local differential operator.

It should be noted here that the transformation connecting the arbitrary variables and the Darboux variables is also representable in the form (A.10). Thus the representation (A.27) can be reformulated in arbitrary variables as follows:

$$\Delta = \exp(-U)\Delta_2\exp(U) + \delta, \qquad \delta = \text{const}, \tag{A.29}$$

with Δ_2 given by the formula (A.2).

If the theory under consideration does not include external Fermions, then the Fermionic constant δ is equal to zero.

Now let us prove the lemma formulated above. First, introduce the operator d, that is

$$d = [\bar{\Delta}_2, \dots]. \tag{A.30}$$

The nilpotency property of the operator $\bar{\Delta}_2$ is inherited by d:

$$d^2 = 0 \tag{A.31}$$

because of the Jacobi identity for supercommutators. In terms of the operator d, equation (A.14) takes the form

$$dX_n = 0. \tag{A.32}$$

Let us represent the result of applying the operator d to X_n in a more explicit form

$$dX_n = \sum_{m=0}^{n}[(\Omega X_{n|m})^{A_{m+1}\cdots A_1}\partial_{A_1}\cdots\partial_{A_{m+1}} + (\bar{\Delta}_2 X_{n|m})^{A_m\cdots A_1}\partial_{A_1}\cdots\partial_{A_m}], \tag{A.33}$$

where

(A.34) $$X_n = \sum_{m=0}^{n} X_{n|m}^{A_m \cdots A_1} \partial_{A_1} \cdots \partial_{A_m},$$

(A.35) $$\varepsilon(X_{n|m}^{A_m \cdots A_1}) = \varepsilon(X_n) + \sum_{k=1}^{m} \varepsilon_{A_k},$$

(A.36) $$(\Omega X_{n|m})^{A_{m+1} \cdots A_1} \equiv \frac{1}{m+1} \Big[(-1)^{(\varepsilon(X_n)+1)\varepsilon_{A_{m+1}}} (\Gamma^{A_{m+1}}, X_{n|m}^{A_m \cdots A_1}) + \sum_{k=1}^{m} (-1)^{(\varepsilon(X_n)+1+\sum_{j=k+1}^{m+1} \varepsilon_{A_j})\varepsilon_{A_k}} \times (\Gamma^{A_k}, X_{n|m}^{A_{m+1} \cdots A_{k+1} A_{k-1} \cdots A_1}) \Big],$$

(A.37) $$(\Gamma^A, F) \equiv \bar{E}^{AB} \partial_B F.$$

It can be checked directly that

(A.38) $$\Omega^2 = 0.$$

In equation (A.32) let us consider the contribution of the highest order $n+1$ to the corresponding operators:

(A.39) $$(\Omega X_{n|n})^{A_{n+1} \cdots A_1} = 0.$$

Multiplying this equation by

(A.40) $$(-1)^{(\varepsilon(X_n)+1)\varepsilon_{A_{n+1}}} \Gamma^A \bar{E}^{-1}_{AA_{n+1}}$$

from the left and then summing over A_{n+1}, we find

(A.41) $$(\Gamma^A \partial_A + n) X_{n|n}^{A_n \cdots A_1} = (\Omega \bar{Y}_{n-1})^{A_n \cdots A_1},$$

where

(A.42) $$\bar{Y}_{n-1}^{A_{n-1} \cdots A_1}(\Gamma) \equiv n \Gamma^A \bar{E}^{-1}_{AB} X_{n|n}^{BA_{n-1} \cdots A_1}(\Gamma)(-1)^{\varepsilon_B \varepsilon(X_n)}.$$

Solving equation (A.41) we obtain for $n > 0$,

(A.43) $$X_{n|n}^{A_n \cdots A_1} = (\Omega Y_{n-1})^{A_n \cdots A_1},$$

where

(A.44) $$Y_{n-1}^{A_{n-1} \cdots A_1}(\Gamma) \equiv \int_0^1 \alpha^{n-2} \bar{Y}_{n-1}^{A_{n-1} \cdots A_1}(\alpha\Gamma)\, d\alpha.$$

When $n = 0$, i.e., when $X_{0|0}$ is a function, equation (A.39) takes the form

(A.45) $$(\Gamma^{A_1}, X_{0|0}) = 0,$$

which gives

(A.46) $$X_{0|0} = \text{const}\,.$$

So, we have proved that the general solution to equation (A.39) for all $n \geqslant 0$ is given by formulas (A.43), (A.46). Thus all the tensor cohomologies of the operator Ω are trivial locally, except for $n = 0$.

Now let us return to equation (A.32) at $n > 0$. Due to the representation (A.43), we can write out the operator (A.34) as follows:

$$X_n = d(Y_{n-1}^{A_{n-1}\cdots A_1}\partial_{A_1}\cdots\partial_{A_{n-1}}) + \widetilde{X}_{n-1}, \tag{A.47}$$

where

$$\widetilde{X}_{n-1} \equiv \sum_{m=0}^{n-1} \widetilde{X}_{n-1|m}^{A_m\cdots A_1}\partial_{A_1}\cdots\partial_{A_m}, \tag{A.48}$$

$$\widetilde{X}_{n-1|m}^{A_m\cdots A_1} \equiv X_{n|m}^{A_m\cdots A_1} - (\bar{\Delta}_2 Y_{n-1}^{A_{n-1}\cdots A_1})\delta_{n-1}^m, \tag{A.49}$$

so that $\widetilde{X}_{n-1}$ is a local differential operator of order $n-1$.

It follows from equations (A.31), (A.32), and (A.47) that

$$d\widetilde{X}_{n-1} = 0. \tag{A.50}$$

So, we have stepped from n in (A.32) to $n-1$ in (A.50).

Repeating the above reasoning $n-1$ times, we finally arrive to the case $n = 0$. As the result, we see that formula (A.15) is the general solution to equation (A.32). Thus the lemma is proved.

Note also that one can regard the expression (A.36) as the natural generalization of the so-called Schouten bracket $[\![\ ,\ ]\!]$.

Acknowledgments. The work of I. A. Batalin is partially supported by the Russian Foundation for Basic Research under grant 96-01-00482 and by the Human Capital and Mobility Program of the European Community under the project INTAS 93-0633, 93-2058. The work of I. V. Tyutin is partially supported by the Russian Foundation for Basic Research under grant 96-02-17314 and by the Human Capital and Mobility Program of the European Community under the project INTAS 93-2068.

References

1. R. P. Feynman, *Quantum theory of gravitation*, Acta Phys. Polon. **24** (1963), no. 6, 697–722.
2. L. D. Faddeev and V. N. Popov, *Feynman diagrams for the Yang–Mills fields*, Phys. Lett. B **25** (1967), no. 1, 29–30.
3. B. S. DeWitt, *Quantum theory of gravity*, I, II, Phys. Rev. **160** (1967), no. 5, 1113–1149; **162** (1967), no. 5, 1195–1239.
4. I. A. Batalin and G. A. Vilkovisky, *Gauge algebra and quantization*, Phys. Lett. B **102** (1981), no. 1, 27–31.
5. ———, *Quantization of gauge theories with linearly dependent generators*, Phys. Rev. D **28** (1983), no. 10, 2567–2582.
6. C. B. Thorn, *Perturbation theory for quantized strings*, Nuclear Phys. B **287** (1987), no. 1, 61–92.
7. ———, *String field theory*, Phys. Rep. **175** (1989), no. 1, 2, 1–101.
8. B. Zwiebach, *Closed string field theory: quantum action and the B–V master equation*, Preprint IASSNS-HEP-92/41.
9. E. Witten, *On background independent open-string field theory*, Preprint IASSNS-HEP-92/53.

10. A. Sen and B. Zwiebach, *Quantum background independence of closed string theory*, Preprint MIT-CTP-2244 (1993).
11. H. Hata, *Theory of theories approach to string theory*, Preprint KUNS-1212 (1993).
12. M. Henneaux, *Lectures on the antifield-BRST formalism for gauge theories*, Nuclear Phys. B Proc. Suppl. **18A** (1990), 47–106.
13. J. Zinn-Justin, *Renormalization of gauge theories*, Trends in Elementary Particle Theory, Lecture Notes in Phys. (H. Rollnik and K. Dietz, eds.), vol. 37, Springer-Verlag, Berlin, 1975, pp. 2–39.
14. R. Kallosh, *On gauge invariance in supergravity*, Pis′ma Zh. Èksper. Teoret. Fiz. **26** (1977), no. 7, 575–578; English transl. in JETP Lett. **26 yr 1977**.
15. B. de Wit and J. van Holten, *Covariant quantization of gauge theories with open gauge algebra*, Phys. Lett. B **79** (1979), no. 4, 5, 389–393.
16. E. Witten, *A note on the antibracket formalism*, Modern Phys. Lett. A **5** (1990), no. 7, 487–494.
17. I. A. Batalin, P. M. Lavrov, and I. V. Tyutin, *Covariant quantization of gauge theories in the framework of extended BRST symmetry*, J. Math. Phys. **31** (1990), no. 6, 1487–1493.
18. ———, *An Sp(2)-covariant quantization of gauge theories with linearly dependent generators*, J. Math. Phys. **32** (1991), no. 2, 532–539.
19. ———, *Remarks on the Sp(2)-covariant Lagrangian quantization of gauge theories*, J. Math. Phys. **32** (1991), no. 9, 2513–2521.
20. M. Henneaux, *Geometric interpretation of the quantum master equation in the BRST-anti-BRST formalism*, Preprint ULB-PHIF-92/01.
21. G. V. Grigoryan, R. P. Grigoryan, and I. V. Tyutin, *Equivalence of Lagrangian and Hamiltonian BRST quantizations: the general case*, Nuclear Phys. B **379** (1992), no. 1, 2, 304–318.
22. I. A. Batalin and E. S. Fradkin, *Operatorial quantization of dynamical systems subject to constraints. A further study of the construction*, Ann. Inst. H. Poincaré Phys. Théor. **49** (1988), no. 2, 145–214.
23. A. Schwarz, *Geometry of Batalin–Vilkovisky quantization*, Comm. Math. Phys. **155** (1993), no. 2, 249–260.
24. ———, *Semiclassical approximation in the Batalin–Vilkovisky formalism*, Comm. Math. Phys. **158** (1993), no. 2, 373–396.
25. M. Penkava and A. Schwarz, *On some algebraic structures arising in string theory*, Perspectives in Mathematical Physics, Conf. Proc. Lecture Notes Math. Phys., vol. III, Internat. Press, Cambridge, MA, 1994, pp. 219–227.
26. O. M. Khudaverdyan, *Geometry of superspace with even and odd brackets*, J. Math. Phys. **32** (1991), no. 7, 1934–1937.
27. O. M. Khudaverdyan and A. P. Nersessyan, *Even and odd symplectic and Kählerian structures on projective superspaces*, Preprint JINR E2-92-411.
28. D. V. Volkov, V. A. Soroka, A. I. Pashnev, and V. I. Tkach, *On Hamiltonian systems with odd and even Poisson brackets, and on duality of their conservation laws*, Pis′ma Zh. Èksper. Teoret. Fiz. **44** (1986), no. 1, 55–57; English transl. in JETP Lett. **44** (1986).
29. D. V. Volkov and V. A. Soroka, *On quantization of dynamical systems with odd Poisson bracket*, Yadernaya Fiz. **46** (1987), no. 1, 110–121; English transl. in Soviet J. Nuclear Phys. **46** (1987).
30. I. A. Batalin and I. V. Tyutin, *On possible generalizations of the field–antifield formalism*, Internat. J. Modern Phys. A **8** (1993), no. 13, 2333–2350.
31. ———, *On the multilevel generalization of field–antifield formalism*, Modern Phys. Lett. A **8** (1993), no. 38, 3673–3681.
32. ———, *On local quantum deformations of antisymplectic differential*, Internat. J. Modern Phys. A **9** (1994), no. 4, 517–525.
33. ———, *On the multilevel field–antifield formalism with the most general Lagrangian hypergauges*, Preprint FIAN/TD/4-94, hep-th/9403180.

LEBEDEV PHYSICAL INSTITUTE, 117924 MOSCOW RUSSIA

Translated by THE AUTHORS

Amer. Math. Soc. Transl.
(2) Vol. **177**, 1996

Semiclassical Integral Equations on the Semiaxis

A. M. Budylin and V. S. Buslaev

ABSTRACT. The asymptotic behavior of solutions on the semiaxis of integral equations of the form

$$\frac{1}{\varepsilon}\int_0^\infty A\Big(\frac{x-y}{\varepsilon};x,y\Big)f(y)\,dy=g(x),\qquad x>0,$$

is described as $\varepsilon\searrow 0$. It is assumed that the symbol

$$a(\xi;x,y)=\int_{-\infty}^{+\infty}e^{-i\xi z}\mathcal{A}(z;x,y)\,dz$$

has a jump with respect to ξ at $\xi=0$. The procedure proposed here can be regarded as an asymptotic generalization of the Wiener–Hopf method to the case in which the symbol $a(\xi;x,y)$ depends nontrivially on x and y.

1. The object of our investigation is the pseudodifferential operator $A[\varepsilon]$ on the semiaxis $(x,y>0)$ with the kernel

$$\frac{1}{\varepsilon}A\Big(\frac{x-y}{\varepsilon};x,y\Big)=\frac{1}{2\pi\varepsilon}\int_{\infty}^{+\infty}e^{(i/\varepsilon)\xi(x-y)}a(\xi;x,y)\,d\xi,$$

where ε is a small positive parameter. The symbol (amplitude) $a=a(\xi;x,y)$ is assumed to be a bounded function $\mathbb{R}\to\mathbb{C}$ smooth with respect to the arguments except for discontinuity in ξ at $\xi=0$. It is also assumed that the function $a=a(\xi;x,y)$ never vanishes and has nonzero limiting values $a=a(\pm\infty;x,y)$ as $\xi\to\pm\infty$ (here the integral is understood in the sense of distributions).

The aim of this article is to study the asymptotics of the operator $(A[\varepsilon])^{-1}$ as $\varepsilon\to 0$. If the symbol a does not depend on x and y, the problem can be solved exactly by the Wiener–Hopf method. Our investigation may be regarded as an asymptotic generalization of this method to the case where such a dependence occurs.

1991 *Mathematics Subject Classification.* Primary 47G10; Secondary 47N20.

This research was partially supported by the Russian Foundation for Basic Research under grant No. 93-011-1697.

The distinctive feature of the problem under investigation is the presence of discontinuities of the operator symbol with respect to both reciprocal variables (at $\xi = 0$ and $x, y = 0$). The smooth symbol a defines a short-range operator (for small ε) whose kernel decreases rapidly as $x - y \to \infty$. The inverse operator can be asymptotically studied as $\varepsilon \to 0$ by using the localization concept. Conversely, in the case under study, the opertator is a long-range operator in both representations.

This investigation is part of a larger work concerned with the calculation of the asymptotics of the solution to the Korteweg–de Vries equation for large times. This work consists of several parts. In our view, the part presented here is of interest by itself.

In a number of applications (see references in [**1**]), one needs the asymptotical inversion of integral operators of the form

$$(\tilde{A}[\varepsilon]f)(x) = \frac{1}{\varepsilon}\int_{-1}^{1} A\Big(\frac{x-y}{\varepsilon}; x, y\Big) f(y)\, dy, \qquad x \in (-1, 1)\,.$$

For such operators (as well as for operators of the type $A[\varepsilon]$) additive perturbation theory cannot be used, since, formally, the limiting operator $\tilde{A}[0]$ is an asymptotically unsatisfactory approximation to the operator $\tilde{A}[\varepsilon]$ as $\varepsilon \to 0$. The case $a(\xi; x, y) = a(\xi)$ was studied in [**1–3**]. An exact solution was found for the problem on the semiaxis. In these papers a general approach was developed, which made it possible to construct the operator $(\tilde{A}[\varepsilon])^{-1}$ and its asymptotics for small ε via $(\tilde{A}[0])^{-1}$ and similar objects used for solving problems on the semiaxes $(-\infty, 1)$ and $(-1, +\infty)$.

This article describes necessary generalizations to the case of the semiaxis in which the symbol $a(\xi; x, y)$ does not depend trivially on x and y. Now the asymptotics of the operator inverse to the operator $\tilde{A}[\varepsilon]$ for $\varepsilon \to 0$ can be calculated by using the scheme described in [**1–3**]. Applications to the Korteweg–de Vries equation will be discussed in another article.

2. First, let us make a few general remarks. Suppose that $\mathcal{X}$ is a complex vector space, $\mathcal{X} = X + X'$ is its direct decomposition, and $I = P + P'$ are the corresponding projectors $X = P\mathcal{X}$ and $X' = P'\mathcal{X}$. Further, let $\mathcal{A}$ be the bijection of $\mathcal{X}$ and $A = P\mathcal{A}|_X$ be its restriction to the subspace X.

Similarly to $\mathcal{X}$ and $\mathcal{A}$, we introduce a space $\mathcal{Y}$, its direct decomposition $\mathcal{Y} = Y + Y'$, the respective projectors $I = Q + Q'$, the bijection $\mathcal{B}\colon \mathcal{Y} \to \mathcal{Y}$, and its restriction to Y, $B = Q\mathcal{B}|_Y$, where B is assumed to be the bijection of Y.

Finally, suppose that the operator $\mathcal{A}$ admits the factorization $\mathcal{A} = \mathcal{Z}\mathcal{B}R$, where $\mathcal{Z}\colon \mathcal{Y} \to \mathcal{X}$ and $R\colon \mathcal{X} \to \mathcal{Y}$ are bijective mappings having the triangle property:

$$P\mathcal{Z}Q' = 0, \quad Q\mathcal{Z}^{-1}P' = 0, \quad P'R^{-1}Q = 0, \quad Q'RP = 0\,.$$

Then the operator A is the bijection of X, and the inverse operator can be represented as $A^{-1} = PR^{-1}B^{-1}Q\mathcal{Z}^{-1}|_X$.

The scheme descibed above is an abstract modification of the Wiener–Hopf method. Its application to the solution of asymptotic problems is described in detail in [**1**]. This scheme will be studied here from a more general point of view.

Suppose there exist triangular bijective mappings $\mathcal{Z}_1 : \mathcal{X} \to \mathcal{Y}$ and $R_1 : \mathcal{Y} \to \mathcal{X}$,

$$Q\mathcal{Z}_1 P' = 0, \quad P\mathcal{Z}_1^{-1}Q' = 0, \quad P'R_1 Q = 0, \quad Q'R_1^{-1}P = 0,$$

such that the operator

$$B_1 = Q\mathcal{Z}_1\mathcal{A}R_1|_Y$$

is a bijection of Y. Then the operator A is a bijection of X and

$$A^{-1} = PR_1B_1^{-1}Q\mathcal{Z}_1|_Y\,.$$

3. In the case of the operator $A[\varepsilon]$, the space $\mathcal{X}$ is formed by the class of functions $\mathbb{R} \to \mathbb{C}$. Its subspaces X and X' are formed by functions whose supports are on the semiaxes $[0,\infty)$ and $(-\infty,0]$, respectively. The space $\mathcal{Y}$ and its subspaces differ from $\mathcal{X}$ and its subspaces by the choice of the norm; the norm depends on the form of the symbol. The operator $\mathcal{A}$ is defined by the kernel $(1/\varepsilon)\mathcal{A}((x-y)/\varepsilon;x,y)$ on the entire axis, and the operator A coincides with $A[\varepsilon]$.

Now let us turn to a more detailed description of admissible symbols in the case $\mathcal{X} = \mathcal{Y} = L_2(\mathbb{R})$.

It is assumed that the symbol a can be expressed in the form

$$a(\xi;x,y) = k(\xi;x,y)\cdot b(\operatorname{sgn}\xi;x,y),$$

where b is a bounded smooth function bounded together with all derivatives, that is the symbol of an operator B (with bounded inverse) on the semiaxis:

$$Bf(x) = \frac{1}{2\pi}\int_{-\infty}^{+\infty} d\xi \int_0^\infty dy\, e^{i\xi(x-y)} b(\operatorname{sgn}\xi;x,y)f(y), \qquad x>0\,.$$

The following conditions are imposed on the function k:

1. $k(\xi;x,y)$ is a continuous function $\mathbb{R}^3 \to \mathbb{C}$ twice continuously differentiable in the half-spaces $[-\infty,0]\times\mathbb{R}^2$ and $[0,+\infty]\times\mathbb{R}^2$;
2. $0 < C_1 \leqslant |k(\xi;x,y)| \leqslant C_2$, $k(0;x,y)=1$;
3. $|k(\xi;x,y)-k(\pm\infty;x,y)| \leqslant C(1+|\xi|)^{-\gamma}$, $\gamma > 1/2$, $\|k^{(j)}(\xi;x,y)\| \leqslant C(1+|\xi|)^{-\gamma}$, $j=1,2$, $\xi \gtrless 0$:
4. $\operatorname{Re}\lambda = 0$, where $\lambda = (-1/2\pi i)\ln k(+\infty;x,x)/k(-\infty;x,x)$; moreover, for all x the index of the function $\xi \mapsto k(\xi;x,x)((\xi-i)/(\xi+i))^\lambda$, which is continuous on the closed axis $\overline{\mathbb{R}}$, vanishes.

These conditions ensure, in particular, the following factorization of the function $k(\xi;x,x)$:

$$k(\xi;x,x) = l(\xi,x)\cdot r(\xi,x),$$

where l (r) admits an analytic continuation with respect to ξ to the upper (lower) complex half-plane. In what follows, it is essential that the functions $1/l$ and $1/r$ are continuous and bounded. As for smoothness, these functions are twice differentiable everywhere, except that the partial derivatives with respect to ξ have a logarithmic singularity at $\xi = 0$. At infinity, for $\xi\to\infty$, the following asymptotic estimates hold:

$$\|(1/l)^{(j)}\| \leqslant C(1+|\xi|)^{-\gamma'}, \quad \|(1/r)^{(j)}\| \leqslant C(1+|\xi|)^{-\gamma'}, \qquad j=1,2,\ \gamma'<\gamma\,.$$

The functions l and r are regarded as normed functions owing to the conditions

$$l(0,x)=1, \qquad r(0,x)=1.$$

Let us define the operators $\mathcal{Z}$ and R by the formulas

$$\mathcal{Z}_1 f(x) = \frac{1}{2\pi\varepsilon}\int_{-\infty}^{+\infty} d\xi \int_{-\infty}^{+\infty} dy\, e^{(i/\varepsilon)\xi(x-y)} \frac{f(y)}{l(\xi,x)},$$
$$R_1 f(x) = \frac{1}{2\pi\varepsilon}\int_{-\infty}^{+\infty} d\xi \int_{-\infty}^{+\infty} dy\, e^{(i/\varepsilon)\xi(x-y)} \frac{f(y)}{r(\xi,y)}.$$

By construction, these operators have bounded inverses in $L_2(\mathbb{R})$ and possess the appropriate triangle properties described in the preceding section.

The symbol $b_1(\xi;x,y)$ of the operator $\mathcal{Z}_1\mathcal{A}R_1$ (and of the operator B at the same time) can be readily shown to be

$$b_1(\xi;x,y) = \frac{1}{4\pi^2\varepsilon^2}\int d\eta \int d\zeta \int du \int dv\, e^{(i/\varepsilon)(u\eta - v\zeta)} \frac{a(\xi;x+v,y+u)}{l(\xi+\zeta,x)r(\xi+\eta,y)}.$$

By applying the standard (for smooth symbols) theory of pseudodifferential operators, we can show that in the L_2 norm, as $\varepsilon \to \infty$, we have

$$\|B_1 - B_2\| = O(\varepsilon \ln \varepsilon),$$

where B_2 is a pseudodifferential operator on the semiaxis $(x, y > 0)$ with the symbol

$$b_2(\xi;x,y) = \frac{k(\xi;x,y)}{l(\xi,x)r(\xi,y)} \cdot b(\operatorname{sgn}\xi;x,y)$$

equal to the product of the symbols of $\mathcal{Z}_1$, $\mathcal{A}$, and R_1. Furthermore, as $\varepsilon \to 0$, we have

$$\|B_2 - B\| = o(1).$$

For a heuristic argument we can use the fact that for $|x-y| > \sqrt{\varepsilon}$ the main contribution to the asymptotics of the operator B is due to the point $\xi = 0$, with $b_2(\pm 0;x,y) \sim b(\pm 0;x,y)$. At the same time, for $|x-y| < \sqrt{\varepsilon}$ the main contribution is defined by the condition $x = y$, with $b_2(\xi;x,x) \sim b(\operatorname{sgn}\xi;x,x)$.

It follows that for small ε the operator B_1, together with B, has a bounded inverse. The operator inverse to B_1 can be expressed as a Neumann series:

$$B_1^{-1} = B^{-1} + B^{-1}(B - B_1)B^{-1} + \dots;$$

moreover, in the L_2 norm, as $\varepsilon \to 0$, we have

$$\|B_1^{-1} - B^{-1}\| = o(1).$$

4. It should be emphasized that the operator B, which is independent of ε, belongs to the class of singular integral operators. To invert such operators, we must often use various weight spaces L_2. Then the results given above can be generalized.

We must also note that condition (4) imposed on the function k can be weakened. Thus, the restriction $\operatorname{Re}\lambda = 0$ can be replaced by $|\operatorname{Re}\lambda| < \gamma$. However, we must then distinguish between the spaces $\mathcal{X}$ and $\mathcal{Y}$ by taking into account Sobolev spaces.

5. As an example important for applications, we consider the operator $A[\varepsilon] = I - Pg[\varepsilon]p\bar{g}[\varepsilon]|_X$ with the symbol

$$a(\xi; x, y) = 1 - g[x, \xi]\,p(\xi)\,\bar{g}[\xi, y],$$

where p is the characteristic function of the semiaxis $[0, \infty)$, and g and $\bar{g}$ are smooth rapidly decreasing functions. Then the symbol of B is defined by

$$b(\operatorname{sgn}\xi; x, y) = 1 - g(x, 0)\,p(\xi)\,\bar{g}(0, y)\,.$$

The operator B can be inverted explicitly by the standard Wiener–Hopf method. Namely, the symbol of B^{-1} is

$$1 - g(x, 0)\,G_+^{-1}(x)\,p(\xi)\,G_-^{-1}(y)\,\bar{g}(0, y),$$

where the functions $G_\pm$ are analytic on $\mathbb{C}_\pm = \{x \mid \operatorname{Im} x \gtrless 0\}$, respectively; they can be determined from the factorization relation

$$G_+(x) \cdot G_-(x) = 1 - g(x, 0)\,\bar{g}(0, x) \cdot p(x)\,.$$

In the case of a small jump (no change of sign) the inversion procedure for the operator $A[\varepsilon]$ as $\varepsilon \to 0$ can be justified in the norm L_2.

References

1. A. M. Budylin and V. S. Buslaev, *Reflection operators and their applications to asymptotic investigations of semiclassical integral equations*, Estimates and Asymptotics for Discrete Spectra (M. Sh. Birman, ed.), Advances in Soviet Math., vol. 7, Amer. Math. Soc., Providence, RI, 1991, pp. 107–157.
2. ______, *Semiclassical integral equations*, Dokl. Akad. Nauk SSSR **319** (1991), no. 3, 527–530; English transl. in Soviet Math. Dokl. **44** (1992).
3. ______, *Semiclassical integral equations with slowly decreasing kernels on bounded domains*, Algebra i Analiz **5** (1993), no. 1, 160–178; English transl. in St. Petersburg Math. J. **5** (1994).

Translated by N. K. KULMAN

Amer. Math. Soc. Transl.
(2) Vol. **177**, 1996

Integrability of $N = 3$ Super Yang–Mills Equations

Ch. Devchand and V. Ogievetsky*

ABSTRACT. We describe the harmonic superspace formulation of the Witten–Manin supertwistor correspondence for $N = 3$ extended super Yang–Mills theories. The essence is that on being sufficiently supersymmetrised (up to the $N = 3$ extension), the Yang–Mills equations of motion can be recast in the form of Cauchy–Riemann-like holomorphicity conditions for a pair of prepotentials in the appropriate harmonic superspace. This formulation makes the explicit construction of solutions a rather more tractable proposition than previous attempts.

§1. Introduction

Alik Berezin was enthusiastic about the possibility of solving the Yang–Mills equations and he frequently discussed this intriguing problem. Recalling these discussions, we feel that he would have enjoyed knowing about our recent work on this topic, which we shall describe here as our contribution to this volume dedicated to his memory. Today the above possibility certainly shows much promise and many existence proofs exist (e.g., [**1**]). An older promise based on the twistor transform [**2–4**] has also been renewed recently [**5**]; and this will be the subject of the present paper. The latter approach was based on the observation that one could approach non-self-dual Yang–Mills fields by combining self-dual (SD) and anti-self-dual (ASD) fields in some way. Specifically, the twistor transform for self-dual Yang–Mills [**6**] establishes a correspondence between self-dual fields and certain holomorphic fields, which effectively linearizes the (nonlinear) self-duality equations, and the observation of [**2, 3**] concerning the intermingling of SD and ASD holomorphic data extends to non-self-dual data. In general, the procedure works in a certain formal neighborhood of the self-dual solution, but Witten [**2**] observed that for the $N = 3$ supersymmetric equations the SD and ASD data actually interlock to give an exact non-self-dual solution. This observation was

1991 *Mathematics Subject Classification.* Primary 81T13.
*Viktor Isaakovich Ogievetsky died on March 23, 1996.

given a global formulation by Manin [**7**], who also discussed some solutions for rather complicated gauge groups.

Restricting ourselves entirely to local considerations, we shall show that this supertwistorial construction possesses a very elegant formulation in the language of "harmonic superspaces", which promises to be very effective for the explicit construction of local solutions. The crux of the formulation is that the $N = 3$ extended supersymmetric Yang–Mills equations can be rewritten as Cauchy–Riemann-like conditions for a pair of prepotentials in an appropriate harmonic superspace. "Holomorphic" prepotentials therefore encode local $N = 3$ super Yang–Mills solutions; leaving only the decoding as the remaining technical problem. The formulation involves a crucial modification of the harmonic (super)space formulation of (anti-)self-duality equations (see, e.g., [**8, 9**] and previous references therein). The latter involve the harmonisation of "half" of the Lorentz group and only allow one to deal with base spaces of signatures $(4, 0)$, $(2, 2)$, or with complexified space. For non-self-dual $N = 3$, however, we need to harmonize the *whole* Lorentz group, allowing us to consider a base space of any signature, including $(3, 1)$, the Minkowski space.

We should mention that harmonics were originally introduced [**10**] in order to construct the first unconstrained off-shell $N = 2$ and $N = 3$ supersymmetric gauge theories. In that case the internal $SU(2)$ and $SU(3)$ groups were harmonized instead of the Lorentz group, which is harmonized here. Harmonisation of the three-dimensional Lorentz group was discussed for $N = 6$, $d = 3$ gauge theories by Zupnik [**11**].

§2. $N = 3$ super Yang–Mills equations

The three times extended Yang–Mills multiplet contains the following fields [**12**]: the gauge vector field represented by its self- and anti-self-dual field-strengths $f_{\dot\alpha\dot\beta}$ and $f_{\alpha\beta}$, defined by

$$[\nabla_{\alpha\dot\alpha}, \nabla_{\beta\dot\beta}] \equiv \varepsilon_{\alpha\beta} f_{\dot\alpha\dot\beta} + \varepsilon_{\dot\alpha\dot\beta} f_{\alpha\beta},$$

spinor singlet and triplet fields $\{\lambda_\alpha, \lambda_{\dot\alpha}, \chi^i_{\dot\alpha}, \chi_{i\alpha}\}$ and two triplets of scalar fields W^i, W_i, where α and $\dot\alpha$ are undotted and dotted Lorentz spinor indices, while $i = 1, 2, 3$ is the $SU(3)$ index. All fields are in the adjoint representation of the gauge group and are Lie algebra valued. The dynamical equations for this supermultiplet are [**13**]

$$\begin{aligned}
\nabla^\alpha{}_{\dot\beta} f_{\beta\alpha} + \nabla_\beta{}^{\dot\alpha} f_{\dot\alpha\dot\beta} &= \{\chi^k_\beta, \chi_{k\dot\beta}\} + \{\lambda_\beta, \lambda_{\dot\beta}\} + [W^i, \nabla_{\beta\dot\beta} W_i] + [W_i, \nabla_{\beta\dot\beta} W^i], \\
\nabla^\beta{}_{\dot\beta} \lambda_\beta &= [\chi_{i\dot\beta}, W^i], \quad \nabla_\alpha{}^{\dot\beta} \lambda_{\dot\beta} = [\chi^i_\alpha, W_i], \\
\nabla^\alpha{}_{\dot\alpha} \chi^j_\alpha &= [\chi_{i\dot\beta}, W_k] \varepsilon^{ijk} - [\lambda_{\dot\beta}, W^j], \quad \nabla_\alpha{}^{\dot\alpha} \chi_{j\dot\alpha} = [\chi^i_\alpha, W^k] \varepsilon_{ijk} - [\lambda_\alpha, W_j], \\
\nabla_{\alpha\dot\alpha} \nabla^{\alpha\dot\alpha} W_j &= -2\,[[W^i, W_j], W_i] + [[W^i, W_i], W_j] \\
&\quad - \{\chi_{j\dot\alpha}, \lambda^{\dot\alpha}\} + \varepsilon_{ijk} \{\chi^i_\alpha, \chi^{k\beta}\}/2, \\
\nabla_{\alpha\dot\alpha} \nabla^{\alpha\dot\alpha} W^j &= -2\,[[W_i, W^j], W_i] + [[W_i, W^i], W_j] \\
&\quad - \{\chi^j_\alpha, \lambda^\alpha\} + \varepsilon^{ijk} \{\chi_{i\dot\alpha}, \chi^{\dot\alpha}_k\}/2\,.
\end{aligned} \tag{1}$$

To describe this theory invariantly in ordinary superspace with coordinates

$$\{x^{\alpha\dot{\alpha}}, \vartheta^{i\alpha}, \bar{\vartheta}^{\dot{\alpha}}_i\}, \tag{2}$$

one introduces gauge-covariant derivatives

$$\mathcal{D}_A \equiv \partial_A + A_A = (\nabla_{\alpha\dot{\beta}}, \mathcal{D}_{i\alpha}, \overline{\mathcal{D}}^j_{\dot{\beta}}), \qquad i, j = 1, 2, 3\,.$$

The superconnections A_A contain the above supermultiplet consistently only if the gauge covariant derivatives are constrained as follows [**14, 15**]:

$$\{\mathcal{D}_{(i\alpha}, \mathcal{D}_{j)\beta}\} = 0, \quad \{\overline{\mathcal{D}}^{(i}_{\dot{\alpha}}, \overline{\mathcal{D}}^{j)}_{\dot{\beta}}\} = 0, \quad \{\mathcal{D}_{i\alpha}, \overline{\mathcal{D}}^j_{\dot{\beta}}\} = 2\delta^j_i \nabla_{\alpha\dot{\beta}}. \tag{3}$$

The crucial message [**13**] is that these constraints turn out to be equivalent to the equations of motion (1). Moreover, they take the form of a Cauchy–Riemann system in an appropriately enlarged base space, as we now explain.

§3. Harmonic superspace

Let us begin with the Euclidean superspace. The coordinates (2) parameterize the cosets of the super Poincaré group by its Lorentz subgroup. In this case the Lorentz group is $SO(4) = SU(2) \times SU(2)$, with independent $SU(2)$ groups. Factoring, instead, by a *subgroup* of the Lorentz group, one obtains a correspondingly larger space, a construction which turns out to be very useful. In particular, factoring by the $U(1) \times U(1)$ subgroup of the Lorentz group, one obtains an enlargement of superspace by a direct product of two 2-spheres, $S^2 = SU(2)/U(1)$. As coordinates for these spheres, we shall use the harmonics $u^+_{\dot{\alpha}}$, $u^-_{\dot{\alpha}}$ and $v^{\oplus}_{\alpha}$, $v^{\ominus}_{\alpha}$ [**8**], defined up to the respective $U(1)$ phases, where $(+,-)$ and $(\oplus,\ominus)$ being the respective $U(1)$ charges; these harmonics obey the constraints

$$u^{+\dot{\alpha}} u^-_{\dot{\alpha}} = 1, \quad v^{\oplus\alpha} v^{\ominus}_{\alpha} = 1,$$

(or, equivalently, satisfy the completeness relations

$$u^{+\dot{\alpha}} u^-_{\dot{\beta}} - u^{-\dot{\alpha}} u^+_{\dot{\beta}} = \delta^{\dot{\alpha}}_{\dot{\beta}}, \quad v^{\oplus\alpha} v^{\ominus}_{\beta} - v^{\ominus\alpha} v^{\oplus}_{\beta} = \delta^{\alpha}_{\beta})\,.$$

This enlarged space, with spinor harmonics as additional coordinates, is our harmonic superspace.

For the signature $(3,1)$ Minkowski space, the Lorentz group is the simple group $SL(2,\mathbb{C})$. In this case a similar construction involves an enlargement by a coset $SL(2,\mathbb{C})/SL(1,\mathbb{C})$, so that the charge of the harmonics becomes complex: the u and v harmonics, and the $(+,-)$ and $(\oplus,\ominus)$ charges becoming complex conjugates of each other. On the other hand, for a signature $(2,2)$ space, the Lorentz group is $SL(2,\mathbb{R}) \times SL(2,\mathbb{R})$, a direct product of two *real* groups. In these noncompact cases there appear richer structures and peculiarities. We shall not go into subtleties in the present paper. It suffices to say that in these cases we understand the construction in the sense of a Wick rotated version of the Euclidean one.

The upshot is that we have additional coordinates and are able to pass to a basis of harmonic superspace with coordinates

$$\{x^{\pm\oplus}, x^{\pm\ominus}, \vartheta^{i\oplus}, \vartheta^{i\ominus}, \bar{\vartheta}^+_i, \bar{\vartheta}^-_i, u^+_{\dot{\alpha}}, u^-_{\dot{\alpha}}, v^{\oplus}_{\alpha}, v^{\ominus}_{\alpha}\} \tag{4}$$

where the x's and ϑ's are related to the usual superspace coordinates (2) by the formulas

$$x^{\pm\oplus} = x^{\alpha\dot\alpha} u^{\pm}_{\dot\alpha} v^{\oplus}_{\alpha}, \quad x^{\pm\ominus} = x^{\alpha\dot\alpha} u^{\pm}_{\dot\alpha} v^{\ominus}_{\alpha},$$
$$\vartheta^{i\oplus} = \vartheta^{i\alpha} v^{\oplus}_{\alpha}, \quad \vartheta^{i\ominus} = \vartheta^{i\alpha} v^{\ominus}_{\alpha}, \quad \bar\vartheta^{\pm}_{i} = \bar\vartheta^{\dot\alpha}_{i} u^{\pm}_{\dot\alpha}.$$

By virtue of these relations, we may recover ordinary superspace fields as coefficients in the double harmonic expansions (in both u and v) of harmonic superspace fields.

§4. The harmonic superconnections

Now we come to the crucial point. In the coordinates (4) the constraints (3) are radically simplified: they turn out to be equivalent to the following set of commutation relations

$$(5) \qquad \{\overline{\mathcal{D}}^{+i}, \overline{\mathcal{D}}^{+j}\} = 0 = \{\mathcal{D}^{\oplus}_{i}, \mathcal{D}^{\oplus}_{j}\}, \quad \{\overline{\mathcal{D}}^{+j}, \mathcal{D}^{\oplus}_{i}\} = 2\nabla^{+\oplus},$$

where $\overline{\mathcal{D}}^{+i}$, $\mathcal{D}^{\oplus}_{i}$ are gauge-covariant spinorial derivatives and $\nabla^{+\oplus} = \partial/\partial x^{-\ominus} + A^{+\oplus}$, together with the conditions

$$(6) \qquad \begin{aligned} &[D^{++}, \overline{\mathcal{D}}^{+j}] = 0 = [D^{++}, \mathcal{D}^{\oplus}_{i}],\\ &[D^{++}, \nabla^{+\oplus}] = 0 = [D^{\oplus\oplus}, \nabla^{+\oplus}],\\ &[D^{\oplus\oplus}, \overline{\mathcal{D}}^{+j}] = 0 = [D^{\oplus\oplus}, \mathcal{D}^{\oplus}_{i}] \end{aligned}$$

and the important consistency relation

$$(7) \qquad [D^{++}, D^{\oplus\oplus}] = 0.$$

Here $D^{\oplus\oplus}$, D^{++} are harmonic space derivatives, which act on the respective negatively charged harmonic space coordinates to yield their positively-charged counterparts, i.e.,

$$D^{++} x^{-\oplus} = x^{+\oplus}, \quad D^{++} x^{-\ominus} = x^{+\ominus}, \quad D^{++} u^{-}_{\dot\alpha} = u^{+}_{\dot\alpha}, \quad D^{++} \bar\vartheta^{-}_{i} = \bar\vartheta^{+}_{i}$$

and

$$D^{\oplus\oplus} x^{\pm\ominus} = x^{\pm\oplus}, \quad D^{\oplus\oplus} v^{\ominus}_{\dot\alpha} = v^{\oplus}_{\dot\alpha}, \quad D^{\oplus\oplus} \vartheta^{i\ominus} = \vartheta^{i\oplus},$$

while giving zero when applied to the respective positively charged coordinates:

$$D^{++} x^{+\oplus} = 0, \quad D^{++} x^{+\ominus} = 0, \quad D^{++} u^{+}_{\dot\alpha} = 0, \quad D^{++} \bar\vartheta^{+}_{i} = 0,$$
$$D^{\oplus\oplus} x^{\pm\oplus} = 0, \quad D^{\oplus\oplus} v^{\oplus}_{\dot\alpha} = 0, \quad D^{\oplus\oplus} \vartheta^{i\oplus} = 0.$$

Correspondingly, the action on derivatives is given by

$$[D^{++}, \partial^{-\oplus}] = \partial^{+\oplus}, \quad [D^{++}, \partial^{-\ominus}] = \partial^{+\ominus}, \quad [D^{++}, \mathcal{D}^{-i}] = \mathcal{D}^{+i},$$
$$[D^{\oplus\oplus}, \partial^{\pm\ominus}] = \partial^{\pm\oplus}, \quad [D^{\oplus\oplus}, \mathcal{D}^{\ominus}_{i}] = \mathcal{D}^{\oplus}_{i},$$
$$[D^{++}, \partial^{+\oplus}] = [D^{++}, \partial^{+\ominus}] = [D^{++}, \mathcal{D}^{+i}] = 0,$$
$$[D^{\oplus\oplus}, \partial^{\pm\oplus}] = [D^{\oplus\oplus}, \mathcal{D}^{\oplus}_{i}] = 0.$$

By virtue of these properties of D^{++} and $D^{\oplus\oplus}$, conditions (6) ensure that in this basis the covariant derivatives $\{\overline{\mathcal{D}}^{+i}, \mathcal{D}^{\oplus}_{i}, \nabla^{+\oplus}\}$ are homogeneous of degree one in the correspondingly charged harmonics. The relations (5) mean that these covariant derivatives take the pure-gauge forms

$$\mathcal{D}^{+i} = D^{+i} - D^{+i}\varphi\varphi^{-1}, \quad \mathcal{D}_i^{\oplus} = D_i^{\oplus} - D_i^{\oplus}\varphi\varphi^{-1}, \quad \nabla^{+\oplus} = \partial^{+\oplus} - \partial^{+\oplus}\varphi\varphi^{-1}; \tag{8}$$

in other words, equations (5) are integrability conditions of the system

$$\mathcal{D}^{+i}\varphi = 0, \quad \mathcal{D}_i^{\oplus}\varphi = 0, \quad \nabla^{+\oplus}\varphi = 0.$$

The matrix function φ takes values in the gauge group and is defined up to the gauge freedom

$$\varphi \mapsto e^{-\tau}\varphi e^{\lambda}, \quad D^{++}\tau = 0 = D^{\oplus\oplus}\tau, \quad D^{+i}\lambda = 0 = D_i^{\oplus}\lambda = \partial^{+\oplus}\lambda,$$

where τ and λ are matrix functions in the gauge algebra.

In contrast to the covariant derivatives (8), the harmonic derivatives D^{++}, $D^{\oplus\oplus}$ are "short" (i.e., have no connection) in this frame. This choice of frame (the "central frame") is actually inherited from the four-dimensional superspace and is not the most natural one for harmonic superspace. However, we may pass to another frame, what we call the "analytic frame", in which the derivatives $\{\mathcal{D}^{+i}, \mathcal{D}_i^{\oplus}, \nabla^{+\oplus}\}$ are "short" and D^{++}, $D^{\oplus\oplus}$ are "long" (i.e., acquire Lie-algebra-valued connections) instead. Namely,

$$D^{++} \mapsto \mathcal{D}^{++} = \varphi^{-1}[D^{++}]\varphi = D^{++} + V^{++},$$
$$D^{\oplus\oplus} \mapsto \mathcal{D}^{\oplus\oplus} = \varphi^{-1}[D^{\oplus\oplus}]\varphi = D^{\oplus\oplus} + V^{\oplus\oplus},$$

with the thus acquired harmonic superconnections given by

$$V^{++} = \varphi^{-1}D^{++}\varphi, \quad V^{\oplus\oplus} = \varphi^{-1}D^{\oplus\oplus}\varphi, \tag{9}$$

while the covariant derivatives $\{\mathcal{D}^{+i}, \mathcal{D}_i^{\oplus}, \nabla^{+\oplus}\}$ lose their connections, i.e., instead of (8) we have, in this analytic frame,

$$\mathcal{D}^{+i} = D^{+i}, \quad \mathcal{D}_i^{\oplus} = D_i^{\oplus}, \quad \nabla^{+\oplus} = \partial^{+\oplus}.$$

§5. The Cauchy–Riemann equivalence

In this analytic frame it is natural to use an analytic basis that manifestly distinguishes the analytic subspace; this basis is analogous to the well-known chiral basis of ordinary superspace, which distinguishes the chiral subspace. Such a basis is defined by the following change of coordinates:

$$x_A^{+\ominus} = x^{+\ominus} + \bar\vartheta_i^{+}\vartheta^{\ominus i}, \quad x_A^{-\oplus} = x^{-\oplus} - \bar\vartheta_i^{-}\vartheta^{\oplus i}, \quad x_A^{-\ominus} = x^{-\ominus}, \quad x_A^{+\oplus} = x^{+\oplus},$$

with all other coordinates remaining unchanged. In such an analytic basis, the derivatives occurring the system (5) take the form

$$(10)\qquad \begin{aligned} D_i^{\oplus} &= -\frac{\partial}{\partial\vartheta^{\ominus i}} + \bar\vartheta_i^{-}\partial_A^{+\oplus}, \quad \overline{D}^{+i} = \frac{\partial}{\partial\bar\vartheta_i^{-}} - \vartheta^{\ominus i}\partial_A^{+\oplus}, \quad \partial_A^{+\oplus} = \frac{\partial}{\partial x_A^{-\ominus}}, \\ D^{++} &= u^{+\dot\alpha}\frac{\partial}{\partial u^{-\dot\alpha}} + x_A^{+(\oplus}\partial_A^{+\ominus)} + \bar\vartheta_i^{+}\frac{\partial}{\partial\bar\vartheta_i^{-}} - \bar\vartheta_i^{+}\vartheta^{(\ominus i}\partial_A^{+\oplus)}, \\ D^{\oplus\oplus} &= v^{\oplus\alpha}\frac{\partial}{\partial v^{\ominus\alpha}} + x_A^{(+\oplus}\partial_A^{-)\oplus} + \vartheta^{\oplus i}\frac{\partial}{\partial\vartheta^{\ominus i}} - \vartheta^{i\oplus}\bar\vartheta_i^{(+}\partial_A^{-)\oplus}, \end{aligned}$$

where brackets denote symmetrizations in the corresponding charges. In the analytic frame, equations (5) become identities for the derivatives (10), whereas equations (6) take the form of generalized Cauchy–Riemann conditions

$$(11)\qquad \begin{aligned} \frac{\partial}{\partial\bar\vartheta_i^{-}}V^{++} &= 0 = \frac{\partial}{\partial\bar\vartheta_i^{-}}V^{\oplus\oplus}, \\ \frac{\partial}{\partial\vartheta^{\ominus i}}V^{++} &= 0 = \frac{\partial}{\partial\vartheta^{\ominus i}}V^{\oplus\oplus}, \\ \frac{\partial}{\partial x_A^{-\ominus}}V^{++} &= 0 = \frac{\partial}{\partial x_A^{-\ominus}}V^{\oplus\oplus} \end{aligned}$$

with the consistency relation (7) taking the form of a zero curvature relation

$$(12)\qquad D^{++}V^{\oplus\oplus} - D^{\oplus\oplus}V^{++} + [V^{++}, V^{\oplus\oplus}] = 0\,.$$

These equations are merely analytic frame versions of the central frame equations (5), (6), and (7); hence, they are equivalent to the constraints (3). The complete dynamical information of the $N = 3$ Yang–Mills system (1) is therefore encoded into the harmonic connections $\{V^{++}, V^{\oplus\oplus}\}$ solving the first order *linear* differential equations (11), (12) in harmonic superspace. The self-dual (respectively anti-self-dual) subsets of solutions simply correspond to v- (resp. u-) independent chiral (i.e., $\bar\vartheta$- (resp. ϑ-) independent) solutions of (11). The condition of v (resp. u) independence being tantamount to the vanishing of one of the harmonic connections, viz. $V^{\oplus\oplus}$ (resp. V^{++}), and chirality implying the independence of an additional x-variable: $x^{-\oplus}$ (resp. $x^{+\ominus}$). This formulation of the super self-duality equations is described in [**9**]. The interlocking of SD and ASD data mentioned at the beginning is clearly manifest in (11), (12).

In order to construct the superconnection satisfying (3), we first need to recover the "bridge" φ connecting the analytic frame to the central frame by solving the linear system of equations (9)

$$(13)\qquad D^{++}\varphi = \varphi V^{++}, \quad D^{\oplus\oplus}\varphi = \varphi V^{\oplus\oplus}$$

for arbitrary holomorphic (i.e., independent of the variables $\{x^{-\ominus}, \vartheta^{\ominus i}, \bar\vartheta_i^{-}\}$) superfields $\{V^{++}, V^{\oplus\oplus}\}$, which enjoy (12) as an integrability condition. For semi-simple gauge groups it follows from (13) that the determinant of the bridge obeys the equations

$$D^{++}\det\varphi = D^{\oplus\oplus}\det\varphi = 0\,.$$

Using a consistent solution φ of this system, we may return to the central basis in which solutions of the constraints (3) take the form

$$A^{+i} = -D^{+i}\varphi\varphi^{-1}, \quad A_i^{\oplus} = -D_i^{\oplus}\varphi\varphi^{-1}, \quad A^{+\oplus} = -\partial^{+\oplus}\varphi\varphi^{-1}.$$

These $\{A^{+i}, A^{+\oplus}\}$ (resp. $\{A_i^{\oplus}, A^{+\oplus}\}$) are guaranteed by (5) to be linear in u (resp. v) in the central frame, so the superconnections satisfying (1) can be immediately extracted from the harmonic expansions

$$A^{+i} = u^{+\dot\alpha} A^i_{\dot\alpha}, \quad A_i^{\oplus} = v^{\oplus\alpha} A_{i\alpha}, \quad A^{+\oplus} = u^{+\dot\alpha} v^{\oplus\alpha} A_{\alpha\dot\alpha}\,.$$

§6. The static limit

The explicit construction is somewhat simpler for the *static* case, which is similar to the three-dimensional $N = 6$ Yang–Mills theory considered in [**11**]. In these cases the spinor indices α and $\dot\alpha$ as well as two sets of harmonics are identified and we have the relations to three-dimensional superspace coordinates:

$$x^{\pm\pm} = x^{\alpha\beta} u^{\pm}_{\alpha} u^{\pm}_{\beta}, \quad x^{+-} = x^{\alpha\beta} u^{+}_{\alpha} u^{-}_{\beta}, \quad \vartheta^{i\pm} = \vartheta^{i\alpha} u^{\pm}_{\alpha}, \quad \bar\vartheta^{\pm}_i = \bar\vartheta^{\dot\alpha}_i u^{\pm}_{\dot\alpha},$$

which supersymmetrize the three dimensional twistor relations of [**16**]. In this reduced harmonic superspace, the harmonic connections V^{++} and $V^{\oplus\oplus}$ become identified and the somewhat difficult consistency relation (12) disappears, leaving the simplified CR system

$$\frac{\partial}{\partial\bar\vartheta^{-}_i} V^{++} = 0, \quad \frac{\partial}{\partial\vartheta^{-i}} V^{++} = 0, \quad \frac{\partial}{\partial x^{--}} V^{++} = 0,$$

in which the imposition of chirality (i.e., independence of $\bar\vartheta^{+}_i$ as well, which implies x^{+-} independence) corresponds to the Bogomolny reduction for self-dual monopoles (supersymmetrising the construction of [**17**]). We have presented explicit solutions elsewhere ([**5**]).

§7. Conclusion

We hope to have convinced the reader that the harmonic superspace approach is an effective framework for discussing integrability properties of the $N = 3$ Yang–Mills equations. We find it remarkable that after sufficient supersymmetrisation, nonintegrable equations become integrable and the corresponding quantum theory becomes ultraviolet finite; and we expect an analogous phenomenon for the higher extended supergravity theories, which we shall discuss elsewhere.

Acknowledgements. V. Ogievetsky would like to gratefully acknowledge the receipt of a Humboldt Forschungspreis that allowed him to carry out this work at Bonn University and to warmly thank the professors of this university, V. Rittenberg, G. von Gehlen, and E. Sokatchev for discussions and hospitality.

References

1. L. Sadun and J. Segert, Comm. Math. Phys. **145** (1992), 363; G. Bor, Comm. Math. Phys. **145** (1992), 393; H.-Y. Wang, J. Differential Geom. **34** (1991), 701; L. Sibner, R. Sibner, and K. Uhlenbeck, Proc. Nat. Acad. Sci. U.S.A. **86** (1989), 860.
2. E. Witten, Phys. Lett. B **77** (1978), 394.

3. P. Green, J. Isenberg, and P. Yasskin, Phys. Lett. B **78** (1978), 462.
4. M. G. Eastwood, Trans. Amer. Math. Soc. **301** (1987), 615; N. P. Buchdahl, Trans. Amer. Math. Soc. **288** (1985), 431;J. Harnad, J. Hurtubise, and S. Shnider, Ann. Phys. (N.Y.) **193** (1989), 40.
5. C. Devchand and V. Ogievetsky, *The structure of all extended supersymmetric self-dual gauge theories*, Nuclear Phys. B **414** (1994), 763–782.
6. R. S. Ward, Phys. Lett. A **61** (1977), 81.
7. Yu. Manin, J. Soviet Math. **2** (1983), 465; Yu. Manin, *Gauge field theory and complex geometry*, "Nauka", Moscow, 1984; English transl., Springer-Verlag, Berlin, 1988.
8. M. Evans, F. Gürsey, and V. Ogievetsky, Phys. Rev. D **47** (1993), 3496.
9. C. Devchand and V. Ogievetsky, *Super-self-duality as analyticity in harmonic superspace*, Phys. Lett. B **297** (1992), 93–98; *The matreoshka of supersymmetric self-dual theories* Bonn preprint, Bonn-HE-93-23, Nuclear Phys. B (to appear).
10. A. Galperin, E. Ivanov, S. Kalitzin, V. Ogievetsky, and E. Sokatchev, Classical Quantum Gravity **1** (1984), 469; **2** (1985), 155.
11. B. Zupnik, Soviet J. Nuclear Phys. **48** (1988), 744.
12. L. Brink, J. Schwarz, and J. Scherk, Nuclear Phys. B **121** (1977), 77.
13. J. Harnad, J. Hurtubise, M. Légaré, and S. Shnider, Nuclear Phys. B **256** (1985), 609; J. Harnad and S. Shnider, Comm. Math. Phys. **106** (1986), 183.
14. R. Grimm, M. Sohnius, and J. Wess, Nuclear Phys. B **133** (1978), 275.
15. M. Sohnius, Nuclear Phys. B **136** (1978), 461.
16. R. S. Ward, J. Math. Phys. **30** (1989), 2246.
17. O. Ogievetsky, in: Group Theoretical Methods in Physics (H.-D. Doebner et al, eds.), Lecture Notes in Phys., vol. 313, Springer-Verlag, Berlin, 1988.

JOINT INSTITUTE FOR NUCLEAR RESEARCH DUBNA, RUSSIA

PHYSIKALISCHES INSTITUT DER UNIVERSITÄT BONN, GERMANY

Amer. Math. Soc. Transl.
(2) Vol. **177**, 1996

Estimates of Semi-invariants for the Ising Model at Low Temperatures

R. L. Dobrushin*

§1. Formulation of the result

Let $V \subset \mathbb{Z}^d$ be a finite subset of the d-dimensional integer lattice $\mathbb{Z}^d$ and $X(V) = \{-1, 1\}^V$ be the set of configurations $x = (x_t,\, t \in V)$, where $x_t = \pm 1$. The (symmetric ferromagnetic) *Ising distribution in the volume V with the plus-boundary condition and inverse temperature $\beta > 0$* is the probability distribution on the set $X(V)$ given by

$$p_V(x) = (Z(V))^{-1} \exp\{-\beta U_V(x)\}, \qquad x \in X(V), \tag{1.1}$$

with the *partition function*

$$Z(V) = \sum_{x \in X(V)} \exp\{-\beta U_V(x)\} \tag{1.2}$$

and the *energy*

$$U_V(x) = -\Biggl(\sum_{\substack{\{s,t\} \subset V \\ |s-t|=1}} x_s x_t + \sum_{\substack{\{s,t\}: t \in V,\, s \in V^c \\ |s-t|=1}} x_t \Biggr), \qquad x \in X(V),\ V^c = \mathbb{Z}^d \setminus V. \tag{1.3}$$

Let $T \subset V$ be a finite set. The *generating function* of the values of the field on the set T is defined by the relation

$$F_T(V; z_t, t \in T) = \sum_{x \in X(V)} p_V(x) \exp\Biggl\{ \sum_{t \in T} z_t x_t \Biggr\}, \tag{1.4}$$

where z_t, $t \in T$, are complex numbers. Let $r = (r_t,\, t \in T)$ be a function with integer values $r_t \geqslant 1$, $t \in T$, and $|r| = \sum_{t \in T} r_t$. The *semi-invariant* (sometimes

---

1991 *Mathematics Subject Classification.* Primary 60K35; Secondary 82B20.
*Roland L'vovich Dobrushin died November 12, 1995.

the terms *cumulant, truncated correlation function* are also used) *of order* r of the values of the field on the set T is defined as the value of the partial derivative

$$(1.5)\qquad s_V(T,r) = \left.\frac{\partial^{|r|} \ln F_T(V; z_t,\, t \in T)}{\prod_{t\in T} \partial z_t^{r_t}}\right|_{z_t\equiv 0,\, t\in T}.$$

We introduce a geometric characteristic of the set T

$$(1.6)\qquad q(T) = \max_{S\subseteq T,\, |S|>1} \frac{|S|}{(\operatorname{diam} S)^{d-1}},$$

which we call the *concentration coefficient* of the set T.[1] This coefficient does not exceed $T^{1/d}$ and takes on values of this order for sets like cubes or balls, but can be arbitrarily small for sets evenly spread on the lattice.

The following estimate is the main result of the paper.

THEOREM 1.1. *For each* $d \geqslant 2$ *there exist constants* $\tilde{\beta} = \tilde{\beta}_d > 0$ *and* $K = K_d < \infty$ *such that for all* $\beta > \tilde{\beta}$, *all finite sets* $T \subset V \subset \mathbb{Z}^d$, *where* $|T| > 1$, *and all* $r = (r_t,\, t \in T)$ *the following estimates for the semi-invariants hold*

$$(1.7)\qquad |s_V(T,r)| \leqslant (K \max(q(T), 1))^{|r|} \exp\{-2(\beta - \tilde{\beta})|\Gamma_T|\},$$

where $|\Gamma_T|$ *is the* $(d-1)$*-dimensional area of the minimal contour containing all the set* T.

REMARK 1.1. See §3 for the definition of the notion of contours usually used in the literature on the Ising model. Observe here only that

$$(1.8)\qquad |\Gamma_T| \geqslant d \operatorname{diam} T.$$

REMARK 1.2. The estimate (1.7) is substantially stronger than the estimates which were obtained earlier ([**Ma**], [**MM,** §6.5]). The factor in (1.7), which exponentially depends on $|\Gamma_T|$, seems natural since it estimates the probability that the set T lies inside a contour separating different phases. Strong tree-like estimates of semi-invariants (see [**MM, DS**]) which are natural for the high-temperature case, are not valid in the low-temperature situation. The improvement in comparison with the previous estimates is connected with the pre-exponential factor, which is larger than $2^{2^{|r|}}$ in [**Ma, MM**].

REMARK 1.3. In the derivation of the theorem formulated above, we use a new variant of the cluster expansion method. Usually these expansions are applied to contours. We apply them to certain more complex objects adequate for the problem under consideration. This yields an improvement of the estimates. This new version of the cluster expansion method is described in §2 of this paper. There are many publications devoted to the development and applications of the powerful cluster expansion method. (See the books and the survey [**Br, GJ, Ma, MM, Se, Sim**] and references there). The general model used in §2 is taken from the paper of Kotecký and Preiss [**KP**] (see also [**HKZ,** Appendix B], and [**MS**]). In contrast with the other approaches, we apply elementary facts of the theory of analytical functions instead of involved combinatorial estimates of the individual terms of the

[1] Here and in the following $|A|$ is the number of elements in a finite set A.

cluster expansion. This simplifies the construction and yields estimates convenient for applications. This approach was used earlier by the author and his coauthors ([**D1, D2, DM, DW**]), but in more specific situations and in more complicated versions.

REMARK 1.4. Without essential changes in the constructions, the main result can be generalized in some directions. Thus instead of semi-invariants of values of the field, it is possible to obtain similar estimates for semi-invariants of functions of finite collections of values of the field. Also instead of the ferromagnetic Ising model, it is possible to consider other spin models which can be studied by the contour method (see [**Sin**], for example).

§2. Cluster expansions in animal models

Let Θ be a countable (or finite) set. Its elements will be called *animals*.[2] Assume that the structure of an undirected graph without loops and multiple edges with vertex set Θ is fixed. We say that elements $\theta_1, \theta_2 \in \Theta$ are *incompatible* if they are connected by an edge of the graph; in this case we write $\theta_1 \not\sim \theta_2$. If the animals $\theta_1, \theta_2 \in \Theta$ are not connected by an edge, we say that they are *compatible* and write $\theta_1 \sim \theta_2$. A finite subset $\pi \subseteq \Theta$ is called a *herd* if any two distinct animals $\theta_1, \theta_2 \in \pi$ are compatible. For any finite subset $\Lambda \subseteq \Theta$, the set of all herds π such that $\pi \subseteq \Lambda$ will be denoted $H(\Lambda)$ and called the set of all *herds in* Λ. (The set $H(\Lambda)$ includes the empty herd, which is denoted by $\varnothing$.)

Assume that a complex-valued function $w(\theta)$, $\theta \in \Theta$, is given. We call the number $w(\theta)$ the *weight* of the animal θ. Let $\Lambda \subseteq \Theta$ be a finite set. The number

$$Z_w(\Lambda) = \sum_{\pi \in H(\Lambda)} \prod_{\theta \in \pi} w(\theta) \tag{2.1}$$

is called *the partition function in* Λ *defined by the weights* $w = w(\theta)$. (In the case $\pi = \varnothing$, the product in (2.1) is equal to one by definition. So if Λ is the empty set, the partition function is $Z_w(\Lambda) = 1$.) Assume that a positive function $b(\theta)$, $\theta \in \Theta$, is given. The value $b(\theta)$ will be called the *might of the animal* θ.

THEOREM 2.1. *Let us suppose that a positive weight function* $w_0 = w_0(\theta) > 0$ *is fixed such that for any animal* $\theta \in \Theta$ *one has*

$$1 - w_0(\theta) \exp\left\{ \sum_{\tilde\theta \in \Theta : \tilde\theta \not\sim \theta} w_0(\tilde\theta) b(\tilde\theta) \right\} \geqslant \exp\{-w_0(\theta) b(\theta)\}. \tag{2.2}$$

In particular, condition (2.2) *implies that*

$$w_0(\theta) \exp\left\{ \sum_{\tilde\theta \in \Theta : \tilde\theta \not\sim \theta} w_0(\tilde\theta) b(\tilde\theta) \right\} < 1. \tag{2.3}$$

[2] In previous papers the elements of this set were called polymers or contours. But in the terminology of statistical mechanics, these words are attached to some concrete objects; in applications (including §4 of this papers) these elements can be of varied and sometimes exotic nature. So we propose the term animal and even develop this "animal terminology" further.

Let W_0 be the set of all weight functions $w = w(\theta)$ of $\theta \in \Theta$ such that

$$|w(\theta)| \leqslant w_0(\theta), \qquad \theta \in \Theta. \tag{2.4}$$

Then for any weight function $w \in W_0$ and any finite set $\Lambda \subseteq \Theta$, the partition function $Z_w(\Lambda)$ is nonzero and for any finite $\Lambda' \subseteq \Lambda$, we have

$$\left| \ln \left| \frac{Z_w(\Lambda)}{Z_w(\Lambda')} \right| \right| \leqslant \sum_{\theta \in \Lambda \setminus \Lambda'} w_0(\theta)\, b(\theta). \tag{2.5}$$

PROOF. We shall use induction on the number of elements $|\Lambda|$ in the set Λ. In the case $|\Lambda| = 0$, i.e., $\Lambda = \varnothing$, the statement of the theorem is evident. So we fix a set Λ and suppose that the estimate (2.3) is valid for all sets with numbers of elements smaller than $|\Lambda|$. We also fix a subset $\Lambda' \subset \Lambda$ and an animal $\theta_0 \in \Lambda \setminus \Lambda'$. (The case of $\Lambda = \Lambda'$ is trivial.) Consider the set $\widehat{\Lambda} = \Lambda \setminus \{\theta_0\}$. Then

$$\frac{Z_w(\Lambda)}{Z_w(\Lambda')} = \frac{Z_w(\Lambda)}{Z_w(\widehat{\Lambda})} \, \frac{Z_w(\widehat{\Lambda})}{Z_w(\Lambda')}. \tag{2.6}$$

It follows from the induction hypothesis that

$$\left| \ln \left| \frac{Z_w(\widehat{\Lambda})}{Z_w(\Lambda')} \right| \right| \leqslant \sum_{\theta \in \widehat{\Lambda} \setminus \Lambda'} w_0(\theta)\, b(\theta). \tag{2.7}$$

So the theorem will be proved if we show that

$$\left| \ln \left| \frac{Z_w(\Lambda)}{Z_w(\widehat{\Lambda})} \right| \right| \leqslant w_0(\theta_0)\, b(\theta_0). \tag{2.8}$$

Let

$$\Lambda_0 = \{\theta \in \widehat{\Lambda} : \theta \sim \theta_0\}. \tag{2.9}$$

It is clear from definition (2.1) that

$$Z_w(\Lambda) = Z_w(\widehat{\Lambda}) + w(\theta_0)\, Z_w(\Lambda_0). \tag{2.10}$$

Since $\Lambda_0 \subseteq \widehat{\Lambda}$ and $|\Lambda_0| < |\Lambda|$, using the induction assumption and definition (2.4), we obtain

$$\left| \frac{w(\theta_0)\, Z_w(\Lambda_0)}{Z_w(\widehat{\Lambda})} \right| \leqslant w_0(\theta_0) \exp\Bigg\{ \sum_{\tilde{\theta} : \tilde{\theta} \in \widehat{\Lambda}, \tilde{\theta} \nsim \theta_0} w_0(\tilde{\theta})\, b(\tilde{\theta}) \Bigg\}. \tag{2.11}$$

Now observe that for any complex z such that $|z| < 1$, we have $1 - |z| \leqslant |1 + z| \leqslant 1 + |z|$ and so

$$|\ln|1 + z|| \leqslant \max(\ln(1 + |z|), -\ln(1 - |z|)) = -\ln(1 - |z|). \tag{2.12}$$

It follows from condition (2.3) that the absolute value of the right-hand side of (2.11) is smaller than 1. So, applying relations (2.10), (2.11), and (2.12) we find that

$$\begin{aligned}\left|\ln\left|\frac{Z_w(\Lambda)}{Z_w(\widehat{\Lambda})}\right|\right| &= \left|\ln\left|1+\frac{w(\theta_0)\,Z_w(\Lambda_0)}{Z_w(\widehat{\Lambda})}\right|\right| \\ &\leqslant -\ln\left(1-w_0(\theta_0)\exp\left\{\sum_{\tilde\theta:\tilde\theta\in\Lambda:\tilde\theta\not\sim\theta_0} w_0(\tilde\theta)\,b(\tilde\theta)\right\}\right).\end{aligned} \tag{2.13}$$

The estimate (2.13) together with the main condition (2.2) implies the desired estimate (2.8). □

REMARK 2.1. Instead of condition (2.2) which appeared in a natural way in the proof of Theorem 2.1, it is convenient to use the following slightly stronger condition: for any $\theta\in\Theta$

$$\exp\left\{\sum_{\tilde\theta\in\Theta:\tilde\theta\not\sim\theta} w_0(\tilde\theta)\,b(\tilde\theta)+w_0(\theta)\,b(\theta)\right\}\leqslant b(\theta)\,. \tag{2.14}$$

Let us verify that condition (2.2) implies condition (2.14). It follows from (2.14) that

$$w_0(\theta)\exp\left\{\sum_{\tilde\theta\in\Theta:\tilde\theta\not\sim\theta} w_0(\tilde\theta)\,b(\tilde\theta)\right\}\leqslant w_0(\theta)\,b(\theta)\exp\{-w_0(\theta)\,b(\theta)\}\,. \tag{2.15}$$

We set $x = w_0(\theta)\,b(\theta)$ and see that condition (2.2) follows from (2.15) and the general inequality

$$1-e^{-x}\geqslant xe^{-x}\quad \text{for } x\geqslant 0\,. \tag{2.16}$$

To derive this inequality, it suffices to check that for $x\geqslant 0$ the function $f(x) = 1-(x+1)\,e^{-x}$ has a nonnegative derivative for $x\geqslant 0$. Condition (2.14) was used by Kotecký and Preiss [**KP**].

Now we intend to construct a cluster expansion for logarithms of partition functions. For any finite set $\Lambda\subseteq\Theta$, any *group of animals* in Λ is any pair $\rho=(\bar\rho,\alpha)$ such that $\bar\rho\subseteq\Lambda$ is a set and $\alpha=\alpha(\theta)\geqslant 1$, $\theta\in\bar\rho$, is an integer-valued function of $\theta\in\bar\rho$. The set $\bar\rho$ is called the *support of the group* and the value $\alpha(\theta)$ is interpreted as the multiplicity of animals of type θ in the group ρ. The set of all groups of animals in Λ is denoted by $D(\Lambda)$. We say that a group $\rho=(\bar\rho,\alpha)$ is the *sum of groups* $\rho_i=(\bar\rho_i,\alpha_i)$, $i=1,\dots,k$, if $\bar\rho=\bigcup_{i=1}^k\bar\rho_i$ and

$$\alpha(\theta)=\sum_{i=1,\dots,k:\theta\in\bar\rho_i}\alpha_i(\theta),\qquad \theta\in\bar\rho\,. \tag{2.17}$$

A *gang of animals in* Λ is a nonempty group of animals $\rho=(\bar\rho,\alpha)\in D(\Lambda)$ such that for any two animals $\theta,\theta'\in\bar\rho$ there is a sequence $\theta=\theta_1,\dots,\theta_n=\theta'$ of animals in $\bar\rho$ such that the animals θ_i and θ_{i+1} are incompatible for all $i=1,\dots,n-1$, i.e., $\bar\rho$ is a connected subset of the graph Θ. The set of all gangs in Λ will be denoted by $G(\Lambda)$. (If $\bar\rho$ is a singleton, then $\rho\in G(\Lambda)$ by definition.)

THEOREM 2.2. *Let the conditions of Theorem 2.1 be fulfilled and a finite set $\Lambda \subseteq \Theta$ be fixed. Consider the closed polydisk*

$$W_0(\Lambda) = \{w = (w(\theta),\, \theta \in \Lambda) : |w(\theta)| \leqslant w_0(\theta),\, \theta \in \Lambda\} \subset \mathbb{C}^\Lambda$$

and its interior $W_0^{\text{in}}(\Lambda)$. For any $w \in W_0^{\text{in}}(\Lambda)$ we have the convergent expansion

$$\ln Z_w(\Lambda) = \sum_{\rho \in G(\Lambda)} q_w(\rho) = \sum_{\rho \in G(\Lambda)} r(\rho) \prod_{\theta \in \bar{\rho}} w(\theta)^{\alpha(\theta)}. \tag{2.18}$$

The coefficients $r(\rho)$ are real numbers depending only on the restriction of the graph structure on Θ to $\bar{\rho}$. For any gang $\rho = (\bar{\rho}, \alpha)$, we have the bound

$$|q_w(\rho)| = \left| r(\rho) \prod_{\theta \in \bar{\rho}} w(\theta)^{\alpha(\theta)} \right| \leqslant \Big(\sum_{\theta \in \bar{\rho}} w_0(\theta) b(\theta) \Big) \Big(\prod_{\theta \in \bar{\rho}} \frac{|w(\theta)|}{w_0(\theta)} \Big)^{\alpha(\theta)}. \tag{2.19}$$

PROOF. For any $w \in W_0(\Lambda)$, let

$$F_\Lambda(w) = \ln Z_w(\Lambda)\,. \tag{2.20}$$

It follows from Theorem 2.1 that $Z_w(\Lambda)$ does not vanish and so the function $F_\Lambda(w)$ is a holomorphic function of $w \in W_0^{\text{in}}(\Lambda)$. Consider the Taylor expansion of this function at the point $w = 0$:

$$F_\Lambda(w) = \sum_{\rho \in D(\Lambda)} r_\Lambda(\rho) \prod_{\theta \in \bar{\rho}} w(\theta)^{\alpha(\theta)}, \tag{2.21}$$

where for $\bar{\rho} = (\theta_1, \dots, \theta_n)$ the coefficients

$$r_\Lambda(\rho) = (\alpha(\theta_1)! \cdots \alpha(\theta_n)!)^{-1} \frac{\partial^{\alpha(\theta_1)+\dots+\alpha(\theta_n)} F_\Lambda(w)}{\partial^{\alpha(\theta_1)} w(\theta_1) \cdots \partial^{\alpha(\theta_n)} w(\theta_n)} \bigg|_{w=0}. \tag{2.22}$$

For any set $\bar{\rho} \subseteq \Lambda$ and any function $w = (w(\theta),\, \theta \in \bar{\rho}) \in W_0(\bar{\rho})$ consider its extension $w_\Lambda = (w_\Lambda(\theta),\, \theta \in \Lambda)$ to Λ such that $w_\Lambda(\theta) = 0$ for $\theta \in \Lambda \setminus \bar{\rho}$. It follows from definitions (2.1) and (2.20) that

$$F_\Lambda(w_\Lambda) = F_{\bar{\rho}}(w) \tag{2.23}$$

and so we see from relation (2.22) that

$$r_\Lambda(\rho) = r_{\bar{\rho}}(\rho)\,. \tag{2.24}$$

Now setting $r(\rho) = r_{\bar{\rho}}(\rho)$, we can rewrite the expansion (2.21) as

$$F_\Lambda(w) = \sum_{\rho \in D(\Lambda)} r(\rho) \prod_{\theta \in \bar{\rho}} w(\theta)^{\alpha(\theta)}. \tag{2.25}$$

The coefficients $r(\rho)$ are real, since the function $F_{\bar{\rho}}(w)$ takes real values for real w.

Since $Z_w(\Lambda) = 1$ for $w = 0$, the coefficient $r(\rho)$ is zero for the empty group of animals $\rho = \varnothing$. Now consider a nonempty group of animals $\rho \in D(\Lambda) \setminus G(\Lambda)$. There exists a representation $\bar{\rho} = \bar{\rho}_1 \cup \bar{\rho}_2$, where the sets $\bar{\rho}_1$, $\bar{\rho}_2$ are nonempty, their intersection $\bar{\rho}_1 \cap \bar{\rho}_2$ is empty and for all $\theta_1 \in \bar{\rho}_1$, $\theta_2 \in \bar{\rho}_2$ the animals θ_1, θ_2 are compatible. Then it follows from definitions (2.1) and (2.20) that

$$F_{\bar{\rho}}(w) = F_{\bar{\rho}_1}(w) + F_{\bar{\rho}_2}(w)\,. \tag{2.26}$$

Differentiating, from (2.22) and (2.26) we obtain

$$r(\rho) = r_{\bar{\rho}}(\rho) = 0 \quad \text{if } \rho \in D(\Lambda) \setminus G(\Lambda), \tag{2.27}$$

and so the expansion (2.25) is reduced to the desired expansion (2.18).

It follows from relation (2.22) and the Cauchy formula that

$$r(\rho) = (2\pi i)^{-|\bar{\rho}|} \oint_{w:|w(\theta)|=w_0(\theta),\theta\in\bar{\rho}} \frac{F_{\bar{\rho}}(w)}{\prod_{\theta\in\bar{\rho}}(w(\theta))^{\alpha(\theta)+1}}\, dw\,. \tag{2.28}$$

The inequality (2.5) applied for $\Lambda' = \varnothing$ yields the bound

$$|F_{\bar{\rho}}(w)| \leqslant \sum_{\theta\in\bar{\rho}} w_0(\theta)\, b(\theta), \qquad w \in W_0(\Lambda)\,. \tag{2.29}$$

The desired estimate (2.19) follows from (2.28) and (2.29). □

§3. Contour representation

The proof of the main theorem will use the well-known contour description of the Ising ferromagnetic model at low temperatures. Following the tradition in the literature on the applications of the contour method, we restrict our exposition to the simplest case, when dimension d is 2. The generalization to the case $d > 2$ is straightforward and sometimes is even discussed in detail (for example in [**Sim**, §V.8] for $d = 3$).

Let $\mathbb{Z}^2$ be the two-dimensional integer lattice. Assume that $\mathbb{Z}^2 \subset \mathbb{R}^2$. Let $(\mathbb{Z}^2)^\star$ be the *dual lattice* with vertices $(n_1 + 1/2, n_2 + 1/2)$, $n_1, n_2 \in \mathbb{Z}^1$. Let $\mathbb{E}$ be the set of *edges of the dual lattice*, i.e., the set of all closed intervals of length 1 connecting adjacent points of this lattice. For each edge $e \in \mathbb{E}$ there are two vertices of the original lattice $\mathbb{Z}^2$ at the distance $1/2$ from e. We say that they are *vertices adjacent to the edge* e. Let $V \subset \mathbb{Z}^2$ be a finite set. The set of edges $e \in \mathbb{E}$ such that at least one of two points to which the edge e is adjacent belongs to V is called the set of *edges in the volume* V and is denoted by $\mathbb{E}(V)$.

Now return to the Ising distribution with the plus-boundary condition described in §1. For each configuration $x \in X(V)$ we define its *boundary* $B(x)$ as the set of all edges $e \in \mathbb{E}(V)$ of the dual lattice such that if t, t' are the points of the original lattice adjacent to e, then $x_t \neq x_{t'}$. (We assume here that $x_t = 1$ for $t \in V^c = \mathbb{Z}^2 \backslash V$, and this means that we introduce plus-boundary conditions.) The contour method for studying the Ising model is based on the possibility of representing the boundary $B(x)$ in a unique way as a union of contours, i.e., nonselfintersecting closed lines consisting of edges $e \in E(V)$, but it is necessary to be careful in the definition of contours. (What is a unique natural way to divide the boundary of the configuration x, $x_t = (-1)^{t_1+t_2}$, $t = (t_1, t_2) \in V$ into contours? See the answer in the right part of Figure 1.)

So we introduce the following convention. Let $e_1, e_2 \in \mathbb{E}$ be two different edges containing a common vertex $t = (t_1, t_2) \in \mathbb{Z}^{2\star}$. We say that these edges make a *legitimate turn*, if either one of these edges connects the vertex t with the vertex $(t_1 + 1, t_2)$ and the other edge connects it with the vertex $(t_1, t_2 + 1)$ or if one of these edges connects the vertex t with the vertex $(t_1 - 1, t_2)$ and the other edge connects it with the vertex $(t_1, t_2 - 1)$. A *contour* is defined as a sequence

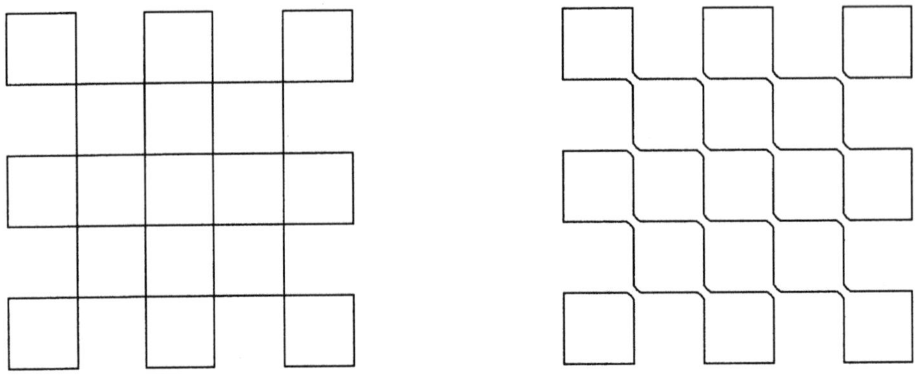

FIGURE 1. Contour representation of the boundary of a configuration.

$e_1, \ldots, e_k$ of mutually distinct edges such that the edges e_i, e_{i+1}, $i = 1, \ldots, k$ (here $k+1 = 1$) have a common vertex and if there is another pair of edges $e_{i'}$, $e_{i'+1}$ of this contour having the same common vertex, then the edges e_i, e_{i+1} make a legitimate turn. In other words, a contour is a closed polygonal line consisting of distinct edges such that a small deformation conserving legitimate turns transforms it into a selfavoiding closed curve. The set of all contours is denoted by $\mathbb{G}$. We say that a contour $\Gamma \in \mathbb{G}$ is a *contour in the volume* V if all its edges belong to $\mathbb{E}(V)$. The set of all such contours is denoted by $\mathbb{G}(V)$. The number of edges in a contour $\Gamma \in \mathbb{G}$ is denoted $|\Gamma|$ and is called the *length* of this contour. The set of all points $t \in \mathbb{Z}^2$ such that there is no continuous curve in $\mathbb{R}^2$ that connects the point $t \in \mathbb{Z}^2 \subset \mathbb{R}^2$ with "infinity" and does not intersect the contour Γ, is called the *interior of the contour* Γ and is denoted by $\operatorname{Int}\Gamma$. We say that the contours Γ_1 and Γ_2 are *compatible* if they have no common edges and at any vertex contained in both of the contours, these contours make legitimate turns.

Let $H(V)$ be the set of all sets $\pi \subseteq \mathbb{G}(V)$ of contours in the volume V such that any two different contours in π are compatible ($H(V)$ includes the empty set of contours). It is easy to understand that for any finite volume V and any configuration $x \in X(V)$ there exists a unique system of contours $\pi(x) \in H(V)$ such that

$$B(x) = \bigcup_{\Gamma \in \pi(x)} \Gamma. \tag{3.1}$$

Further, for any system of contours $\pi \in H(V)$ there is a unique configuration $x(\pi) = (x_t(\pi), t \in V)$ such that

$$\pi(x(\pi)) = \pi\,. \tag{3.2}$$

Actually, this configuration can be defined by the following construction. For any point $t \in \mathbb{Z}^2$ we denote by $O(t)$ the set of all contours $\Gamma \in \mathbb{G}$ such that $t \in \operatorname{Int}\Gamma$. Then

$$x_t(\pi) = (-1)^{|\pi \cap O(t)|}, \qquad t \in V,\ \pi \in H(V)\,. \tag{3.3}$$

The definition (1.3) implies that the energy is given by

$$U_V(x) = 2|\pi(x)| - K_V, \qquad x \in X(V), \tag{3.4}$$

where we set

$$|\pi| = \sum_{\Gamma\in\pi} |\Gamma|, \qquad \pi \in H(V), \tag{3.5}$$

and where the constant K_V is equal to the number of terms in the sums in (1.3) and does not depend on x. Hence, recalling definition (1.1), we find that

$$p_V(x) = \tilde{p}_V(\pi(x)), \qquad x \in X(V), \tag{3.6}$$

where the *contour probability distribution* is

$$\tilde{p}_V(\pi) = (\tilde{Z}(V))^{-1} \exp\{-2\beta|\pi|\}, \qquad \pi \in H(V), \tag{3.7}$$

and the *contour partition function* is

$$\tilde{Z}(V) = \sum_{\pi\in H(V)} \exp\Big\{ - 2\beta \sum_{\Gamma\in\pi} |\Gamma| \Big\}. \tag{3.8}$$

This is the well-known reduction of the ferromagnetic Ising model to the contour Ising model. This reduction is very helpful in the case of large β.

Comparing definition (1.4) and relation (3.6), we find that the generating function will be

$$F_T(V; z_t,\, t \in T) = \frac{Z_T(V; z_t,\, t \in V)}{\tilde{Z}(V)}, \tag{3.9}$$

where the partition function

$$Z_T(V; z_t,\, t \in V) = \sum_{\pi\in H(V)} \exp\Big\{ - 2\beta \sum_{\Gamma\in\pi} |\Gamma| + \sum_{t\in T} z_t x_t(\pi) \Big\}. \tag{3.10}$$

So we can rewrite definition (1.5) as

$$s_V(T, r) = \frac{\partial^{|r|} \ln Z_T(V; z_t,\, t \in T)}{\prod_{t\in T} \partial z_t^{r_t}} \bigg|_{z_t\equiv 0, t\in T}. \tag{3.11}$$

This contour representation of the semi-invariants is the starting point of the following constructions.

We begin with a simple statement about contours to be used later. For any set $V_0 \subset \mathbb{Z}^2$ we denote by $O(V_0)$ the system of all contours which contain in their interior at least one of the points of the set V_0. Clearly,

$$O(V_0) = \bigcup_{t\in V_0} O(t)\,. \tag{3.12}$$

LEMMA 3.1. *There exists a constant $\beta_1 > 0$ such that for any finite set $V_0 \subset \mathbb{Z}^2$ we have*

$$\sum_{\Gamma\in O(V_0)} \exp\{-2\beta_1|\Gamma|\} \leqslant \frac{1}{2}\,|V_0|\,. \tag{3.13}$$

PROOF. Let $t_0 = (t_0^1, t_0^2) \in V_0$ and e_{a,t_0}, where $a = 0, 1, \dots,$ be the edge of the lattice connecting the vertices $(t_0^1 + a + 1/2, t_0^2 + 1/2)$ and $(t_1^0 + a + 1/2, t_0^2 - 1/2)$ of the dual lattice. It is clear that any contour $\Gamma \in O(t_0)$ contains at least one of the edges e_{a,t_0} with $a \leqslant |\Gamma|$. The number of different contours of length d containing a fixed edge does not exceed 3^d and so

$$\sum_{\Gamma \in O(t_0)} \exp\{-2\beta|\Gamma|\} \leqslant \sum_{a=0}^{\infty} \sum_{d=a}^{\infty} \exp\{-(2\beta - \ln 3)\, d\}. \tag{3.14}$$

The right-hand side of (3.14) tends to 0 as $\beta \to \infty$, and thus it follows from the definition (3.12) that the desired estimate (3.13) holds for β_1 large enough. □

§4. Animal model representation

The derivation of the main estimate (1.7) for semi-invariants uses a special representation of the partition function (3.10) as the partition function of a special animal model (see §2). Fixing a finite set $T \subset \mathbb{Z}^2$, we now describe this animal model. The set of animals $\Theta = \Theta(T)$ consists of elements of two types. First of all, it contains all contours $\mathbb{G} \subset \Theta(T)$. Besides this, the set of animals $\Theta(T)$ contains some additional elements, which we call *amoebas*. Amoeba is a pair $A = (S, \pi_S)$, where $S \subseteq T$ is a finite nonempty set whose elements will be called *nuclei* of the amoeba A, and π_S is a finite system of pairwise compatible contours such that $\pi_S \subset O(S)$ (see (3.12)) and if $|S| > 1$, then there is a contour $\Gamma \in \pi_S$ containing the entire set S in its interior: $S \subseteq \operatorname{Int}\Gamma$. The largest (in the sense of the number of $|\operatorname{Int}\Gamma|$) of contours Γ containing S is called the *membrane* of the amoeba A and is denoted by $M(A)$. (In the case $|S| = 1$, the system π_S can be empty and then the membrane is absent.) (See Figure 2.) The set of all amoebas will be denoted by $\mathbb{A}(T)$.

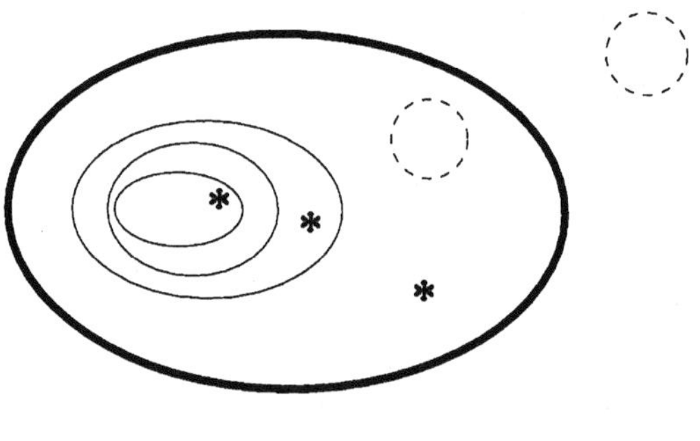

FIGURE 2. An amoeba with three nuclei and two (dashed) contours compatible with it.

We must describe the incompatibility relation ($\nsim$) in $\Theta(T)$. The contours $\Gamma_1, \Gamma_2 \in \mathbb{G} \subset \Theta(T)$ are *incompatible* if they are incompatible in the sense described in §3. A contour $\Gamma \in \mathbb{G}$ and an amoeba $A = (S, \pi_S) \in \mathbb{A}(T)$ are *incompatible* if either one of the contours $\Gamma' \in \pi_S$ is incompatible with the contour Γ or if the contour Γ belongs to $O(S)$. Two amoebas $A_1 = (S_1, \pi_{S_1}^1)$, $A_2 = (S_2, \pi_{S_2}^2) \in \mathbb{A}(T)$ are *incompatible* if either there is a pair of incompatible contours $\Gamma^1 \in \pi_{S_1}^1$ and

$\Gamma^2 \in \pi^2_{S_2}$, or if the intersection $S_1 \cap S_2$ is not empty, or if finally the set of contours $(\pi^1_{S_1} \cap O(S_2)) \cup (\pi^2_{S_2} \cap O(S_1))$ is not empty.

Fixing $\beta > 0$ and complex numbers z_t, $t \in T$, we define the weight function

(4.1)
$$w(\theta) = \begin{cases} \exp\{-2\beta|\Gamma|\}, & \text{if } \theta = \Gamma \in \mathbb{G}, \\ \exp\{-2\beta|\pi_S|\} \prod_{t \in S} (\exp\{z_t x_t(\pi_S)\} - 1), & \text{if } \theta = (S, \pi_S) \in \mathbb{A}(T). \end{cases}$$

PROPOSITION 4.1. *Let $\Lambda(T,V)$, where $T \subseteq V \subset \mathbb{Z}^2$ are finite sets, be the set of animals $\theta \in \Theta(T)$ such that either $\theta = \Gamma \in \mathbb{G}(V)$ or $\theta = (S, \pi_S) \in \mathbb{A}(T)$, where $\pi_S \in H(V)$ (see §3). Then (see definition (2.1)) the partition function is*

$$Z_T(V; z_t, t \in T) = Z_w(\Lambda(T,V)). \tag{4.2}$$

Let

$$w_0(\theta) = \begin{cases} \exp\{-2\beta_0|\Gamma|\}, & \text{if } \theta = \Gamma \in \mathbb{G}, \\ (v(T))^{|S|} \exp\{-2\beta_0|\pi_S|\}, & \text{if } \theta = (S, \pi_S) \in \mathbb{A}(T), \end{cases} \tag{4.3}$$

where (recall (1.6))

$$v(T) = \frac{1}{2} e^{-3/2} \min((q(T))^{-1}, 1), \tag{4.4}$$

$$b(\theta) = \begin{cases} \exp\{2|\Gamma|\}, & \text{if } \theta = \Gamma \in \mathbb{G}, \\ \exp\{2|\pi_S| + |S|\}, & \text{if } \theta = (S, \pi_S) \in \mathbb{A}(T). \end{cases} \tag{4.5}$$

Then there exists a value $\beta_0 > 0$ such that the main condition (2.2) of Theorem 2.1 is fulfilled.

PROOF. We can rewrite the partition function (3.10) as

$$\begin{aligned} Z_T(V; z_t, t \in V) &= \sum_{\pi \in H(V)} \exp\Big\{ -2\beta \sum_{\Gamma \in \pi} |\Gamma| \Big\} \prod_{t \in T} (\exp\{z_t x_t(\pi)\} - 1 + 1) \\ &= \sum_{S \subseteq T,\, \pi \in H(V)} \exp\Big\{ -2\beta \sum_{\Gamma \in \pi} |\Gamma| \Big\} \prod_{t \in S} (\exp\{z_t x_t(\pi)\} - 1). \end{aligned} \tag{4.6}$$

Let us show that for any fixed pair (S, π), where $S \subseteq T$ and $\pi \in H(V)$, there exists a unique representation

$$S = S_1 \cup \cdots \cup S_k, \qquad \pi \cap O(S) = \pi_{S_1} \cup \cdots \cup \pi_{S_k}, \tag{4.7}$$

with some integer k, $1 \leqslant k \leqslant |S|$ such that the amoebas $A_1 = (S_1, \pi_{S_1})$, $A_2 = (S_2, \pi_{S_2}), \ldots, A_k = (S_k, \pi_{S_k})$ are mutually compatible. Indeed, the partition $S = S_1 \cup \cdots \cup S_k$ is uniquely generated by the following construction. We say that points $s_1, s_2 \in S$ are *connected* if there is a contour $\Gamma \in \pi$ such that both points $s_1, s_2 \in \operatorname{Int} \Gamma$ are inside this contour. Observe now that if two compatible contours Γ_1 and Γ_2 and two sets $S_1, S_2 \subset \mathbb{Z}^2$ satisfy $S_1 \subseteq \operatorname{Int} \Gamma_1$, $S_2 \subseteq \operatorname{Int} \Gamma_2$ and the intersection $S_1 \cap S_2$ is not empty, then at least one of these two contours contains the sum $S_1 \cup S_2$ in its interior. So if we choose the sets S_i, $i = 1, \ldots, k$, as maximal connected components of S, we find that there are contours $\Gamma_i \in \pi$, $i = 1, \ldots, k$, such that $S_i \subseteq \operatorname{Int} \Gamma_i$, if $|S_i| > 1$, and no contour $\Gamma \in \pi$ can contain in its interior points of two different sets S_i, S_j, $i \neq j$, $i,j = 1,\dots,k$. So we obtain the representation (4.7) if we set

$$\pi_{S_i} = \pi \cap O(S_i), \qquad i = 1,\dots,k. \tag{4.8}$$

To each pair (S,π) let us assign a herd of animals $\rho(\pi)$:

$$\rho(\pi) = \{\Gamma \in \pi \setminus O(S),\, A_1 = (S_1, \pi_{S_1}), \dots, A_k = (S_k, \pi_{S_k})\} \in H(\Lambda(T,V)).$$

We obtain a one-to-one correspondence between the set of all pairs (S,π), where $S \subseteq T$ and $\pi \in H(V)$, and the set of all herds $\rho \in H(\Lambda(T,V))$. The weights (4.1) were chosen in such a way that for any pair (S,π) we have

$$\prod_{\theta\in\rho(\pi)} w(\theta) = \exp\Big\{ -2\beta \sum_{\Gamma\in\pi} |\Gamma| \Big\} \prod_{t\in S} (\exp\{z_t x_t(\pi)\} - 1). \tag{4.9}$$

So the desired identity (4.2) follows from (4.6).

Instead of condition (2.2), we verify the stronger condition (2.14), which states that for any animal $\theta \in \Theta(T)$ we have

$$\sum_{\tilde\theta\in\Theta:\tilde\theta\nsim\theta} w_0(\tilde\theta)\, b(\tilde\theta) + w_0(\theta)\, b(\theta) \leqslant \ln b(\theta). \tag{4.10}$$

To this end, we need certain estimates of sums of terms $w_0(A)\, b(A)$, where $A \in \mathbb{A}(T)$. Let $\mathbb{A}_S(T)$ be the set of all amoebas with a fixed set of nuclei $S \subseteq T$. In the first place, using the definitions (4.3), (4.4), and (4.5) and Lemma 3.1, we observe that for any finite set $S \subseteq T$, any number $a \geqslant 0$, and any β_1 we have

(4.11)

$$\begin{aligned}
&\sum_{A=(S,\pi_S)\in\mathbb{A}_S(T):|\pi_S|\geqslant a} w_0(A)\, b(A) \\
&\quad = (v(T))^{|S|} \sum_{A=(S,\pi_S)\in\mathbb{A}_S(T):|\pi_S|\geqslant a} \exp\{-2\,(\beta_0 - 1)|\pi_S| + |S|\} \\
&\quad \leqslant (v(T))^{|S|} e^{|S|} \exp\{-2\,(\beta_0 - \beta_1 - 1)\, a\} \sum_{A=(S,\pi_S)\in\mathbb{A}_S(T)} \exp\{-2\beta_1 |\pi_S|\} \\
&\quad \leqslant (v(T))^{|S|} e^{|S|} \exp\{-2\,(\beta_0 - \beta_1 - 1)\, a\} \prod_{\Gamma\in O(S)} (1 + \exp\{-2\beta_1|\Gamma|\}) \\
&\quad \leqslant (v(T))^{|S|} e^{|S|} \exp\{-2\,(\beta_0 - \beta_1 - 1)\, a\} \exp\Big\{ \sum_{\Gamma\in O(S)} \exp\{-2\beta_1|\Gamma|\} \Big\} \\
&\quad \leqslant (1/2)^{|S|} \min\big((q(T))^{-|S|}, 1\big) \exp\{-2\,(\beta_0 - \beta_1 - 1)\, a\}.
\end{aligned}$$

Using the estimate (4.11) and definition (1.6), we see that for any fixed contour $\Gamma^0 \in \mathbb{G}$ and any number $a \geqslant 0$,

$$
\begin{aligned}
&\sum_{A=(S,\pi_S)\in\mathbb{A}_S(T):S\subseteq\operatorname{Int}\Gamma^0,\,|\pi_S|\geqslant a} w_0(A)\,b(A) \\
&\qquad\leqslant \exp\{-2(\beta_0-\beta_1-1)\,a\} \sum_{S:S\subseteq\operatorname{Int}\Gamma^0\cap T} (2q(T))^{-|S|} \\
&\qquad\leqslant \exp\{-2(\beta_0-\beta_1-1)\,a\}(1+(2q(T))^{-1})^{|\operatorname{Int}\Gamma^0\cap T|} \\
&\qquad\leqslant \exp\{-2(\beta_0-\beta_1-1)\,a+(2q(T))^{-1}|\operatorname{Int}\Gamma^0\cap T|\} \\
&\qquad\leqslant \exp\{-2(\beta_0-\beta_1-1)\,a+\tfrac{1}{2}\operatorname{diam}(\operatorname{Int}\Gamma^0\cap T)\} \\
&\qquad\leqslant \exp\{-2(\beta_0-\beta_1-1)\,a+\tfrac{1}{4}|\Gamma^0|\}.
\end{aligned}
\tag{4.12}
$$

Similarly, fixing a point $t \in \operatorname{Int}\Gamma^0\cap T$ and observing that $\operatorname{diam}(\operatorname{Int}\Gamma^0\setminus\{t\}) \leqslant |\Gamma^0|/2$, we find that

$$
\begin{aligned}
&\sum_{A=(S,\pi_S)\in\mathbb{A}_S(T):S\subseteq\operatorname{Int}\Gamma^0,t\in S,|\pi_S|\geqslant a} w_0(A)\,b(A) \\
&\qquad\leqslant \frac{1}{2}\min((q(T))^{-1},1) \\
&\qquad\qquad\times \exp\{-2(\beta_0-\beta_1-1)\,a\} \sum_{S'\subseteq(\operatorname{Int}\Gamma^0\cap T)\setminus\{t\}} (2q(T))^{-|S'|} \\
&\qquad\leqslant \frac{1}{2}\min((q(T))^{-1},1) \\
&\qquad\qquad\times \exp\{-2(\beta_0-\beta_1-1)\,a\}(1+(2q(T))^{-1})^{|\operatorname{Int}\Gamma^0\cap T|\setminus\{t\}|} \\
&\qquad\leqslant \frac{1}{2}\min((q(T))^{-1},1) \\
&\qquad\qquad\times \exp\{-2(\beta_0-\beta_1-1)\,a+\frac{1}{4}|\Gamma^0|\}\,.
\end{aligned}
\tag{4.13}
$$

Fix a point $t \in \mathbb{Z}^2$ and a contour $\Gamma^0 \in O(t)$ and let $\mathbb{A}_{t,\Gamma^0}(T)$ be the set of all amoebas $A=(S,\pi_S)\in\mathbb{A}(T)$ such that $t\in S$ and the membrane $M(A)$ is Γ^0. Since $|\pi(S)| \geqslant |\Gamma^0|$ for $A=(S,\pi_S)\in\mathbb{A}_{t,\Gamma^0}(T)$, it follows from the estimate (4.13) that for $\beta_2 = 2\beta_1+9/8$ we have

$$
\begin{aligned}
&\sum_{A=(S,\pi_S)\in\mathbb{A}_{t,\Gamma^0}(T):|\pi_S|\geqslant a} w_0(A)\,b(A) \\
&\qquad\leqslant \tfrac{1}{2}\min((q(T))^{-1},1) \\
&\qquad\qquad\times \exp\{-2(\beta_0-\beta_1-1)\max(a,|\Gamma^0|)+\frac{1}{4}|\Gamma^0|\} \\
&\qquad= \tfrac{1}{2}\min((q(T))^{-1},1) \\
&\qquad\qquad\times \exp\{-2(\beta_0-\beta_2)\max(a,|\Gamma^0|)\} \\
&\qquad\qquad\times \exp\{-2(\beta_1+\tfrac{1}{8})\max(a,|\Gamma^0|)+\tfrac{1}{4}|\Gamma^0|\} \\
&\qquad\leqslant \tfrac{1}{2}\min((q(T))^{-1},1) \\
&\qquad\qquad\times \exp\{-2(\beta_0-\beta_2)\,a\}\exp\{-2\beta_1|\Gamma^0|\}\,.
\end{aligned}
\tag{4.14}
$$

Using the estimate (4.11) for $S = \{t\}$, the estimate (4.14) and Lemma 3.1, and recalling that $\beta_2 \geqslant \beta_1 + 1$, we find that for any fixed point $t \in \mathbb{Z}^2$ we have

(4.15)
$$\begin{aligned}
&\sum_{A=(S,\pi_S)\in\mathbb{A}(T):t\in S,|\pi_S|\geqslant a} w_0(A)\,b(A) \\
&\quad\leqslant \sum_{A=(\{t\},\pi_{\{t\}})\in\mathbb{A}(T):|\pi_{\{t\}}|\geqslant a} w_0(A)\,b(A) \\
&\qquad + \sum_{\Gamma^0\in O(t)} \sum_{\substack{A=(S,\pi_S)\in\mathbb{A}(T):|S|>1,\\ t\in S,|\pi_S|\geqslant a, M(A)=\Gamma^0}} w_0(A)\,b(A) \\
&\quad\leqslant \frac{1}{2}\min\left((q(T))^{-1},1\right)\Big(\exp\{-2(\beta_0-\beta_1-1)\,a\} \\
&\qquad\qquad + \exp\{-2(\beta_0-\beta_2)\,a\}\sum_{\Gamma^0\in O(t)}\exp\{-2\beta_1|\Gamma^0|\}\Big) \\
&\quad\leqslant \frac{3}{4}\min\left((q(T))^{-1},1\right)\exp\{-2(\beta_0-\beta_2)\,a\}\,.
\end{aligned}$$

Now from (4.15) and (1.6) we obtain that for any contour $\Gamma_0 \in \mathbb{G}$ and $\beta_3 = \beta_2 + 1/2$, we have

$$\begin{aligned}
&\sum_{A=(S,\pi_S)\in\mathbb{A}(T):\Gamma_0\in\pi_S} w_0(A)\,b(A) \\
&\quad\leqslant \sum_{t\in \operatorname{Int}\Gamma^0\cap T}\ \sum_{A=(S,\pi_S)\in\mathbb{A}(T):t\in S,|\pi_S|\geqslant|\Gamma^0|} w_0(A)\,b(A) \\
&\quad\leqslant \exp\{-2(\beta_0-\beta_2)|\Gamma^0|\}(q(T))^{-1}|\operatorname{Int}\Gamma^0\cap T| \\
&\quad\leqslant \exp\{-2(\beta_0-\beta_2)|\Gamma^0|\}\operatorname{diam}(\operatorname{Int}\Gamma^0\cap T) \\
&\quad\leqslant \exp\{-2(\beta_0-\beta_2)|\Gamma^0|\}|\Gamma^0| \\
&\quad\leqslant \exp\{-2(\beta_0-\beta_3)|\Gamma^0|\}\,.
\end{aligned} \tag{4.16}$$

In a similar way, applying (4.15) for $a = 0$, we find that for any contour $\Gamma_0 \in \mathbb{G}$ we have

$$\begin{aligned}
&\sum_{A=(S,\pi_S)\in\mathbb{A}(T),\Gamma_0\in O(S)} w_0(A)\,b(A) \\
&\quad\leqslant \sum_{t\in \operatorname{Int}\Gamma^0\cap T}\ \sum_{A=(S,\pi_S)\in\mathbb{A}(T):t\in S} w_0(A)\,b(A) \\
&\quad\leqslant (q(T))^{-1}|\operatorname{Int}\Gamma^0\cap T| \\
&\quad\leqslant \operatorname{diam}(\operatorname{Int}\Gamma^0\cap T) \leqslant |\Gamma^0|/2\,.
\end{aligned} \tag{4.17}$$

Now we start the proof of the desired estimate (4.10). We begin with the case $\theta = \Gamma \in \mathbb{G}$. Let $R(\Gamma)$ be the set consisting of all contours $\widetilde{\Gamma} \in \mathbb{G}$ that are incompatible with the contour Γ and the contour Γ itself. It follows from the definition of incompatibility that for $\theta = \Gamma$ one has

$$
\begin{aligned}
&\sum_{\tilde{\theta}\in\Theta:\tilde{\theta}\not\sim\theta} w_0(\tilde{\theta})\, b(\tilde{\theta}) + w_0(\theta)\, b(\theta) \\
&\qquad \leqslant \sum_{\widetilde{\Gamma}\in R(\Gamma)} \Big(w_0(\widetilde{\Gamma})\, b(\widetilde{\Gamma}) + \sum_{A=(S,\pi_S)\in\mathbb{A}(T):\widetilde{\Gamma}\in\pi_S} w_0(A)\, b(A) \Big) \\
&\qquad\quad + \sum_{A=(S,\pi_S)\in\mathbb{A}(T):\Gamma\in O(S)} w_0(A)\, b(A)\,.
\end{aligned}
\tag{4.18}
$$

It follows from the estimate (4.16) applied for $\Gamma^0 = \widetilde{\Gamma}$ that

$$
\sum_{A=(S,\pi_S)\in\mathbb{A}(T):\widetilde{\Gamma}\in\pi_S} w_0(A)\, b(A) \leqslant \exp\{-2\,(\beta_0 - \beta_3)|\widetilde{\Gamma}|\}\,.
\tag{4.19}
$$

Hence, also recalling definitions (4.3) and (4.5), we find that

$$
\begin{aligned}
&\sum_{\widetilde{\Gamma}\in R(\Gamma)} \Big(w_0(\widetilde{\Gamma})\, b(\widetilde{\Gamma}) + \sum_{A=(S,\pi_S)\in\mathbb{A}(T):\widetilde{\Gamma}\in\pi_S} w_0(A)\, b(A) \Big) \\
&\qquad \leqslant \sum_{\widetilde{\Gamma}\in R(\Gamma)} \big(\exp\{-2\,(\beta_0 - 1)|\widetilde{\Gamma}|\} + \exp\{-2\,(\beta_0 - \beta_3)|\widetilde{\Gamma}|\}\big)\,.
\end{aligned}
\tag{4.20}
$$

Since each of the contours $\widetilde{\Gamma} \in R(\Gamma)$ contains a vertex of the dual lattice which belongs to Γ and the number of different contours $\widetilde{\Gamma}$ of length d containing a fixed vertex does not exceed 3^d, it follows from the estimate (4.20) that for a large enough $\beta_0 > 0$ we have

$$
\sum_{\widetilde{\Gamma}\in R(\Gamma)} \Big(w_0(\widetilde{\Gamma})\, b(\widetilde{\Gamma}) + \sum_{A=(S,\pi_S)\in\mathbb{A}(T):\widetilde{\Gamma}\in\pi_S} w_0(A)\, b(A) \Big) \leqslant |\Gamma|\,.
\tag{4.21}
$$

The estimate (4.17) shows that

$$
\sum_{A=(S,\pi_S)\in\mathbb{A}(T):\Gamma\in O(S)} w_0(A)\, b(A) \leqslant \frac{1}{2}\,|\Gamma|\,.
\tag{4.22}
$$

The estimates (4.18), (4.21), and (4.22) imply that for $\theta = \Gamma$ we have

$$
\sum_{\tilde{\theta}\in\Theta:\tilde{\theta}\not\sim\theta} w_0(\tilde{\theta})\, b(\tilde{\theta}) + w_0(\theta)\, b(\theta) \leqslant 2\,|\Gamma|\,.
\tag{4.23}
$$

This is the desired estimate (4.10) for the case under consideration.

In the remaining case we have $\theta = A = (S, \pi_S) \in \mathbb{A}(T)$. Using the definition of incompatibility, we find that

$$\begin{aligned}
&\sum_{\tilde\theta \in \Theta : \tilde\theta \nsim \theta} w_0(\tilde\theta)\, b(\tilde\theta) + w_0(\theta)\, b(\theta) \\
&\quad \leqslant \sum_{\Gamma \in \pi_S} \sum_{\widetilde\Gamma \in R(\Gamma)} \Bigl(w_0(\widetilde\Gamma)\, b(\widetilde\Gamma) + \sum_{\tilde A = (\tilde S, \tilde\pi_{\tilde S}) \in \mathbb{A}(T) : \widetilde\Gamma \in \tilde\pi_{\tilde S}} w_0(\tilde A)\, b(\tilde A) \Bigr) \\
&\qquad + \sum_{\tilde A = (\tilde S, \tilde\pi_{\tilde S}) \in \mathbb{A}(T) : \tilde S \cap S \neq \varnothing} w_0(\tilde A)\, b(\tilde A) \\
&\qquad + \sum_{\Gamma \in \pi_S} \sum_{\tilde A = (\tilde S, \tilde\pi_{\tilde S}) \in \mathbb{A}(T) : \Gamma \in O(\tilde S)} w_0(\tilde A)\, b(\tilde A) \\
&\qquad + \sum_{\bar\Gamma \in O(S)} \Bigl(\sum_{\tilde A = (\tilde S, \tilde\pi_{\tilde S}) \in \mathbb{A}(T) : \bar\Gamma \in \tilde\pi_{\tilde S}} w_0(\tilde A)\, b(\tilde A) + w_0(\bar\Gamma)\, b(\bar\Gamma) \Bigr).
\end{aligned} \tag{4.24}$$

We shall estimate the separate terms in the right-hand side of estimate (4.24). First, using estimate (4.21), we observe that

$$\sum_{\Gamma \in \pi_S} \sum_{\widetilde\Gamma \in R(\Gamma)} \Bigl(w_0(\widetilde\Gamma)\, b(\widetilde\Gamma) + \sum_{\tilde A = (\tilde S, \tilde\pi_{\tilde S}) \in \mathbb{A}(T) : \widetilde\Gamma \in \tilde\pi_{\tilde S}} w_0(\tilde A)\, b(\tilde A) \Bigr) \leqslant |\pi_S| . \tag{4.25}$$

Further, using estimate (4.15) for $a = 0$, we find

$$\sum_{\tilde A = (\tilde S, \tilde\pi_{\tilde S}) \in \mathbb{A}(T) : \tilde S \cap S \neq \varnothing} w_0(\tilde A)\, b(\tilde A) \leqslant \sum_{t \in S} \sum_{\tilde A = (\tilde S, \tilde\pi_{\tilde S}) \in \mathbb{A}(T) : t \in \tilde S} w_0(\tilde A)\, b(\tilde A) \leqslant \frac{3}{4} |S| . \tag{4.26}$$

It follows from estimate (4.17) that

$$\sum_{\Gamma \in \pi_S} \sum_{\tilde A = (\tilde S, \tilde\pi_{\tilde S}) \in \mathbb{A}(T) : \Gamma \in O(\tilde S)} w_0(\tilde A)\, b(\tilde A) \leqslant \frac{1}{2} |\pi_S| . \tag{4.27}$$

Finally, using estimate (4.16) and then Lemma 3.1, and the estimate $\min |\bar\Gamma| \geqslant 4$, for some large enough β_0 we get

$$\begin{aligned}
&\sum_{\bar\Gamma \in O(S)} \Bigl(\sum_{\tilde A = (\tilde S, \tilde\pi_{\tilde S}) \in \mathbb{A}(T) : \bar\Gamma \in \tilde\pi_{\tilde S}} w_0(\tilde A)\, b(\tilde A) + w_0(\bar\Gamma)\, b(\bar\Gamma) \Bigr) \\
&\quad \leqslant \sum_{\bar\Gamma \in O(S)} \bigl(\exp\{-2(\beta_0 - \beta_3)|\bar\Gamma|\} + \exp\{-2(\beta_0 - 1)|\bar\Gamma|\} \bigr) \\
&\quad \leqslant 2 \exp\{-8(\beta_0 - \beta_3 - \beta_1 - 1)\} \sum_{\bar\Gamma \in O(S)} \exp\{-\beta_1 |\bar\Gamma|\} \leqslant \frac{1}{4} |S| .
\end{aligned} \tag{4.28}$$

Gathering the estimates (4.24)–(4.28) together and recalling definition (4.5), for any $\theta = A = (S, \pi_S) \in \mathbb{A}(T)$ we obtain

$$\sum_{\tilde\theta \in \Theta : \tilde\theta \nsim \theta} w_0(\tilde\theta)\, b(\tilde\theta) + w_0(\theta)\, b(\theta) \leqslant 2|\pi_S| + |S| = \ln b(A) . \tag{4.29}$$

We have checked condition (4.10) in all cases. $\square$

§5. Estimate of semi-invariants

Now, using the animal representation constructed in §4, we can prove the main Theorem 1.1. To this end we need the following lemma.

LEMMA 5.1. *Let $G(\mathbb{G})$ be the set of all gangs of animals consisting only of contours $\Gamma \in \mathbb{G}$. There exists a constant $\hat{\beta} > 0$ such that for any edge $e \in \mathbb{E}$ we have*

$$\sum_{\rho \in G(\mathbb{G}): e \in \cup_{\Gamma \in \bar{\rho}} \Gamma} \exp\Big\{ - 2\hat{\beta} \sum_{\Gamma \in \bar{\rho}} \alpha(\Gamma)|\Gamma| \Big\} \leqslant 1 . \tag{5.1}$$

PROOF. Fix an edge $e \in \mathbb{E}$. By $R(e)$ we denote the set of all gangs $\rho \in G(\mathbb{G})$ such that for all $\Gamma \in \bar{\rho}$ the edge e belongs to Γ. Observe that for any $\beta > 0$ we have

$$\begin{aligned}
&\sum_{\rho \in R(e)} \exp\Big\{ - 2\beta \sum_{\Gamma \in \bar{\rho}} \alpha(\Gamma)|\Gamma| \Big\} \\
&\quad = \prod_{\Gamma \in \mathbb{G}: e \in \Gamma} \Big(1 + \sum_{a=1}^{\infty} \exp\{-2\beta a |\Gamma|\} \Big) - 1 \\
&\quad = \prod_{\Gamma \in \mathbb{G}: e \in \Gamma} (1 + \exp\{-2\beta|\Gamma|\}(1 - \exp\{-2\beta|\Gamma|\})^{-1}) - 1 \\
&\quad \leqslant \exp\Big\{ \sum_{\Gamma \in \mathbb{G}: e \in \Gamma} \exp\{-2\beta|\Gamma|\}(1 - \exp\{-2\beta|\Gamma|\})^{-1} \Big\} - 1 .
\end{aligned} \tag{5.2}$$

(The product in the second line of this relation is sum of terms corresponding to any subset of the set of all contours containing the edge e. The term -1 arises since we need to exclude the empty subset.) Since the number of different contours Γ such that $|\Gamma| = d$ and $e \in \Gamma$ does not exceed 3^{d-1}, the series in the exponent in the right-hand side of (5.2) converges for $\beta > \ln 3$ and vanishes as $\beta \to \infty$. Hence, there exists a value $\hat{\beta}$ so large that

$$\sum_{\rho \in R(e)} \exp\Big\{ - (2\hat{\beta} - \ln 2) \sum_{\Gamma \in \bar{\rho}} \alpha(\Gamma)|\Gamma| \Big\} \leqslant 1 . \tag{5.3}$$

We prove below that, for this value of $\hat{\beta}$, the statement of Lemma 5.1 holds.

Consider a finite set $C \subset \mathbb{E}$ and an edge $e \in C$. The set of all gangs $\rho \in G(\mathbb{G})$ such that $e \in \bigcup_{\Gamma \in \bar{\rho}} \Gamma \subseteq C$ will be denoted by $R(e, C)$. Instead of (5.1), it suffices to prove that for any finite $C \subset \mathbb{E}$ and any edge $e \in C$ one has

$$\sum_{\rho \in R(e,C)} \exp\Big\{ - 2\hat{\beta} \sum_{\Gamma \in \bar{\rho}} \alpha(\Gamma)|\Gamma| \Big\} \leqslant 1 . \tag{5.4}$$

We prove this estimate by induction on the number of elements $|C|$. In the case $|C| = 1$, the set of gangs $R(e, C)$ is empty and so the estimate (5.4) is evident.

We use it as the initial step of our induction. Suppose now that for any edge $\tilde{e} \in C$, $\tilde{e} \neq e$ one has

$$\sum_{\rho \in R(\tilde{e}, C \setminus \{e\})} \exp\left\{ - 2\hat{\beta} \sum_{\Gamma \in \bar{\rho}} \alpha(\Gamma)|\Gamma| \right\} \leqslant 1. \tag{5.5}$$

Any gang of contours $\rho \in R(e, C)$ can be represented (perhaps in a nonunique way) as the sum (see the definition of sums of groups of animals in §2)

$$\rho = \rho_0 + \sum_{(e_1, \ldots, e_k)} \rho_{e_i}, \tag{5.6}$$

where the gang ρ_0 belongs to $R(e)$, the gangs ρ_{e_i} belong to $R(e_i, C \setminus \{e\})$ and the summation is taken over all subsets of edges $(e_1, \ldots, e_k) \subseteq \bigcup_{\Gamma \in \bar{\rho}_0} \Gamma \setminus \{e\}$ (including the empty subset with $k = 0$). Let $R(e, C, \rho_0)$ be the set of gangs $\rho \in R(e, C)$ that can be represented in the form (5.6) with a fixed gang ρ_0. Then, using the induction assumption (5.5), we find that

$$\begin{aligned}
&\sum_{\rho \in R(e, C, \rho_0)} \exp\left\{ - 2\hat{\beta} \sum_{\Gamma \in \bar{\rho}} \alpha(\Gamma)|\Gamma| \right\} \\
&\qquad \leqslant \exp\left\{ - 2\hat{\beta} \sum_{\Gamma \in \bar{\rho}_0} \alpha(\Gamma)|\Gamma| \right\} \\
&\qquad\quad \times \left(\sum_{(e_1, \ldots, e_k)} \prod_{i=1}^{k} \left(\sum_{\rho \in R(e_i, C \setminus \{e\})} \exp\left\{ - 2\hat{\beta} \sum_{\Gamma \in \bar{\rho}} \alpha(\Gamma)|\Gamma| \right\} \right) \right) \\
&\qquad \leqslant \exp\left\{ - 2\hat{\beta} \sum_{\Gamma \in \bar{\rho}_0} \alpha(\Gamma)|\Gamma| \right\} \\
&\qquad\quad \times \prod_{\tilde{e} \in \bigcup_{\Gamma \in \bar{\rho}_0} \Gamma} \left(1 + \sum_{\rho \in R(\tilde{e}, C \setminus \{e\})} \exp\left\{ - 2\hat{\beta} \sum_{\Gamma \in \bar{\rho}} \alpha(\Gamma)|\Gamma| \right\} \right) \\
&\qquad \leqslant \exp\left\{ - 2\hat{\beta} \sum_{\Gamma \in \bar{\rho}_0} \alpha(\Gamma)|\Gamma| \right\} 2^{\sum_{\Gamma \in \bar{\rho}_0} |\Gamma|} \\
&\qquad \leqslant \exp\left\{ - (2\hat{\beta} - \ln 2) \sum_{\Gamma \in \bar{\rho}_0} \alpha(\Gamma)|\Gamma| \right\}.
\end{aligned} \tag{5.7}$$

Returning to the representation (5.6), we see that

$$\sum_{\rho \in R(e, C)} \exp\left\{ - 2\hat{\beta} \sum_{\Gamma \in \bar{\rho}} \alpha(\Gamma)|\Gamma| \right\} \leqslant \sum_{\rho_0 \in R(e)} \exp\left\{ - (2\hat{\beta} - \ln 2) \sum_{\Gamma \in \bar{\rho}_0} \alpha(\Gamma)|\Gamma| \right\} \tag{5.8}$$

and the desired induction estimate (5.4) follows from the estimate (5.3) proved above. □

PROOF OF THEOREM 1.1. Proposition 4.1 allows us to apply the cluster expansion (2.18) to the logarithm of the partition functions $Z_T(V; z_t,\, t \in T) = Z_w(\Lambda(T,V))$ (see (4.2)). So if (see (4.1) and (4.3))

$$|w(\theta)| < w_0(\theta), \qquad \theta \in \Theta(T), \tag{5.9}$$

then we have the expansion

$$\ln Z_T(V; z_t,\, t \in T) = \sum_{\rho \in G(\Lambda(T,V))} q_w(\rho)\,. \tag{5.10}$$

Condition (5.9) is fulfilled if $\beta > \beta_0$ and the values of the variables $(z_t,\, t \in T)$ belong to the polydisk

$$P_T = \{(z_t,\, t \in T) : |z_t| < \ln(1 + v(T))\} \subset \mathbb{C}^T. \tag{5.11}$$

So $\ln Z_T(V; z_t,\, t \in T)$ and the terms $q_w(\rho)$ of the series (5.10) are functions of the variables z_t, $t \in T$, holomorphic in this polydisk, continuous in its closure, and we can differentiate it in z_t term by term. Let $G_T(V)$ be the set of all gangs $\rho \in G(\Lambda(T,V))$ consisting of animals $\theta \in \Lambda(T,V)$ such that the sum of nuclei of the amoebas included in ρ is equal to T. The derivatives of $q_w(\rho)$ with respect to all the variables z_t, $t \in T$, vanish if $\rho \notin G_T(V)$. So from the relations (3.11) and (5.10) we obtain

$$s_V(T,r) = \sum_{\rho \in G_T(V)} \left.\frac{\partial^{|r|} q_w(\rho)}{\prod_{t \in T} \partial z_t^{r_t}}\right|_{z_t \equiv 0,\, t \in T}. \tag{5.12}$$

The derivatives included in (5.12) will be estimated by the Cauchy formula using integration over the boundary of the polydisk (5.11). It follows from estimate (2.19) and definitions (4.1) and (4.3) that for $|z_t| = \ln(1 + v(t))$, $t \in T$, one has

$$\begin{aligned} |q_w(\rho)| \leqslant \Bigl(\sum_{\theta \in \bar\rho} w_0(\theta)\, b(\theta)\Bigr) \\ \times \exp\Bigl\{ - 2(\beta - \beta_0) \\ \times \Bigl(\sum_{\Gamma \in \mathbb{G}(\bar\rho)} \alpha(\Gamma)|\Gamma| + \sum_{A = (S,\pi_S) \in \mathbb{A}(\bar\rho)} \alpha(A)|\pi_S|\Bigr)\Bigr\}. \end{aligned} \tag{5.13}$$

Here we denote the set of all contours $\Gamma \in \mathbb{G}$ such that $\Gamma \in \bar\rho$ by $\mathbb{G}(\bar\rho)$ and the set of all amoebas $A \in \mathbb{A}$ such that $A \in \bar\rho$ by $\mathbb{A}(\bar\rho)$. It follows from (5.12) and (5.13), the Cauchy formula, and the estimate $\ln(1+x) \leqslant x$ for $x \geqslant 0$, that

(5.14)
$$\begin{aligned}|s_V(T,r)| \leqslant &(\ln(1+v(t)))^{-|r|} \\ &\times \sum_{\rho\in G_T(V)} \Big[\sum_{\theta\in\bar\rho} w_0(\theta)\, b(\theta) \\ &\qquad \times \exp\Big\{-2(\beta-\beta_0) \\ &\qquad\qquad \times \Big(\sum_{\Gamma\in\mathbb{G}(\bar\rho)} \alpha(\Gamma)|\Gamma| + \sum_{A=(S,\pi_S)\in\mathbb{A}(\bar\rho)} \alpha(A)|\pi_S|\Big)\Big\}\Big].\end{aligned}$$

Under the condition $\beta_0 > 1$ it follows from definitions (4.3), (4.4), and (4.5) that for any gang ρ,

$$\sum_{\theta\in\bar\rho} w_0(\theta)\, b(\theta) \leqslant |\bar\rho| \leqslant \exp\Big\{\sum_{\Gamma\in\mathbb{G}(\bar\rho)} \alpha(\Gamma)|\Gamma| + \sum_{A=(S,\pi_S)\in\mathbb{A}(\bar\rho)} \alpha(A)|\pi_S|\Big\}. \tag{5.15}$$

So we can rewrite the estimate (5.14) as

$$\begin{aligned}|s_V(T,r)| \leqslant &(\ln(1+v(t)))^{-|r|} \sum_{\rho\in G_T(V)} \\ &\times \exp\Big\{-2(\beta-\beta_0-1) \\ &\qquad \times \Big(\sum_{\Gamma\in\mathbb{G}(\bar\rho)} \alpha(\Gamma)|\Gamma| + \sum_{A=(S,\pi_S)\in\mathbb{A}(\bar\rho)} \alpha(A)|\pi_S|\Big)\Big\}.\end{aligned} \tag{5.16}$$

This formula is the starting point for the following estimates.

Fix a gang of animals $\rho \in G_T(V)$. We must describe the structure of its support $\bar\rho$. It follows from the definition of $G_T(V)$ that the set $\mathbb{A}(\bar\rho)$ is not empty and

$$\bigcup_{A=(S,\pi_S)\in\mathbb{A}(\bar\rho)} S = T. \tag{5.17}$$

Consider the set of contours

$$K(\bar\rho) = \mathbb{G}(\bar\rho) \cup \Big(\bigcup_{A=(S,\pi_S):A\in\mathbb{A}(\bar\rho)} \pi_S\Big). \tag{5.18}$$

Treating $K(\bar\rho)$ as a subgraph of the graph of all contours $\mathbb{G}$ with the edges connecting incompatible contours, we consider the system of all connected components of this subgraph: $K_1(\bar\rho), \dots, K_N(\bar\rho)$, where N is an integer, $N = N(K(\bar\rho))$. Observe that each component $K_i(\bar\rho)$ contains a contour $\Gamma_i \in O(T)$. Indeed, in the opposite case, $K_i(\bar\rho) \subseteq \mathbb{G}(\bar\rho)$ and there is no path along incompatible animals belonging to $\bar\rho$ which connects $K_i(\bar\rho)$ with the amoebas from $\mathbb{A}(\bar\rho)$. This contradicts the definition of gangs. Denote by $Q_i(\bar\rho)$ the set of all $t \in T$ such that the intersection $K_i(\bar\rho) \cap O(t)$

is nonempty. Since we assumed that $|T| > 1$, the definition of compatibility of animals and condition (5.17) exclude situations when $O(t) \cap K(\bar{\rho}) = \varnothing$ for some $t \in T$. So,

$$\bigcup_{i=1}^{N} Q_i(\bar{\rho}) = T\,. \tag{5.19}$$

Observe also that if $Q_i(\bar{\rho}) \cap Q_j(\bar{\rho}) \neq \varnothing$, then either

$$Q_i(\bar{\rho}) \subseteq Q_j(\bar{\rho}) \quad \text{or} \quad Q_j(\bar{\rho}) \subseteq Q_i(\bar{\rho})\,.$$

Indeed, if

$$t \in Q_i(\bar{\rho}) \cap Q_j(\bar{\rho}), \quad \Gamma_i \in K_i(\bar{\rho}) \cap O(t), \quad \Gamma_j \in K_j(\bar{\rho}) \cap O(t),$$

then either

$$\operatorname{Int}\Gamma_i \supseteq \bigcup_{\Gamma \in K_j(\bar{\rho})} \operatorname{Int}\Gamma \quad \text{or} \quad \operatorname{Int}\Gamma_j \supseteq \bigcup_{\Gamma \in K_i(\bar{\rho})} \operatorname{Int}\Gamma.$$

Finally the situation when there are two nonempty disjoint sets $\widehat{T}_1$ and $\widehat{T}_2$ such that each of the sets $Q_i(\bar{\rho})$ is a subset of either $\widehat{T}_1$ or $\widehat{T}_2$ is also impossible. Indeed, in the opposite case there is no path along incompatible animals belonging to $\bar{\rho}$ which connect amoebas having nuclei belonging to $\widehat{T}_1$ with amoebas having nuclei belonging to $\widehat{T}_2$; again, this contradicts the definition of gangs. So the only remaining possibility is that $Q_{i_0}(\bar{\rho}) = T$ for some i_0. But then

$$\sum_{\Gamma \in K(\bar{\rho})} |\Gamma| \geqslant \sum_{\Gamma \in K_{i_0}(\bar{\rho})} |\Gamma| \geqslant |\Gamma_T|, \tag{5.20}$$

where the contour Γ_T was introduced in the formulation of Theorem 1.1. This estimate is the main conclusion of the last paragraph.

Fix a sufficiently large constant $\bar{\beta}$, which will be specified below. Observe, using the estimates (5.16) and (5.20), that

$$\begin{aligned} |s_V(T, r)| &\leqslant \exp\{-2(\beta - \bar{\beta} - \beta_0 - 1)|\Gamma_T|\}(\ln(1 + v(t)))^{-|r|} \\ &\times \sum_{\rho \in G_T(V)} \exp\Big\{ - 2\bar{\beta}\Big(\sum_{\Gamma \in \mathbb{G}(\bar{\rho})} \alpha(\Gamma)|\Gamma| + \sum_{A=(S,\pi_S) \in \mathbb{A}(\bar{\rho})} \alpha(A)|\pi_S|\Big)\Big\}. \end{aligned} \tag{5.21}$$

We must estimate the sum over $\rho \in G_T(V)$ in (5.21). Let $\mathbb{A}(\rho) = (\mathbb{A}(\bar{\rho}), \alpha)$ be the group of animals consisting of all amoebas A included in the gang ρ with the same multiplicity $\alpha(A)$ that they have in $\bar{\rho}$. It follows from the description of connected components of the set $K(\bar{\rho})$ given above that the gang ρ can be represented (perhaps in a nonunique way) as the sum (see (2.17))

$$\rho = \mathbb{A}(\rho) + \sum_{e \in E(\mathbb{A}(\bar{\rho}))} \rho_e, \tag{5.22}$$

where $E(\mathbb{A}(\rho))$ is the set of all edges of the dual lattice having nonempty intersection with contours included in $\mathbb{A}(\bar{\rho})$, and ρ_e is either a gang of contours such that some of its contours contain e or the empty group of contours. Assuming that $\bar{\beta} \geqslant \hat{\beta}$, we can estimate the contributions of all possible gangs ρ_e with a fixed e by using Lemma 5.1. Thus we find

$$\begin{aligned}\sum_{\rho\in G_T(V)} &\exp\Big\{-2\bar\beta\Big(\sum_{\Gamma\in\mathbb{G}(\bar\rho)}\alpha(\Gamma)|\Gamma| + \sum_{A=(S,\pi_S)\in\mathbb{A}(\bar\rho)}\alpha(A)|\pi_S|\Big)\Big\}\\ &\leqslant \sum_{\rho\in\mathcal{A}_T}\exp\Big\{-(2\bar\beta-\ln 6)\Big(\sum_{A=(S,\pi_S)\in\bar\rho}\alpha(A)|\pi_S|\Big)\Big\},\end{aligned} \tag{5.23}$$

where $\mathcal{A}_T$ is the set of all groups of amoebas $\rho = (\bar\rho, \alpha)$, $\bar\rho \in \mathbb{A}(T)$. (The extra $\ln 6$ term in the right-hand side of (5.23) arises in the following way. In the representation $6 = 3(1+1)$, the first 1 originates in estimate (5.1), the second 1 originates from the possibility that ρ_e is the empty set, and the factor 3 originates from the following evident estimate: the number of edges e having nonempty intersection with the contour Γ does not exceed $3|\Gamma|$.) Let $\mathcal{A}_t$, $t\in T$, be the set of all groups of amoebas $\rho = (\bar\rho,\alpha)\in\mathcal{A}_T$ such that $t\in S$ for all $A=(S,\pi_S)\in\bar\rho$. It is clear that

$$\begin{aligned}\sum_{\rho\in\mathcal{A}_T}&\exp\Big\{-(2\bar\beta-\ln 6)\Big(\sum_{A=(S,\pi_S)\in\bar\rho}\alpha(A)|\pi_S|\Big)\Big\}\\ &\leqslant\prod_{t\in T}\Big(1+\sum_{\rho\in\mathcal{A}_t}\exp\Big\{-(2\bar\beta-\ln 6)\Big(\sum_{A=(S,\pi_S)\in\bar\rho}\alpha(A)|\pi_S|\Big)\Big\}\Big).\end{aligned} \tag{5.24}$$

Further, for any $t\in T$, we have

$$\begin{aligned}\sum_{\rho\in\mathcal{A}_t}&\exp\Big\{-(2\bar\beta-\ln 6)\Big(\sum_{A=(S,\pi_S)\in\bar\rho}\alpha(A)|\pi_S|\Big)\Big\}\\ &\leqslant \prod_{A=(S,\pi_S)\in\mathbb{A}(T):t\in S,\alpha=1,2,\dots}\big(1+\exp\{-(2\bar\beta-\ln 6)\,\alpha|\pi_S|\}\big)\\ &\leqslant\exp\Big\{\sum_{A=(S,\pi_S)\in\mathbb{A}(T):t\in S,\alpha=1,2,\dots}\exp\{-(2\bar\beta-\ln 6)\,\alpha|\pi_S|\}\Big\}.\end{aligned} \tag{5.25}$$

Finally, we need the following statement. For a sufficiently large $\bar\beta$ and all T and $t\in T$, we have

$$\sum_{A=(S,\pi_S)\in\mathbb{A}(T):t\in S,\alpha=1,2,\dots}\exp\{-(2\bar\beta-\ln 6)\alpha|\pi_S|\}\leqslant\ln 2\,. \tag{5.26}$$

Indeed, summing the geometrical series with the quotient $\exp\{-(2\bar\beta-\ln 6)|\pi_S|\}$, for $\bar\beta$ large enough, we find

$$\begin{aligned}\sum_{A=(S,\pi_S)\in\mathbb{A}(T):t\in S,\alpha=1,2,\dots}&\exp\{-(2\bar\beta-\ln 6)\,\alpha|\pi_S|\}\\ &\leqslant 2\sum_{A=(S,\pi_S)\in\mathbb{A}(T):t\in S}\exp\{-(2\bar\beta-\ln 6)|\pi_S|\}\,.\end{aligned} \tag{5.27}$$

After this we can almost literally repeat the derivation of relation (4.15) for $a=0$ and find that the right-hand side of (5.27) vanishes as $\bar\beta\to\infty$. Gathering the estimates (5.21), (5.23), (5.24), (5.25), and (5.26), we obtain

$$|s_V(T,R)| \leqslant 3^{|T|} \exp\{-2(\beta - \bar{\beta} - \beta_0 - 1)|\Gamma_T|\}(\ln(1+v(t)))^{-|r|}. \tag{5.28}$$

Observing that

$$\ln(1+x) \geqslant x - \frac{x^2}{2} \geqslant \frac{1}{2}x \quad \text{for } 0 \leqslant x \leqslant 1 \tag{5.29}$$

and that $|T| \leqslant |r|$, we finally obtain the desired estimate (1.7) from (5.28) and definition (4.4). □

References

[Br] D. Brydges, *A short course on cluster expansions*, Critical Phenomena, Random Systems, Gauge Theory (Les Houches, 1984) (K. Ostewalder and R. Stora, eds.), North-Holland, Amsterdam–New York, 1986, pp. 129–183.

[Do1] R. Dobrushin, *Induction on volume and cluster expansions*, Proc. VIIIth Intern. Congress on Math Phys. (M. Mebkhout and B. Seneor, eds.), World Scientific, Singapore, 1987, pp. 73–91.

[Do2] ———, *A new approach to the analysis of Gibbs perturbations of Gaussian fields*, Selecta Math. Soviet. **7** (1988), 221–277.

[DM] R. Dobrushin and M. Martirosjan, *Non-finite perturbation of Gibbsian fields*, Teoret. Mat. Fiz. **74** (1988), no. 1, 16–28; English transl., Theoret. and Math. Phys. **74** (1988), 10–20.

[DS] R. L. Dobrushin and S. B. Shlosman, *Completely analytical interactions: constructive description*, J. Statist. Phys. **46** (1987), 983–1014.

[DW] R. Dobrushin and V. Warstat, *Completely analytical interactions with infinite values*, Probab. Theory Related Fields **84** (1990), 335–359.

[GJ] J. Glimm and A. Jaffe, *Quantum physics. A functional integral point of view*, Springer-Verlag, New York–Berlin, 1987.

[HKZ] P. Holický, R. Kotecký, and M. Zahradník, *Rigid interfaces for lattice models at low temperatures*, J. Statist. Phys. **50** (1988), 755–812.

[KP] R. Kotecký and D. Preiss, *Cluster expansion for abstract polymer models*, Comm. Math. Phys. **103** (1986), 491–498.

[Ma] V. A. Malyshev, *Cluster expansions in lattice models of statistical physics and the quantum field theory*, Uspekhi Mat. Nauk **35** (1980), no. 2, 3–53; English transl., Russian Math. Surveys **35** (1980), no. 2, 1–62.

[MM] V. A. Malyshev and R. A. Minlos, *Gibbs random fields. Cluster expansions*, Kluwer, Dordrecht, 1991.

[MS] A. E. Mazel and Y. M. Suhov, *Ground states of a boson quantum lattice model*, Sinaĭ's Moscow Seminar on Dynamical Systems, Amer. Math. Soc. Transl. Ser. 2, vol. 171, Amer. Math. Soc., Providence, RI, 1996, pp. 185–226.

[Se] E. Seiler, *Gauge theories as a problem of constructive quantum field theory and statistical mechanics*, Lecture Notes in Phys., vol. 159, Springer-Verlag, Berlin–New York, 1982.

[Sim] B. Simon, *The statistical mechanics of lattice gases,* Vol. 1, Princeton University Press, Princeton, NJ, 1993.

[Sin] Ya. G. Sinai, *The theory of phase transitions*: *Rigorous results*, Pergamon, London, 1981.

Institute for Problems of Information Transmission, Moscow, Russia

The Erwin Schrödinger International Institute for Mathematical Physics, Wien, Austria

Amer. Math. Soc. Transl.
(2) Vol. **177**, 1996

Pseudoclassical Theory of a Relativistic Spinning Particle

Dmitri M. Gitman

ABSTRACT. A short survey of the pseudoclassical theory of relativistic spin one half particle and its recent development is given. In particular, the canonical quantization of the pseudoclassical model is described for the massive and massless cases. Different representations of the Dirac propagator in terms of path integrals are derived. Introduction of the anomalous magnetic moment in the model is presented, as well as the pseudoclassical description of Weyl particles.

§1. Introduction

The systematic utilization of the algebra and analysis of anticommuting variables, initiated to a great extent by Felix Berezin ([**1**]), has undoubtedly exerted considerable influence on the development of theoretical physics. Among numerous examples of the application of these methods is the so-called pseudoclassical theory of relativistic spin one half particle. Felix Berezin was also one of the originators of this theory.

Classical and pseudoclassical models of relativistic particles and their quantization have been discussed lately in different contexts. One of the reasons is that these simple examples may be used to learn how to solve some typical problems that also arise in string theory, gravity, and so on. On the other hand, it is an important question in itself whether there exist classical models for any relativistic particles (with any spin) whose quantization reproduces, in some sense, the corresponding field theory or one particle sector of the corresponding quantum field theory.

In this paper I would like to give a short survey of the pseudoclassical theory of spin one half relativistic particle and recent developments in this direction, since that can serve as a convincing demonstration of how fruitful the methods and ideas introduced by Berezin still are.

1991 *Mathematics Subject Classification.* Primary 81Q20.

§2. Action of a relativistic spin one half particle in an external electromagnetic field

The action of spin one half relativistic particle with spinning degrees of freedom described by Grassmannian (odd) variables was first proposed by Berezin and Marinov ([**2**]) and immediately was discussed and investigated in the papers [**3**–**7**]. In the most symmetric form, the action of a spinning particle in an external electromagnetic field can be written as [**4, 6, 7**]

$$(1)\quad S = \int_0^1 \left[-\frac{\dot{x}^2}{2e} - e\,\frac{m^2}{2} - g\dot{x}^\alpha A_\alpha + igeF_{\alpha\beta}\psi^\alpha\psi^\beta + i\left(\frac{\dot{x}_\alpha\psi^\alpha}{e} - m\psi^5\right)\chi - i\psi_n\dot{\psi}^n \right] d\tau,$$

where x^α, e are ordinary (bosonic or even) variables and ψ^n, χ are odd variables depending on the parameter τ, $\tau \in [0,1]$, the latter playing the role of time in this theory, $A_\alpha(x)$ is the external electromagnetic field potential, $F_{\alpha\beta}(x)$ is the Maxwell strength tensor, and g the electrical charge. Greek indices run over $0, 1, 2, 3$ and Latin indices n, m run over $0, 1, 2, 3, 5$. The metric tensors are $\eta_{\alpha\beta} = \operatorname{diag}(1, -1, -1, -1)$ and $\eta_{mn} = \operatorname{diag}(1, -1, -1, -1, -1)$. The spinning degrees of freedom in such a model are described by the odd variables ψ^n, which is why the model is called pseudoclassical. There are two type of gauge transformations in the theory with the action (1): reparametrizations,

$$(2)\qquad \delta x = \dot{x}\xi, \quad \delta e = \frac{d}{d\tau}(e\xi), \quad \delta\psi^n = \dot{\psi}^n\xi, \quad \delta\chi = \frac{d}{d\tau}(\chi\xi),$$

and supertransformations,

$$(3)\quad \delta x^\alpha = i\psi^\alpha\epsilon, \quad \delta e = i\chi\epsilon, \quad \delta\chi = \dot{\epsilon}, \quad \delta\psi^\alpha = \frac{1}{2e}(\dot{x}^\alpha - i\chi\psi^\alpha)\epsilon, \quad \delta\psi^5 = \frac{m}{2}\epsilon,$$

where ξ is an even and ϵ is an odd τ-dependent parameter. The Lagrangian equations of motion have the form

$$\frac{d}{d\tau}\left[\frac{1}{e}(\dot{x}_\alpha - i\psi_\alpha\chi)\right] + g\dot{x}^\beta F_{\beta\alpha} + iegF_{\beta\gamma,\alpha}\psi^\beta\psi^\gamma = 0,$$

$$(4)\qquad \frac{1}{2e^2}(\dot{x}^\alpha - i\psi^\alpha\chi)^2 - \frac{m^2}{2} + igF_{\alpha\beta}\psi^\alpha\psi^\beta = 0, \quad \dot{x}_\alpha\psi^\alpha - me\psi^5 = 0,$$

$$2\dot{\psi}_\alpha + 2egF_{\beta\alpha}\psi^\beta - \frac{\dot{x}_\alpha}{e}\chi = 0, \quad 2\dot{\psi}^5 - m\chi = 0\,.$$

Calculating the total angular momentum corresponding to the action (1), we obtain

$$M_{\mu\nu} = L_{\mu\nu} + S_{\mu\nu}, \quad L_{\mu\nu} = x_\mu p_\nu - x_\nu p_\mu, \quad S_{\mu\nu} = i(\psi_\mu\psi_\nu - \psi_\nu\psi_\mu)\,.$$

The spatial part of $S_{\mu\nu}$ forms a tree-dimensional spin vector $\mathbf{s} = (s_k)$

$$(5)\qquad s_k = \epsilon_{kjl}S_{ij}/2 = i\epsilon_{kjl}\psi_l\psi_j,$$

where ϵ_{kjl} is the tree-dimensional Levi–Civita symbol. To demonstrate that this vector really behaves like a spin, one can use, for example, the nonrelativistic approximation $dx^i/dx^0 \ll 1$ and consider the case of a magnetic field only. In this case $F_{0i} = 0$, and $F_{ij} = -\epsilon_{ijk}B_k$, where B_k are the components of a magnetic field

B. Selecting the gauge $\chi = 0$ and $x^0 = \tau$, the latter leads effectively to $e = 1/m$ in the approximation in question, and one can find from the equations (4)

$$\frac{d\mathbf{s}}{dt} = \frac{g}{m}\,[\mathbf{s}\times\mathbf{B}], \qquad m\,\frac{d^2\mathbf{x}}{dt^2} = g\left[\frac{d\mathbf{x}}{dt}\times\mathbf{B}\right] + \frac{g}{m}\,\nabla(\mathbf{s}\cdot\mathbf{B})\,. \tag{6}$$

The equations (6) describe ([**8**]) a nonrelativistic motion of a particle with total spin momentum $\mathbf{s}$ and total magnetic momentum $g\mathbf{s}/m$, which confirms the interpretation of the action (1).

Passing to the Hamiltonian formulation, we introduce the canonical momenta

$$\begin{gathered} p_\alpha = \frac{\partial L}{\partial \dot{x}^\alpha} = -\frac{1}{e}\,(\dot{x}_\alpha - i\psi_\alpha\chi) - gA_\alpha, \\ P_e = \frac{\partial L}{\partial \dot{e}} = 0, \quad P_\chi = \frac{\partial_r L}{\partial \dot{\chi}} = 0, \quad P_n = \frac{\partial_r L}{\partial \dot{\psi}^n} = -i\psi_n\,. \end{gathered} \tag{7}$$

It follows from equation (7) that there exist primary constraints $\Phi^{(1)} = 0$,

$$\Phi^{(1)} = (\Phi^{(1)}_1 = P_\chi,\ \Phi^{(1)}_2 = P_e,\ \Phi^{(1)}_{3n} = P_n + i\psi_n)\,. \tag{8}$$

We construct the Hamiltonian $H^{(1)}$ according to the standard procedure ([**9, 10**]) (we use the notations of the book [**10**]),

$$H^{(1)} = H + \lambda_a\Phi^{(1)}_a, \quad H = -\frac{e}{2}\,(\mathcal{P}^2 + 2igF_{\alpha\beta}\psi^\alpha\psi^\beta - m^2) - i\,(\mathcal{P}_\alpha\psi^\alpha - m\psi^5)\,\chi,$$

where $\mathcal{P} = -p_\mu - gA_\mu(x)$. From the condition of conservation of primary constraints $\Phi^{(1)}_{1,2}$ in time τ, $\dot{\Phi}^{(1)}_{1,2} = \{\Phi^{(1)}_{1,2}, H^{(1)}\} = 0$, we find the secondary constraints $\Phi^{(2)} = 0$,

$$\Phi^2 = (\Phi^{(2)}_1 = \mathcal{P}_\alpha\psi^\alpha - m\psi^5,\ \Phi^{(2)}_2 = \mathcal{P}^2 + 2igF_{\alpha\beta}\psi^\alpha\psi^\beta - m^2), \tag{9}$$

and the same conditions for the constraints $\Phi^{(1)}_{3n}$ give equations for the determination of λ_{3n}. Thus, the Hamiltonian H turns out to be proportional to the constraints, as one can expect in the case of reparametrization invariant theory, $H = i\chi\Phi^{(2)}_1 - (e/2)\,\Phi^{(2)}_2$. No more secondary constraints arise from the Dirac procedure, and the Lagrange multipliers λ_1 and λ_2 remain undetermined, in accordance with the fact that the number of gauge transformations parameters equals two for the theory in question ([**10**]). One can pass from the initial set of constraints $(\Phi^{(1)}, \Phi^{(2)})$ to the equivalent one $(\Phi^{(1)}, T)$, where $T = \Phi^{(2)} - (i/2)\,\Phi^{(1)}_{3n}\partial_r\Phi^2/\partial\psi_n$. The new set of constraints can be explicitly divided into a set of first-class constraints, which is $(\Phi^{(1)}_{1,2}, T)$, and into a set of second-class constraints, which is $\Phi^{(1)}_{3n}$.

§3. Canonical quantization

Here we consider the canonical operator quantization of the theory in question. Because it contains first-class constraints, well-known problems arise in the quantization procedure. Moreover, in this particular case, due to the reparametrization invariance, an additional problem appears; namely, the Hamiltonian vanishes on the constraints surface. Usually one tries to avoid this difficulty by using the so-called Dirac method of quantization ([**9**]), in which first-class constraints in the sense of restrictions on the state vectors are considered. In the problem in question, the Dirac wave equation arises precisely in this way ([**3–7**]). Unfortunately, this scheme of quantization creates many questions, e.g., how to construct the appropriate Hilbert space, what the Schrödinger equation is, and so on. A consistent, but technically more complicated way is to work in the physical sector; namely, first, on the classical level, one has to impose gauge conditions to reduce the theory to one with second-class constraints only, and then to quantize it using Dirac brackets. Here we present this method of quantization ([**10, 11**]), which gives the Dirac equation as a Schrödinger one. For simplicity, we restrict ourselves to the free particle case.

We fix a gauge, preliminarily imposing three additional conditions

$$\Phi_1^G = x_0 - \zeta\tau = 0, \qquad \Phi_2^G = \chi = 0, \qquad \Phi_3^G = \psi^5 = 0,$$

where $\zeta = -\operatorname{sign} p_0$. The gauge $x_0 - \zeta\tau = 0$ was first proposed in [**11**] as a conjugated gauge condition for the constraint $p^2 = m^2$ in the case of scalar and spinning particles. In contrast with the gauge $x_0 = \tau$, which, together with continuous reparametrization symmetry, breaks time reflection symmetry and therefore fixes the variables ζ, the former gauge breaks only the continuous symmetry, so that the variable ζ remains in the theory to describe states of particles $\zeta = +1$ and states of antiparticles $\zeta = -1$. It is precisely this circumstance that allows one to get Klein–Gordon and Dirac equations as Schrödinger equations in the course of canonical quantization. From the consistency condition $\dot{\Phi}^G = 0$, we find an additional condition

$$\Phi_4^G = e - |p_0|^{-1}.$$

The total set of constraints $\Phi = (\Phi^1, \Phi^2, \Phi^G)$ is already a second-class one. However, now we are dealing with constraints that depend on time. In this case the canonical method of quantization by means of Dirac brackets, generally speaking, has to be modified ([**10, 12**]). To avoid this new problem, one can pass to a time-independent set of constraints by performing the canonical transformation $x_0' = x_0 - \zeta\tau$, $x'^i = x^i$, $p_\mu' = p_\mu$, with a generating function of the form $W = x^\mu p_\mu' + \tau|p_0'| + W_0$, where W_0 is the generating function of the identity transformation with respect to all the variables except x_0, p_0. We change, in fact, only the coordinate x_0, and therefore the primes on the other variables are henceforth omitted. The transformed Hamiltonian $H'^{(1)}$ is of the form

$$H'^{(1)} = H^{(1)} + \partial W/\partial\tau = H + \{\Phi\},$$

where H is the physical Hamiltonian,

$$H = \omega = (\mathbf{p}^2 + m^2)^{1/2}, \qquad \mathbf{p} = (p_k), \tag{10}$$

and $\{\Phi\}$ are terms proportional to the constraints Φ. We present the constraints Φ in an equivalent form, dividing them into two groups K and ϕ, each being a set of second-class constraints

$$K = (e - \omega^{-1}, P_e, \chi, P_\chi, \psi^5, P_5, x_0', |p_0| - \omega), \qquad \phi = (P_\mu + i\psi_\mu, p\psi)\,. \tag{11}$$

In (11) and below $p_0 = -\zeta\omega$. Next we eliminate the variables e, P_e, χ, P_χ, ψ^5, P_5, x_0', and $|p_0|$ from our considerations by using the constraints K. These constraints have a certain special form ([**10**]) such that for the rest of the variables

$\eta = (x^k, p_k, \zeta, \psi^\mu, P_\mu)$ the Dirac brackets with respect to all the constraints Φ reduce to brackets with respect to the constraints ϕ only. Calculating the Dirac brackets between the variables η, we obtain[1]

$$\begin{aligned}
&\{x^j, x^k\}_{D(\phi)} = -(i/m^2)[R^j, R^k]_-, \qquad &&\{x^j, p_k\}_{D(\phi)} = \delta^j_k,\\
(12)\quad &\{\psi^\mu, \psi^\nu\}_{D(\phi)} = (i/2)(\eta^{\mu\nu} - p^\mu p^\nu m^{-2}), &&\{x^j, \psi^\mu\}_{D(\phi)} = -R^j p^\mu m^{-2},\\
&\{p_j, p_k\}_{D(\phi)} = \{p_j, \psi^\mu\}_{D(\phi)} = \{\zeta, \eta\}_{D(\phi)} = 0, &&R^j = \psi^j - \psi^0 p^j p_0^{-1}.
\end{aligned}$$

According to the recipes of quantization of theories with second-class constraints ([**9, 10**]), the operators $\hat{\eta}$ corresponding to the variables η must satisfy the relations[2]

$$(13)\qquad [\hat{\eta}, \hat{\eta}'\} = i\{\eta, \eta'\}_{D(\phi)}\big|_{\eta\to\hat{\eta}}, \quad \widehat{P}_\mu + i\hat{\psi}_\mu = 0, \quad \hat{p}\hat{\psi} = 0.$$

Besides, we assume the operator $\hat{\zeta}$ to have the eigenvalues $\zeta = \pm 1$ by analogy with the classical theory, so that $\hat{\zeta}^2 = 1$. We can construct the realization of the algebra (13) of the operators $\hat{\eta}$ in the Hilbert state space $\mathcal{R}$ whose elements $\mathbf{f} \in \mathcal{R}$ are four-component columns

$$\mathbf{f} = \begin{pmatrix} f_1(\mathbf{x}) \\ f_2(\mathbf{x}) \end{pmatrix},$$

where $f_1(\mathbf{x})$ and $f_2(\mathbf{x})$ are two component columns. We seek all the operators in block-diagonal form, in particular, we immediately choose the operators $\hat{\zeta}$ and $\hat{p}_k$ in the form[3]

$$(14)\qquad \hat{\zeta} = \gamma^0, \qquad \hat{p}_k = -i\partial_k \mathbf{I},$$

where I and $\mathbf{I}$ are 2×2 and 4×4 identity matrices, respectively. By the assumption concerning block diagonality, the general form of the operators $\hat{\psi}^\mu$ will be $\hat{\psi}^\mu = A^\mu \mathbf{I} + B^\mu_k \Sigma^k$, where $\mathbf{\Sigma} = \operatorname{diag}(\sigma, \sigma)$ and σ^k, $k = 1, 2, 3$ are Pauli matrices. The commutation relations involving $\hat{\psi}^\mu$ imply that for nonzero B^μ_k the equation $A^\mu = 0$ must hold and that B^μ_k must depend on the operators $\hat{p}_k$ and $\hat{\zeta}$ only. Assuming that under spatial rotations $\hat{\psi}^0$ behaves as a scalar and $\hat{\psi}^k$ as a 3-vector, one can write

$$\hat{\psi}^0 = a\mathbf{\Sigma}\hat{\mathbf{p}}, \qquad \hat{\psi}^k = b\Sigma^k + c\hat{p}^k\mathbf{\Sigma}\hat{\mathbf{p}} + d\varepsilon^{kjl}\hat{p}^j\Sigma^l,$$

where a, b, c, and d depend on $\hat{\mathbf{p}}^2$ and $\hat{\zeta}$ only. From (13) and (12) it follows that $(\hat{\psi}^0)^2 = \hat{\mathbf{p}}^2/4m^2$. Hence, we can choose $a = 1/2m$. Besides, we immediately set $d = 0$, since this coefficient can always be made equal to zero by using the similarity transformation $\exp(r\hat{\mathbf{p}}\mathbf{\Sigma})\hat{\psi}^\mu \exp(-r\hat{\mathbf{p}}\mathbf{\Sigma})$ for a certain r. The coefficients b and c are uniquely determined (up to the similarity transformation mentioned above) from the relations (13) and (12). We are finally led to

$$(15)\qquad \hat{\psi}^0 = \frac{1}{2m}\mathbf{\Sigma}\hat{\mathbf{p}}, \quad \hat{\psi}^k = \frac{\gamma^0}{2}\left(\Sigma^k + \frac{\hat{p}^k\mathbf{\Sigma}\hat{\mathbf{p}}}{m(\widehat{\omega} + m)}\right), \quad \widehat{\omega} = (-\partial_k^2 + m^2)^{1/2}.$$

---

[1]We mean a generalized Dirac bracket whose form depends on the parities of the variables involved ([**10**]).

[2]In (13) $[\cdot,\cdot\}$ stands for the generalized commutator which is either a commutator or an anticommutator depending on the parities of the corresponding operators.

[3]We use the standard representation for the γ matrices, $[\gamma^\mu, \gamma^\nu]_+ = 2\eta^{\mu\nu}$.

Similarly one can find the operator $\hat{x}^k$,

$$\hat{x}^k = x^k\mathbf{I} + \frac{\epsilon^{kij}\Sigma^i\hat{p}^j}{2m(\widehat{\omega}+m)}\,.$$

The evolution within the time τ of the state vectors from $\mathcal{R}$ is described by the Schrödinger equation

$$(i\partial/\partial\tau - \hat{H})\mathbf{f} = 0,$$

where according to (10) and (14), $\hat{H} = \widehat{\omega}\mathbf{I}$. Passing to physical time $x_0 = \zeta\tau$ in this equation, we obtain ([**11**]),

$$i\,\frac{\partial}{\partial x_0}\,f = \gamma^0\widehat{\omega}f\,. \tag{16}$$

We interpret $f_+(x) = f_1(x)$ as the wavefunction of a particle and $f_-(x) = \sigma^2 f_2^*(x)$ as that of an antiparticle and define the scalar product in $\mathcal{R}$ accordingly,

$$(f,g) = \int [f_1^+ g_1 + g_2^+ f_2]\,d\mathbf{x} = \int f_\zeta^+ g_\zeta d\mathbf{x}, \qquad \zeta = \pm\,.$$

The operators $\hat{H}$, $\hat{\psi}^\mu$, and $\hat{x}^k$ are selfconjugate with respect to this scalar product. The equation for $f_\zeta(x)$ follows from (16), $(i\partial/\partial x_0 - \widehat{\omega})\,f_\zeta = 0$. In this case, the equations for the wavefunctions of a particle and antiparticle have an equivalent form due to the absence of an external electromagnetic field. In the rest frame, the spin operator $\hat{s}_k$ acts upon the wavefunctions as $\sigma^k/2$ and coincides with $\hat{\psi}^k$.

The quantum mechanics constructed is completely equivalent to the one obtained from Dirac theory. Indeed, (16) is simply the Dirac equation in the Foldy–Wouthuysen (FW) representation ([**13**]). Carrying out the unitary FW transformation, we come to the usual Dirac equation,

$$f = U\varphi, \qquad U = \frac{\widehat{\omega}+m+\gamma\hat{\mathbf{p}}}{[2\widehat{\omega}(\widehat{\omega}+m)]^{1/2}}, \qquad (i\gamma^\mu\partial_\mu - m)\varphi = 0\,.$$

Applying the FW transformation to the operators $\hat{\psi}^\mu$ and $\hat{x}^k$, we obtain

$$U^+\hat{\psi}^0U = \hat{\Psi}^0 = \frac{1}{2m}\gamma^5\hat{p}_k\sigma^{k0}, \qquad U^+\hat{\psi}^kU = \hat{\Psi}^k = \frac{1}{2m}\gamma^5[\hat{p}_j\sigma^{jk} + H_D\sigma^{0k}],$$
$$U^+\hat{x}^kU = \widehat{X}^k = x^k\mathbf{I} + \frac{i}{2m}(\gamma^k - \hat{p}^k\gamma\hat{\mathbf{p}}\widehat{\omega}^{-2})\,.$$

The operators $\hat{x}^k$ and $\widehat{X}^k$ are the operators of middle position in the FW and Dirac pictures, respectively. The expression for these operators were obtained from covariance considerations by Pryce ([**14**]).

Let us discuss the quantization of a massless spinning particle ([**15, 16**]). In this connection, one can consider the limit $m = 0$ of the massive case and compare it with an independent quantization of the classical action describing a massless particle in the beginning. To consider the limit, it is convenient to use a different gauge at $m \neq 0$; namely, to use, instead of the gauge condition $\psi^5 = 0$, the condition $\psi_0 = 0$. After fixing the gauge and eliminating some variables, we arrive at the theory with the variables x^i, p_i, ζ, ψ^l, P_l, $l = (i, 5)$, and second-class constraints $\phi = 0$,

$$\phi = (p_i \psi^i + m\psi^5, P_l + i\psi_l) . \tag{17}$$

It is useful to introduce the transversal and longitudinal

$$\psi^{i\perp} = \Pi^i_j \psi^j, \quad \Pi^i_j = \delta^i_j - p^{-2} p_i p_j, \quad p = |\mathbf{p}|, \quad \psi^{\|} = p_i \psi^i$$

parts of ψ^i, because of it is convenient to treat cases $m \neq 0$ and $m = 0$ in these variables in a similar way. The first constraint (17) is, in fact, a relation between $\psi^{\|}$ and ψ^5, $\psi^{\|} = -m\psi^5$, whereas $\psi^{i\perp}$ are not constrained. Nonzero Dirac brackets between all the variables have the form

$$\begin{aligned} \{x^k, x^j\}_{D(\phi)} &= \frac{i}{\omega^2} [\psi^{k\perp}, \psi^{j\perp}]_- + \frac{im}{\omega^2 p^2} (p_k [\psi^{j\perp}, \psi^5]_- - p_j [\psi^{k\perp}, \psi^5]_-), \\ \{x^i, \psi^{j\perp}\}_{D(\phi)} &= -\frac{\psi^{i\perp} p_j}{p^2} + \frac{m}{p^2} \Pi^i_j \psi^5, \quad \{x^i, \psi^5\}_{D(\phi)} = -\frac{m}{\omega^2} \psi^{i\perp} + \frac{m^2}{\omega^2 p^2} p_i \psi^5, \\ \{\psi^{i\perp}, \psi^{j\perp}\}_{D(\phi)} &= -\frac{i}{2} \Pi^i_j, \quad \{\psi^5, \psi^5\}_{D(\phi)} = -\frac{i}{2} \frac{p^2}{\omega^2}, \quad \{x^k, p_j\}_{D(\phi)} = \delta^k_j . \end{aligned} \tag{18}$$

One can introduce new variables θ^i and X^k instead of x^k and ψ^μ, ψ^5,

$$\begin{aligned} X^k &= x^k - \frac{i}{\omega + m} [\psi^{k\perp}, \psi^5]_-, \quad \theta^i = \psi^{i\perp} - \frac{\omega}{p^2} p_i \psi^5, \\ x^i &= X^i - \frac{i}{\omega(\omega + m)} [\theta^i, \theta^{\|}], \quad \psi^{i\perp} = \theta^{i\perp}, \quad \psi^5 = -\frac{1}{\omega} \theta^{\|}, \end{aligned} \tag{19}$$

which are independent with respect to the second-class constraints (17). Using (18), for the nonzero Dirac brackets we obtain

$$\{X^k, p_j\}_{D(\phi)} = \delta^k_j, \qquad \{\theta^k, \theta^j\}_{D(\phi)} = -(i/2)\, \delta_{kj} . \tag{20}$$

Now the commutation relations between the corresponding operators $\widehat{X}^i$, $\hat{p}_i$, $\hat{\zeta}$, $\hat{\theta}^k$ must be calculated by means of Dirac brackets (20), so that the nonzero commutators are

$$[\widehat{X}^k, \hat{p}_j]_- = i\delta^k_j, \qquad [\hat{\theta}^k, \hat{\theta}^j]_+ = \delta_{kj}/2 . \tag{21}$$

One can construct a realization of the algebra (21) and relation $\hat{\zeta}^2 = 1$ in the same Hilbert space $\mathcal{R}$ as before. The realization for $\hat{\zeta}$ and $\hat{p}_k$ remains the same and

$$\widehat{X}^k = X^k \mathbf{I}, \qquad \hat{\theta}^k = \frac{1}{2} \Sigma^k. \tag{22}$$

In the realization, the operators of angular momentum $\widehat{M}_{\mu\nu}$ and the spin $\hat{s}^k$ have the form

$$\begin{aligned} \widehat{M}_{0j} &= \widehat{X}_0 \hat{p}_j - \widehat{X}_j \hat{p}_0 - \frac{i}{2} \frac{\hat{p}_j}{\hat{p}_0} + \frac{\hat{p}_0}{2\hat{\omega}(\hat{\omega} + m)} \epsilon_{jkl} \hat{p}_k \Sigma^l, \\ \widehat{M}_{ij} &= \widehat{X}_i \hat{p}_j - \widehat{X}_j \hat{p}_i - \frac{1}{2} \epsilon_{ijk} \Sigma^k, \qquad \hat{s}^k = i\epsilon_{kjl} \hat{\psi}^j \hat{\psi}^l = \frac{1}{2} \Sigma^k. \end{aligned} \tag{23}$$

It is known that the square of the Pauli–Lubanski vector $\widehat{W}^\mu = 1/2\epsilon^{\mu\nu\lambda\sigma}\widehat{M}_{\mu\nu}\hat{p}_\sigma$ is a Casimir operator for the Poincaré algebra. For this realization and in the center of mass system we have

$$\begin{aligned}
\widehat{W}^0 &= 0, \\
\widehat{W}^k &= m\,\frac{\hat{p}_0}{\widehat{\omega}}\,\hat{s}^k, \\
\widehat{W}^2 &= -(\widehat{W}^i)^2 = -\frac{3}{4}m^2.
\end{aligned}$$

The latter confirms that the system in question has spin one half.

Since the Schrödinger equation has the same form (16), the quantum mechanics constructed in this gauge is completely equivalent to the one from the standard Dirac theory; namely, it is connected with the latter by the same Foldy–Wouthuysen transformation. Moreover, applying the same transformation to the operators (23), we get the operators of the angular momentum in the Dirac theory.

$$\mathcal{U}^+\widehat{M}_{\mu\nu}\mathcal{U} = \widehat{X}_\mu\hat{p}_\nu - \widehat{X}_\nu\hat{p}_\mu - \frac{1}{2}\sigma_{\mu\nu}, \qquad \sigma_{\mu\nu} = \frac{i}{2}\left[\gamma_\mu, \gamma_\nu\right]_- .$$

Considering the limit $m = 0$, one can remark that all formulas are nonsingular in the mass and admit such a limit. On the classical level, after the gauge fixing, it is possible to use both the variables x^i, p_i, ζ, $\psi^{i\perp}$, ψ^5 or the variables X^i, p_i, ζ, θ^i, the Dirac brackets of the latter do not contain mass at all and expressions of the former via the latter are nonsingular in mass. The first set of variables at $m = 0$ splits into two (anti) commuting variables with other groups x^i, p_i, $\psi^{i\perp}$, and ψ^5. The Poincaré generators are only expressed via the first group of variables and commute with ψ^5. Instead of the Casimir operator W^2, which vanishes at $m = 0$, a new one appears: helicity $\Lambda = \hat{p}^{-1}\hat{p}_k\hat{s}^k$. It turns out that at $m = 0$ the variable ψ^5_a can be omitted from the action (1). The quantization of such a modified action reproduces the physical sector of the limit of the massive quantum mechanics. As we mentioned above, the Dirac brackets for the variables X^i, p_i, ζ, θ^k do not depend on the mass, which means that the realization (22) remains in the limit $m = 0$. It is clear that the realization does not depend on the presence of the operator $\hat{\psi}^5$. In the limit we have $\psi^5 = \Lambda$. The Schrödinger equation (16) with $m = 0$ gives the Dirac equation with $m = 0$ after the corresponding FW transformation. The total Hilbert space now forms a reducible representation of the Poincaré group (right and left neutrinos). It follows from the described structure of the quantum mechanics that in the limit $m = 0$ one does not need the variable ψ^5 in the theory. Indeed, one can take the action (1) at $m = 0$ and omit ψ^5 in the beginning. In such a theory, after the same gauge fixing (in particular, $\psi_0 = 0$), we have only the variables x^i, p_i, ζ, $\psi^{i\perp}$ on the constraint surface. Their Dirac brackets and the expressions of the Poincaré generators coincide with the corresponding expressions of the massive theory at $m = 0$. The same realization is available. If one introduces the operator $i\hat{p}^{-1}\hat{p}_k\epsilon_{kjl}\hat{\psi}^{l\perp}\hat{\psi}^{j\perp}$, which is in fact the operator $\hat{\psi}^5$ of the massive case, then the theory literally coincides with the limit of the massive case. In this connection one can remark that the dimension of the Hilbert space in the realization discussed does not depend on the presence of the variable ψ^5 at $m = 0$ and coincides with the dimension in the massive case.

§4. Path integral representation for Dirac propagator in external electromagnetic field

Here we are going to discuss path integral representations for the propagator of the relativistic spinning particle in an external electromagnetic field. We shall demonstrate that in such representations integrands have the form $\exp(iS_{\text{eff}})$, where the effective actions S_{eff} are very closely related to the action (1). We should mention that different kinds of such representations were derived and discussed in [**17–24, 44**]. Below we mainly follow the paper [**24**].

It is well known that the propagator of a relativistic spinning particle is the casual Green function $S^c(x,y)$ of the Dirac equation. For our purpose, it is convenient to deal with the transformed by $\gamma^5 = \gamma^0\gamma^1\gamma^2\gamma^3$ function $\tilde{S}^c(x,y) = S^c(x,y)\gamma^5$, which obeys the properly transformed Dirac equation for the Green function

$$(\widehat{\mathcal{P}}_\nu\tilde{\gamma}^\nu - m\gamma^5)\,\tilde{S}^c(x,y) = \delta^4(x-y), \tag{24}$$

where $\widehat{\mathcal{P}}_\mu = i\partial_\mu - gA_\mu(x)$ and $\tilde{\gamma}^\mu = \gamma^5\gamma^\mu$. The matrices $\tilde{\gamma}^\nu$, have the same commutation relations as the initial ones γ^ν, $[\tilde{\gamma}^\mu, \tilde{\gamma}^\nu]_+ = 2\eta^{\mu\nu}$, so that the tilde sign will be omitted hereafter. For all the γ-matrices we have $[\gamma^m, \gamma^n]_+ = 2\eta^{mn}$, $m,n = 0,1,2,3,5$, $\eta^{mn} = \operatorname{diag}(1,-1,-1,-1,-1)$.

Following Schwinger ([**25**]), we present $\tilde{S}^c_{\alpha\beta}(x,y)$ as a matrix element of an operator $\widehat{S}^c_{\alpha\beta}$, but in the coordinate space only,

$$\tilde{S}^c_{\alpha\beta}(x,y) = \langle x|\widehat{S}^c_{\alpha\beta}|y\rangle, \tag{25}$$

where the spinor indices are written explicitly for clarity only once and will be omitted hereafter; $|x\rangle$ are eigenvectors for some Hermitian operators of coordinates X^μ, the corresponding canonically conjugated operators of momenta are P_μ, so that:

$$\begin{gathered}
X^\mu|x\rangle = x^\mu|x\rangle, \quad \langle x|y\rangle = \delta^4(x-y), \quad \int |x\rangle\langle x|\,dx = I,\\
[P_\mu, X^\nu]_- = -i\delta^\nu_\mu, \quad P_\mu|p\rangle = p_\mu|p\rangle, \quad \langle p|p'\rangle = \delta^4(p-p'),\\
\int |p\rangle\langle p|\,dp = I, \quad \langle x|P_\mu|y\rangle = -i\partial_\mu\delta^4(x-y), \quad \langle x|p\rangle = (2\pi)^{-2}e^{ipx},\\
[\Pi_\mu, \Pi_\nu]_- = -igF_{\mu\nu}(X), \quad \Pi_\mu = -P_\mu - gA_\mu(X)\,.
\end{gathered} \tag{26}$$

The equation (24) implies the formal solution for the operator $\widehat{S}^c$, $\widehat{S}^c = [\Pi_\nu\gamma^\nu - m\gamma^5]^{-1}$. The operator in the square brackets is a pure Fermi operator, if one regards the γ-matrices as Fermi operators. In the general case, the inverse operator to a Fermi operator F can be presented by an integral over the superproper time (λ, χ) of an exponential with an even exponent ([**24**]),

$$F^{-1} = \int_0^\infty d\lambda \int e^{i[\lambda(F^2+i\epsilon)+\chi F]}\,d\chi, \tag{27}$$

where λ is an even variable and χ is an odd one, the latter anticommutes with F by definition. Here and below integrals over odd variables are understood as Berezin integrals ([**1**]). The representation (27) is an analog of the Schwinger proper-time representation for an inverse operator, convenient in the Fermi case. Using (27) and taking into account the fact that

$$(\Pi_\nu\gamma^\nu - m\gamma^5)^2 = \Pi^2 - m^2,$$

for the operator $\widehat{S}^c$ one can write

$$\widehat{S}^c = \int_0^\infty d\lambda \int e^{-i\widehat{\mathcal{H}}(\lambda,\chi)} d\chi,$$
$$\widehat{\mathcal{H}}(\lambda,\chi) = \lambda(m^2 - \Pi^2 + (ig/2)F_{\alpha\beta}\gamma^\alpha\gamma^\beta) + (\Pi_\nu\gamma^\nu - m\gamma^5)\chi .$$

Thus, the Green function (25) takes the form

$$\tilde{S}^c = \tilde{S}^c(x_{\text{out}}, x_{\text{in}}) = \int_0^\infty d\lambda \int \langle x_{\text{out}}|e^{-i\hat{\mathcal{H}}(\lambda,\chi)}|x_{\text{in}}\rangle\, d\chi . \tag{28}$$

Now one can present the matrix element appearing in the expression (28) by means of a path integral. Despite the fact that the operator $\widehat{\mathcal{H}}(\lambda,\chi)$ has a γ-matrix structure, it is possible to do this as usual. Namely, first we write $\exp(-i\widehat{\mathcal{H}}) = (\exp(-i\widehat{\mathcal{H}}/N))^N$ and then insert $N-1$ resolutions of identity $\int |x\rangle\langle x|\, dx = I$ between all the operators $\exp(-i\widehat{\mathcal{H}}/N)$. Besides, we introduce N additional integrations over λ and χ so as to transform the ordinary integrals over these variables into the corresponding path-integrals,

$$\begin{aligned}\tilde{S}^c = \lim_{N\to\infty} &\int_0^\infty d\lambda_0 \int d\chi_0 \int_{-\infty}^\infty dx_1 \cdots \int_{-\infty}^\infty dx_{N-1} \\ &\times \int_{-\infty}^\infty d\lambda_1 \cdots \int_{-\infty}^\infty d\lambda_N \int_{-\infty}^\infty d\chi_1 \cdots \int_{-\infty}^\infty d\chi_N \\ &\times \prod_{k=1}^N \langle x_k|e^{-i\widehat{\mathcal{H}}(\lambda_k,\chi_k)\Delta\tau}|x_{k-1}\rangle\, \delta(\lambda_k - \lambda_{k-1})\,\delta(\chi_k - \chi_{k-1}),\end{aligned} \tag{29}$$

where $\Delta\tau = 1/N$, $x_0 = x_{\text{in}}$, $x_N = x_{\text{out}}$. Bearing in mind the limiting process, one can calculate the matrix elements from (29) approximately,

$$\langle x_k|e^{-i\widehat{\mathcal{H}}(\lambda_k,\chi_k)\Delta\tau}|x_{k-1}\rangle \approx \langle x_k|1 - i\widehat{\mathcal{H}}(\lambda_k,\chi_k)\,\Delta\tau|x_{k-1}\rangle, \tag{30}$$

using the resolutions of the identity $\int |p\rangle\langle p|\, dp$. In this connection it is important to notice that the operator $\widehat{\mathcal{H}}(\lambda_k,\chi_k)$ was originally in symmetric form with respect to the operators $\hat{x}$ and $\hat{p}$. Indeed, the only term in $\widehat{\mathcal{H}}(\lambda_k,\chi_k)$ that contains products of these operators is $[P_\alpha, A^\alpha(X)]_+$. One can verify that this is the maximally symmetrized expression that can be combined from operators appearing here (see the remark in [**26**]). Thus, one can write

$$\widehat{\mathcal{H}}(\lambda,\chi) = \text{Sym}_{(\hat{x},\hat{p})}\, \mathcal{H}(\lambda,\chi,\hat{x},\hat{p}),$$

where $\mathcal{H}(\lambda,\chi,x,p)$ is the Weyl symbol of the operator $\widehat{\mathcal{H}}(\lambda,\chi)$ in the sector of coordinates and momenta,

$$\mathcal{H}(\lambda,\chi,x,p) = \lambda(m^2 - \mathcal{P}^2 + (ig/2)F_{\alpha\beta}\gamma^\alpha\gamma^\beta) + (\mathcal{P}_\nu\gamma^\nu - m\gamma^5)\chi,$$

and $\mathcal{P}_\nu = -p_\nu - gA_\nu(x)$. It is a general statement ([**27**]), which can be easily verified in this concrete case by direct calculations, that the matrix elements (30) are expressed in terms of the Weyl symbols at the middle point $\overline{x}_k = (x_k + x_{k-1})/2$. Taking all that into account, one can see that in the limiting process the matrix elements (30) can be replaced by the expressions

$$\int_{-\infty}^{\infty} \frac{dp_k}{(2\pi)^4} \exp i\left[p_k \frac{x_k - x_{k-1}}{\Delta\tau} - \mathcal{H}(\lambda_k, \chi_k, \overline{x}_k, p_k)\right]\Delta\tau, \tag{31}$$

which are noncommuting due to the γ-matrix structure and are situated in (29) so that the numbers k increase from right to left. For the two δ-functions accompanying each matrix element (30) in the expression (29), we use the integral representations

$$\delta(\lambda_k - \lambda_{k-1})\,\delta(\chi_k - \chi_{k-1}) = \frac{i}{2\pi}\int_{-\infty}^{\infty} d\pi_k \int d\nu_k\, e^{i[\pi_k(\lambda_k - \lambda_{k-1}) + \nu_k(\chi_k - \chi_{k-1})]},$$

where ν_k are odd variables. Then we formally attribute to the γ-matrices entering into (31) the index k, and then we attribute to all quantities the "time" τ_k corresponding to the index k, i.e., $\tau_k = k\Delta\tau$, so that $\tau \in [0,1]$. Introducing the T-product that acts on γ-matrices, it is possible to gather all the expressions in (29) in one exponent and then deal with the γ-matrices like with odd variables. Thus, for the right side of (29) we get

$$\begin{aligned}\tilde{S}^c = \mathrm{T}\int_0^\infty d\lambda_0 \int d\chi_0 \int_{x_{\text{in}}}^{x_{\text{out}}} Dx \int Dp \int_{\lambda_0} D\lambda \int_{\chi_0} D\chi \int D\pi \int D\nu \\ \times \exp\left\{ i\int_0^1 \left[\lambda\left(\mathcal{P}^2 - m^2 - \frac{ig}{2}F_{\alpha\beta}\gamma^\alpha\gamma^\beta\right)\right.\right. \\ \left.\left. + (m\gamma^5 - \mathcal{P}_\nu\gamma^\nu)\chi + p\dot{x} + \pi\dot{\lambda} + \nu\dot{\chi}\right]d\tau\right\},\end{aligned} \tag{32}$$

where x, p, λ, π, are even and χ, ν are odd trajectories obeying the boundary conditions $x(0) = x_{\text{in}}$, $x(1) = x_{\text{out}}$, $\lambda(0) = \lambda_0$, $\chi(0) = \chi_0$. The operation of T-ordering acts on the γ-matrices, which are assumed to formally depend on the time τ. The expression (32) can be reduced to

$$\begin{aligned}\tilde{S}^c = \int_0^\infty d\lambda_0 \int d\chi_0 \int_{x_{\text{in}}}^{x_{\text{out}}} Dx \int Dp \int_{\lambda_0} D\lambda \int_{\chi_0} D\chi \int D\pi \int D\nu \\ \times \exp\left\{ i\int_0^1 \left[\lambda\left(\mathcal{P}^2 - m^2 - \frac{ig}{2}F_{\alpha\beta}\frac{\delta_l}{\delta\rho_\alpha}\frac{\delta_l}{\delta\rho_\beta}\right)\right.\right. \\ \left.\left. + \left(m\frac{\delta_l}{\delta\rho_5} - \mathcal{P}_\nu\frac{\delta_l}{\delta\rho_\nu}\right)\chi + p\dot{x} + \pi\dot{\lambda} + \nu\dot{\chi}\right]d\tau\right\} \\ \times \mathrm{T}\exp\int_0^1 \rho_n(\tau)\gamma^n\,d\tau\bigg|_{\rho=0},\end{aligned}$$

where five odd sources $\rho_n(\tau)$ are introduced, which anticommute with the γ-matrices by definition. One can present the quantity

$$\mathrm{T}\exp\int_0^1 \rho_n(\tau)\gamma^n\,d\tau$$

as a path integral over odd trajectories ([**24**]),

$$\mathrm{T}\exp\int_0^1 \rho_n(\tau)\gamma^n d\tau = \exp\left(i\gamma^n\frac{\partial_l}{\partial\theta^n}\right)\int_{\psi(0)+\psi(1)=\theta} \times\exp\left[\int_0^1 (\psi_n\dot{\psi}^n - 2i\rho_n\psi^n)\,d\tau + \psi_n(1)\psi^n(0)\right]\mathcal{D}\psi\Bigg|_{\theta=0},$$

(33)
$$\mathcal{D}\psi = D\psi\left[\int_{\psi(0)+\psi(1)=0} D\psi\exp\left\{\int_0^1 \psi_n\dot{\psi}^n d\tau\right\}\right]^{-1},$$

where θ^n are odd variables anticommuting with the γ-matrices and $\psi^n(\tau)$ are odd trajectories of integration obeying the boundary conditions written out under the integration signs. Using (33), we get the Hamiltonian path integral representation for the Green function in question:

$$\begin{aligned}\tilde{S}^c =&\exp\left(i\gamma^n\frac{\partial_l}{\partial\theta^n}\right)\int_0^\infty d\lambda_0\int d\chi_0\int_{\lambda_0} D\lambda\int_{\chi_0} D\chi\int_{x_{\mathrm{in}}}^{x_{\mathrm{out}}} Dx\\ &\times\int Dp\int D\pi\int D\nu\int_{\psi(0)+\psi(1)=\theta}\mathcal{D}\psi\\ &\times\exp\left\{i\int_0^1 \left[\lambda(\mathcal{P}^2 - m^2 + 2igeF_{\alpha\beta}\psi^\alpha\psi^\beta)\right.\right.\\ &\qquad\left. + 2i\,(\mathcal{P}_\alpha\psi^\alpha - m\psi^5)\chi - i\psi_n\dot{\psi}^n + p\dot{x} + \pi\dot{\lambda} + \nu\dot{\chi}\right]d\tau\\ &\qquad\left.+\psi_n(1)\,\psi^n(0)\right\}\Bigg|_{\theta=0}.\end{aligned}$$

Integrating over momenta, we get a path integral in Lagrangian form,

(34)
$$\begin{aligned}\tilde{S}^c =&\exp\left(i\gamma^n\frac{\partial_\ell}{\partial\theta^n}\right)\int_0^\infty de_0\int d\chi_0\int_{e_0} De\int_{\chi_0} D\chi\int_{x_{\mathrm{in}}}^{x_{\mathrm{out}}} Dx\\ &\times\int D\pi\int D\nu\int_{\psi(0)+\psi(1)=\theta}\mathcal{D}\psi M(e)\\ &\times\exp\left\{i\int_0^1\left[-\frac{\dot{x}^2}{2e} - \frac{e}{2}m^2 - g\dot{x}^\mu A_\mu(x) + iegF_{\mu\nu}(x)\,\psi^\mu\psi^\nu\right.\right.\\ &\qquad\left. + i\left(\frac{\dot{x}_\mu\psi^\mu}{e} - m\psi^5\right)\chi - i\psi_n\dot{\psi}^n + \pi\dot{e} + \nu\dot{\chi}\right]d\tau\\ &\qquad\left.+\psi_n(1)\,\psi^n(0)\right\}\Bigg|_{\theta=0},\end{aligned}$$

where

(35)
$$M(e) = \int Dp\exp\left\{\frac{i}{2}\int_0^1 ep^2 d\tau\right\}.$$

One can find a discussion of the role of the measure (35) in [**24**].

The exponential in the integrand (34) can be regarded as an effective and nondegenerate Lagrangian action of a spinning particle in an external field. It consists of two principal parts. The first one, which contains two summands with the derivatives of e and χ, can be treated as a gauge fixing term and corresponds to the gauge conditions $\dot{e} = \dot{\chi} = 0$. The remaining part of the effective action, in fact, coincides with the gauge invariant action (1) of a spinning particle.

§5. Spinor structure of Dirac propagator

In this section we shall demonstrate that the Dirac propagator can be expressed using a bosonic path integral over coordinates only; the integrand of this path integral differs from the corresponding expression in the scalar case by a spin factor, whose spinor structure is completely described. This problem has attracted the attention of researchers for a long time. Thus, Feynman, who had first written his path integral for the probability amplitude in nonrelativistic quantum mechanics ([**28**]) and then the path integral for the causal Green function of the Klein–Gordon equation (propagator of a scalar particle, [**29**]), had also attempted to write a representation for the Dirac propagator via a bosonic path integral ([**30**]). After Berezin had introduced his integral over Grassmannian variables, it turned out to be possible to represent this propagator using both bosonic and Grassmannian path integrals, as demonstrated in the previous section. Nevertheless, attempts to write the Dirac propagator via a bosonic path integral only continued. Thus, Polyakov ([**31**]) assumed that the propagator of the free Dirac electron in $D = 3$ Euclidean space-time can be presented by means of a bosonic path integral, similar to the scalar particle, modified by a so-called spinor factor. This idea was developed in [**32**] to write a spinor factor for Dirac fermions interacting with a non-Abelian gauge field in D-dimensional Euclidean space-time. Unfortunately, in that representation the spinor factor itself was presented using certain additional bosonic path integrals, and its γ-matrix structure was not defined explicitly. As was shown in [**33**], the problem can be solved directly by all the Grassmannian integrations in expression (34). Let us consider this solution.

We start with representation (34) for the Dirac propagator, in which we first integrate over π and ν, and then use the arising δ-functions to remove the functional integration over e and χ,

$$
\begin{aligned}
\tilde{S}^c = &- \exp\left\{ i\gamma^n \frac{\partial_\ell}{\partial\theta^n} \right\} \int_0^\infty de_0 \int_{x_{\text{in}}}^{x_{\text{out}}} Dx \\
&\times \int_{\psi(0)+\psi(1)=\theta} \mathcal{D}\psi M(e_0) \int_0^1 \left(\frac{\dot{x}_\mu \psi^\mu}{e_0} - m\psi^5 \right) d\tau \\
&\times \exp\left\{ i \int_0^1 \left[-\frac{\dot{x}^2}{2e_0} - \frac{e_0}{2} m^2 - g\dot{x}^\mu A_\mu(x) \right.\right. \\
&\qquad \left. + ige_0 F_{\mu\nu}(x)\psi^\mu\psi^\nu - i\psi_n\dot{\psi}^n \right] d\tau \\
&\qquad \left.\left. + \psi_n(1)\psi^n(0) \right\} \right|_{\theta=0} .
\end{aligned} \tag{36}
$$

Then it is convenient to replace the integration over ψ by one over related odd velocities ω,

$$\psi(\tau)=\frac{1}{2}\int_0^1 \varepsilon(\tau-\tau')\omega(\tau')\,d\tau'+\frac{1}{2}\theta,\quad \omega(\tau)=\dot{\psi}(\tau),\quad \varepsilon(\tau)=\operatorname{sign}\tau\,. \tag{37}$$

There are no more restrictions on ω; because of (37), the boundary conditions for ψ are fulfilled automatically. The corresponding Jacobian does not depend on the variables and cancels with the one from the measure (32). Thus,[4]

$$\begin{aligned}\tilde{S}^c=&-\frac{1}{2}\exp\left\{i\gamma^n\frac{\partial_\ell}{\partial\theta^n}\right\}\int_0^\infty de_0\int_{x_{\rm in}}^{x_{\rm out}}Dx\\&\times\int \mathcal{D}\omega M(e_0)\left[\frac{\dot{x}_\mu}{e_0}(\varepsilon\omega^\mu+\theta^\mu)-m(\varepsilon\omega^5+\theta^5)\right]\\&\times\exp\left\{i\left[-\frac{\dot{x}^2}{2e_0}-\frac{e_0}{2}m^2-g\dot{x}A(x)\right.\right.\\&\qquad\left.\left.-\frac{ie_0g}{4}(\omega^\mu\varepsilon-\theta^\mu)F_{\mu\nu}(x)(\varepsilon\omega^\nu+\theta^\nu)+\frac{i}{2}\omega_n\varepsilon\omega^n\right]\right\}\Bigg|_{\theta=0},\end{aligned}$$

where

$$\mathcal{D}\omega=D\omega\left[\int D\omega\exp\left\{-\frac{1}{2}\omega^n\varepsilon\omega_n\right\}\right]^{-1}. \tag{38}$$

One can prove that for a function $f(\theta)$ from the Grassmann algebra the following identity holds

(39)
$$\begin{aligned}\exp\left\{i\gamma^n\frac{\partial_\ell}{\partial\theta^n}\right\}f(\theta)\Big|_{\theta=0}&=f\left(\frac{\partial_\ell}{\partial\zeta}\right)\exp\{i\zeta_n\gamma^n\}\Big|_{\zeta=0}\\&=\sum_{k=0}^{4}\sum_{n_1\cdots n_k}f_{n_1\cdots n_k}\frac{\partial_\ell}{\partial\zeta_{n_1}}\cdots\frac{\partial_\ell}{\partial\zeta_{n_k}}\sum_{l=0}^{4}\frac{i^l}{l!}(\zeta_n\gamma^n)^l\Big|_{\zeta=0},\end{aligned}$$

where ζ_n are odd variables. Using (39), we get

$$\begin{aligned}\tilde{S}^c=&-\frac{1}{2}\int_0^\infty de_0\int_{x_{\rm in}}^{x_{\rm out}}DxM(e_0)\left[\frac{\dot{x}_\mu}{e_0}\left(\varepsilon\frac{\delta_\ell}{\delta\rho_\mu}+\frac{\partial_\ell}{\partial\zeta_\mu}\right)-m\left(\varepsilon\frac{\delta}{\delta\rho_5}+i\gamma^5\right)\right]\\&\times\exp\left\{i\left[-\frac{\dot{x}^2}{2e_0}-\frac{e_0}{2}m^2-g\dot{x}A(x)+\frac{ie_0g}{4}F_{\mu\nu}(x)\frac{\partial_\ell}{\partial\zeta_\mu}\frac{\partial_\ell}{\partial\zeta_\nu}\right]\right\}\\&\times R\left[x,\rho,\frac{\partial_\ell}{\partial\zeta}\right]\exp\{i\zeta_\mu\gamma^\mu\}\Big|_{\rho,\zeta=0}\end{aligned}\tag{40}$$

[4]Here and further, we use short notation, e.g., $\omega\varepsilon\omega=\int_0^1 d\tau d\tau'\omega(\tau)\varepsilon(\tau-\tau')\omega(\tau')$, and so on.

with

$$R\left[x,\rho,\frac{\partial_\ell}{\partial\zeta}\right]=\int\mathcal{D}\omega\exp\left\{-\frac{1}{2}\omega^n T_{nk}(x|g)\,\omega^k+I_n\omega^n\right\},\tag{41}$$
$$I_\mu=\rho_\mu-\frac{e_0 g}{2}\frac{\partial_\ell}{\partial\zeta_\nu}F_{\nu\mu}(x)\,\varepsilon,\qquad I_5=\rho_5,$$

$$T_{nk}(x|g)=\begin{pmatrix}\Lambda_{\mu\nu}(x|g) & 0\\ 0 & -\varepsilon\end{pmatrix},\qquad \Lambda_{\mu\nu}(x|g)=\eta_{\mu\nu}\,\varepsilon-\frac{e_0}{2}\varepsilon g F_{\mu\nu}(x)\,\varepsilon,\tag{42}$$

where $\rho_n(\tau)$ are odd sources for $\omega^n(\tau)$. The integral in (41) is Gaussian. It can be easily found ([**1**]) from its original definition, which gives

$$R\left[x,\rho,\frac{\partial_\ell}{\partial\zeta}\right]=\left[\frac{\operatorname{Det}T(x|g)}{\operatorname{Det}T(x|0)}\right]^{1/2}\exp\left\{-\frac{1}{2}I_n[T^{-1}(x|g)]^{nk}I_k\right\}.\tag{43}$$

The ratio $\operatorname{Det}T(x|g)/\operatorname{Det}T(x|0)$ in (43) can be replaced by $\operatorname{Det}\Lambda(x|g)/\operatorname{Det}\Lambda(x|0)$, due to the structure (42) of the matrix $T(x|g)$, and the latter can be presented in a convenient form, which allows one to avoid the problems of calculating the determinants of matrices with continuous indices,

$$\frac{\operatorname{Det}\Lambda(x|g)}{\operatorname{Det}\Lambda(x|0)}=\exp\left\{-e_0\int_0^g dg'\operatorname{Tr}\mathcal{G}(x|g')\,F(x)\right\},$$
$$\mathcal{G}^{\mu\nu}(x|g)=\frac{1}{2}\,\varepsilon[\Lambda^{-1}(x|g)]^{\mu\nu}\varepsilon\,.\tag{44}$$

Substituting (44) into (40), and performing functional differentiations with respect to ρ_μ, we get

$$\begin{aligned}\tilde{S}^c=&-\frac{1}{2}\int_0^\infty de_0\int_{x_{\rm in}}^{x_{\rm out}}DxM(e_0)\left[\frac{\dot{x}^\mu}{e_0}K_{\mu\nu}(x)\frac{\partial_\ell}{\partial\zeta_\nu}-im\gamma^5\right]\\ &\times\exp\left\{i\left[-\frac{\dot{x}^2}{2e_0}-\frac{e_0}{2}m^2-g\dot{x}A(x)\right.\right.\\ &\qquad+\frac{ie_0}{2}\int_0^g dg'\operatorname{Tr}\mathcal{G}(x|g')\,F(x)\\ &\qquad\left.\left.+\frac{ie_0 g}{4}(F(x)\,K(x))_{\mu\nu}\frac{\partial_\ell}{\partial\zeta_\mu}\frac{\partial_\ell}{\partial\zeta_\nu}\right]\right\}\exp\{i\zeta_\mu\gamma^\mu\}\Big|_{\zeta=0},\\ K_{\mu\nu}(x)=&\,\eta_{\mu\nu}+e_0 g(\mathcal{G}(x|g)\,F(x))_{\mu\nu}\,.\end{aligned}$$

Now the differentiation over ζ can be performed explicitly, using (39). Thus, finally

$$\begin{aligned}S^c=&\,\frac{i}{2}\int_0^\infty de_0\int_{x_{\rm in}}^{x_{\rm out}}DxM(e_0)\,\Phi(x,e_0)\\ &\times\exp\left\{i\left[-\frac{\dot{x}^2}{2e_0}-\frac{e_0}{2}m^2-g\dot{x}A(x)\right]\right\},\end{aligned}\tag{45}$$

$$
\begin{aligned}
\Phi(x, e_0) = \Big[& m + (2e_0)^{-1} \dot{x} K(x)(2 - g e_0 F(x) K(x))\gamma \\
& + im \frac{e_0 g}{4} (F(x) K(x))_{\mu\nu} \sigma^{\mu\nu} \\
& + i \frac{g}{4} (\dot{x} K(x) \gamma)(F(x) K(x))_{\mu\nu} \sigma^{\mu\nu} \\
& + m \frac{e_0^2 g^2}{16} (F(x) K(x))^*_{\mu\nu} (F(x) K(x))^{\mu\nu} \gamma^5 \Big] \\
& \times \exp\left\{ - \frac{e_0}{2} \int_0^g dg' \operatorname{Tr} \mathcal{G}(x|g') F(x) \right\},
\end{aligned}
\tag{46}
$$

where

$$
(F(x) K(x))^*_{\mu\nu} = \frac{1}{2} \epsilon_{\mu\nu\alpha\beta} (F(x) K(x))^{\alpha\beta},
$$

and $\epsilon_{\mu\nu\alpha\beta}$ is the Levi–Civita symbol.

Thus we get a representation for the Dirac propagator as a path integral over bosonic trajectories of a functional whose spinor structure is found explicitly; namely, its decomposition in all the independent γ-structures is given. The functional $\Phi(x, e_0)$ can be called the *spin factor*, and in fact it distinguishes the Dirac propagator from the scalar one. One needs to stress that the spin factor is gauge invariant, because of it depends on $F_{\mu\nu}(x)$ only. In the same manner one can describe the isospinor structure of relativistic particle propagators ([**33**]). An application of the spin factor to concrete calculations of the propagator in external electromagnetic fields can be found in [**45**].

§6. Spinning particle with anomalous magnetic momentum

It is possible to get a generalization of the action (1) of a spinning particle in the presence of an anomalous magnetic momentum. The relativistic quantum theory of a spinning particle with anomalous magnetic momentum μ was formulated by Pauli ([**35**]). In this case he generalized the Dirac equation to the following form:

$$
(\widehat{\mathcal{P}}_\nu \gamma^\nu - m - (\mu/2) \sigma^{\alpha\beta} F_{\alpha\beta}) \Psi(x) = 0, \qquad \widehat{\mathcal{P}}_\nu = i\partial_\nu - g A_\nu(x) . \tag{47}
$$

An analog of the action (1), which after quantization reproduces the Dirac–Pauli theory, can be written in the following form ([**34**]):

$$
\begin{aligned}
S = \int_0^1 \Big[& - \frac{\dot{x}^2}{2e} - e \frac{M^2}{2} - \dot{x}^\alpha (g A_\alpha + 4 i \mu \psi^5 F_{\alpha\beta} \psi^\beta) \\
& + i g e F_{\alpha\beta} \psi^\alpha \psi^\beta + i \Big(\frac{\dot{x}_\alpha \psi^\alpha}{e} - M^* \psi^5 \Big) \chi - i \psi_n \dot{\psi}^n \Big] d\tau,
\end{aligned}
\tag{48}
$$

where $M = m - 2i\mu F_{\alpha\beta} \psi^\alpha \psi^\beta$, and $M^* = m + 2i\mu F_{\alpha\beta} \psi^\alpha \psi^\beta$. Other possibilities were discussed in [**46–48**], The symmetry (2) and (3) remains for the action (48). One can derive equations of motion and check that they indeed describe a classical particle with anomalous magnetic momentum μ. Performing the Dirac quantization of the theory, we arrive at the equation (47) as a condition on the physical state vectors ([**36**]).

Similarly to §4, one can find a path integral representation for the causal Green function $\tilde{S}^c = \tilde{S}^c(x_{\text{in}}, y_{\text{out}})$ of the Dirac–Pauli equation (47) transformed by γ^5,

$$
\begin{aligned}
\tilde{S}^c = & \exp\left(i\gamma^n \frac{\partial_l}{\partial\theta^n}\right) \int_0^\infty de_0 \int d\chi_0 \int_{e_0} De \int_{\chi_0} D\chi \\
& \times \int_{x_{\text{in}}}^{x_{\text{out}}} Dx \int D\pi \int D\nu \int_{\psi(0)+\psi(1)=\theta} \mathcal{D}\psi M(e) \\
& \times \exp\left\{ i\int_0^1 \left[-\frac{\dot{x}^2}{2e} - e\frac{M^2}{2} - \dot{x}^\alpha(gA_\alpha + 4i\mu\psi^5 F_{\alpha\beta}\psi^\beta) \right.\right. \\
& \qquad + igeF_{\alpha\beta}\psi^\alpha\psi^\beta + i\left(\frac{\dot{x}_\alpha\psi^\alpha}{e} - M^*\psi^5\right)\chi \\
& \qquad \left.\left. - i\psi_n\dot{\psi}^n + \pi\dot{e} + \nu\dot{\chi} \right] d\tau + \psi_n(1)\psi^n(0) \right\}\Bigg|_{\theta=0},
\end{aligned} \tag{49}
$$

where $M^* = m + 2i\mu F_{\alpha\beta}\psi^\alpha\psi^\beta$, and the measure $\mathcal{M}(e)$ has the form (35).

One can treat the exponent in the integrand (49) as an effective and nondegenerate Lagrangian action of a spinning particle with an anomalous magnetic momentum; the gauge invariant part of this action coincides with (48).

§7. Pseudoclassical models for Weyl particles

As we have seen from the discussion presented in §3, the quantum mechanics constructed from the Berezin–Marinov action (1) admits the limit $m = 0$. As a result, one gets the quantum theory of a massless particle described by the Dirac equation with $m = 0$, without any additional restrictions on the four-spinor $\Psi(x)$,

$$
i\partial_\mu\gamma^\mu\Psi(x) = 0\,. \tag{50}
$$

As we explained, the variable ψ^5 can be omitted from the action (1) at $m = 0$. The quantization of such a modified action reproduces the physical sector in the limit $m = 0$ of the massive quantum mechanics. Unfortunately, such a quantum theory describes massless spin one half particle with the all possible values of helicity (right and left neutrinos). It is known that the right (left) neutrino is described by a four-spinor, which obeys, in addition to the Dirac equation (50), the Weyl condition as well,

$$
(\gamma^5 - \alpha)\Psi(x) = 0, \quad \alpha = 1\,(-1), \quad \gamma^5 = i\gamma^0\gamma^1\gamma^2\gamma^3\,. \tag{51}
$$

There were several attempts to modify the action (1) at $m = 0$ so that in the course of quantization one would get a quantum mechanics with wave functions obeying both equations (50) and (51) at the same time. So, in [**37**] the authors proposed the following action,

$$
S = \int_0^1 \left[-\frac{1}{2e}(\dot{x}^\mu + g_1 Q^\mu - i\psi^\mu\chi)^2 - i\psi_\mu\dot{\psi}^\mu + g_2\left(\Lambda - \frac{\alpha}{2}\right)\right] d\tau, \tag{52}
$$

where g_a, $a = 1, 2$, are Lagrange multipliers and $\Lambda = i\epsilon^{\mu\nu\rho\varsigma}\psi_\mu\psi_\nu\psi_\rho\psi_\varsigma/3$, $Q^\mu = \epsilon^{\mu\nu\rho\varsigma}\psi_\nu\psi_\rho\psi_\varsigma$. Quantization by means of the Dirac method gives both equations (50) and (51) as restrictions on states vectors. In particular, the theory involves second-class constraints $P_\mu + i\psi_\mu = 0$, where P_μ are momenta conjugate to ψ^μ, and first-class constraints

$$\pi^2 = 0, \quad \pi_\mu \psi^\mu = 0, \quad \pi_\mu Q^\mu = 0, \quad \Lambda - \alpha/2 = 0,$$

where π_μ are momenta conjugate to x^μ. Calculating the Dirac brackets with respect to the second-class constraints only, in the course of quantization one can find the following realization for the essential variables $\hat{\pi}_\mu = -i\partial_\mu$, $\hat{\psi}^\mu = (i/2)\gamma^\mu$, in the x-representation. Applying the first-class constraints operators to the state vector, following Dirac, one gets only two independent equations

$$\hat{\pi}_\mu \hat{\psi}^\mu \Psi(x) = 0, \qquad (\hat{\Lambda} - \alpha/2)\,\Psi(x) = 0,$$

which are precisely equations (50) and (51). As we mentioned before, this method of quantization is not well founded. Moreover, attempts to quantize this action canonically fail, since as soon as one chooses any gauge condition linear in ψ, the combination Λ vanishes and only the Dirac equation remains after quantization. Another possibility was discussed in [**38**]. The authors considered the theory with the action

$$S = \int_0^1 \left\{ -\frac{1}{2e}\left[\dot{x}^\mu - i\left(\psi^\mu - \frac{2i\alpha}{3}\epsilon^{\mu\nu\rho\varsigma}\psi_\nu\psi_\rho\psi_\varsigma\right)\chi\right]^2 - i\psi_\mu\dot{\psi}^\mu\right\} d\tau\,. \tag{53}$$

A formal quantization of the theory following the Dirac method leads to the equation $i\partial_\mu\gamma^\mu(\gamma^5 - \alpha)\,\Psi(x) = 0$ for state vectors, which is not equivalent to the two equations (50) and (51). Canonical quantization gives the Dirac equation (50), but without any additional restrictions for helicity. This is in agreement with the fact that classically the actions (53) and (1) are equivalent at $m = 0$ ([**38**]).

In the paper [**39**], we proposed a new pseudoclassical action to describe Weyl particles,

$$S = \int_0^1 \left[-\frac{1}{2e}\left(\dot{x}^\mu - i\psi^\mu\chi + i\epsilon^{\mu\nu\rho\varsigma}b_\nu\psi_\rho\psi_\varsigma + \frac{\alpha}{2}b^\mu\right)^2 - i\psi_\mu\dot{\psi}^\mu\right] d\tau, \tag{54}$$

where x^μ, e, ψ^μ, χ have the same meaning as in (1), the variables b^μ form an even four-vector, and α is an even constant. The action admits both quasi-canonical quantization (with fixation of the gauge freedom, which corresponds to two types of gauge transformations of the three existing ones) and the Dirac quantization. Both of them lead to the theory of a Weyl particle.

§8. Discussion

Thus, the model (1) of Berezin–Marinov and the model (54) completely solve the problem of pseudoclassical description of massive and massless spin one half particles in $3+1$ dimensions. Possible generalizations may be related to the construction of the corresponding models in arbitrary dimensions and for all spins available in these dimensions.

It is easy to see ([**15**]) that the action (1) gives an adequate description of the massive spin one half particles in all even dimensions $D = 2n$ if one replaces all four-dimensional quantities in (1) by D-dimensional ones.

The action (54) can also be generalized to any even dimension ([**49, 50**]),

$$S = \int_0^1 \left[-\frac{1}{2e}\left(\dot{x}^\mu - i\psi^\mu\chi + \frac{(2i)^{(D-2)/2}}{(D-2)!}\,\epsilon^{\mu\nu\rho_2\ldots\rho_{D-1}} \times b_\nu\psi_{\rho_2}\cdots\psi_{\rho_{D-1}} + \frac{\alpha}{2^{(D-2)/2}}\,b^\mu\right)^2 - i\psi_\mu\dot{\psi}^\mu\right] d\tau, \quad \mu = 0,\ldots,D-1. \tag{55}$$

In the case of odd dimensions, and, in particular, in the case $D = 2+1$, the action (1) has to be modified essentially. An adequate model, which describes spin one half particles in $2+1$ dimensions, was proposed in [**51**],

$$S = \int_0^1 \left[-\frac{z^2}{2e} - e\frac{m^2}{2} - im\psi^3\chi - \frac{1}{2}\,sm\kappa - i\psi_a\dot{\psi}^a\right] d\tau, \qquad z = \dot{x}^\mu - i\psi^\mu\chi + i\epsilon^{\mu\nu\lambda}\psi_\nu\psi_\lambda\kappa, \tag{56}$$

where $\mu = 0,1,2$, $a = 0,1,2,3$, and κ is a new even variable, whereas s is an even constant. That model can be generalized to arbitrary odd $2n+1$ dimensions ([**53**]),

$$S = \int_0^1 \left[-\frac{z^2}{2e} - e\frac{m^2}{2} - im\psi^{2n+1}\chi - \frac{s}{2^n}\,m\kappa - i\psi_a\dot{\psi}^a\right] d\tau, \qquad z^\mu = \dot{x}^\mu - i\psi^\mu\chi + \frac{(2i)^n}{(2n)!}\,\epsilon^{\mu\rho_1\ldots\rho_{2n}}\psi_{\rho_1}\cdots\psi_{\rho_{2n}}\kappa, \tag{57}$$

where $\mu = 0,\ldots,2n$, $a = 0,\ldots,2n+1$.

One possible way to describe pseudoclassically higher spins (integer and half integer) is to multiply the variables, which are introduced to describe spin one-half. For example, a corresponding model for massive spin $N/2$ particles has the following form in 3+1 dimensions ([**40, 41**])

$$S = \int_0^1 \left[-\frac{1}{2e}\left(\dot{x}^\mu - i\sum_{a=1}^N \psi_a^\mu\chi_a\right)^2 - \frac{e}{2}m^2 - im\sum_{a=1}^N \psi_a^5\chi_a + \frac{1}{2}\sum_{a,b=1}^N f_{ab}(i[\psi_{an},\psi_b^n]_- + \alpha_{ab}) - i\sum_{a=1}^N \psi_{an}\dot{\psi}_a^n\right] d\tau, \tag{58}$$

where f_{ab} are even and antisymmetric, $\mu = 0,1,2,3$, $a,b = 1,\ldots,N$, $n = (\mu,5)$, $\eta_{\mu\nu} = \operatorname{diag}(1,-1,-1,-1)$, $\eta_{mn} = \operatorname{diag}(1,-1,-1,-1,-1)$. The summand

$$\frac{1}{2}\alpha_{ab}\int_0^1 f_{ab}\,d\tau$$

with even coefficients α_{ab} plays the role of a Chern–Simons term and can be added only in case $N = 2$ without breaking the rotational gauge symmetry ([**37**]). Thus, $\alpha_{ab} = \alpha\epsilon_{ab}\delta_{N,2}$ with an even constant α and a two-dimensional Levi–Civita symbol ϵ_{ab}. Dirac quantization of theories with this action, in particular for $N = 2$, was considered in [**42, 43, 37**] and the canonical one, in the case $N = 2$ in [**16**].

A pseudoclassical model for massive higher spins (integer and half-integer) in 2+1 dimensions was proposed in [**52**],

$$
\begin{aligned}
S &= \int_0^1 \left\{ -\frac{z^2}{2e} - \frac{e}{2}m^2 - \sum_{a=1}^{N} \left[\alpha m \left(\frac{\kappa_a}{2} + i\psi_a^3 \chi_a \right) - i\psi_{an}\dot{\psi}_a^n \right] \right\} d\tau, \\
z^\mu &= \dot{x}^\mu + i \sum_{a=1}^{N} (\epsilon^{\mu\nu\lambda}\psi_{a\nu}\psi_{a\lambda}\kappa_a - \psi_a^\mu \chi_a) \,.
\end{aligned}
\tag{59}
$$

In particular, it describes Chern–Simons particles at $N = 2$.

A generalization of the models (58) and (59) to arbitrary even and odd dimensions can be done similarly to the spin one-half case.

One ought to mention a remarkable property of the dimensional duality between the pseudoclassical models of massive and massless particles. Namely, the pseudoclassical models of massive particles in odd $2n+1$ dimensions can be derived by means of a dimensional reduction from the corresponding models of Weyl particles in even $2n+2$ dimensions ([**51, 53**]).

Finally, it is known that in any odd dimension, in particular in 2+1 dimensions, it is possible to have particles with fractional spins, the so-called anyons. The corresponding irreducible representations of the Lorentz group are infinite-dimensional, which implies that these spins must be described on the classical level by means of bosonic variables. Classical models for anyons were recently discussed in the literature (see, for example, [**54, 55**]).

References

1. F. A. Berezin, *The method of second quantization*, "Nauka", Moscow, 1965 (Russian); English transl., Academic Press, New York, 1966; *Introduction to algebra and analysis with anticommuting variables*, Moscow State University Press, Moscow, 1983; A. A. Kirillov (ed.), *Introduction to superanalysis*, D. Reidel, Dordrecht, 1987.
2. F. A. Berezin and M. S. Marinov, *Classical spin and Grassmann algebra*, Pis′ma Zh. Èksper. Teoret. Fiz. **21** (1975), 678–680; English transl., JETP Lett. **21** (1975), 320–321; *Particle spin dynamics as the Grassmann variant of classical mechanics*, Ann. Physics **104** (1977), 336–362.
3. R. Casalbuoni, *On the quantization of systems with anticommuting variables*, Nuovo Cimento A **33** (1976), 115–125; *The classical mechanics for Bose–Fermi systems*, Nuovo Cimento A **33** (1976), 389–431.
4. A. Barducci, R. Casalbuoni, and L. Lusanna, *Supersymmetry and the pseudoclassical relativistic electron*, Nuovo Cimento A **35** (1976), 377–399.
5. L. Brink, S. Deser, B. Zumino, P. di Vecchia, and P. Howe, *Local supersymmetry for spinning particles*, Phys. Lett. B **64** (1976), 435–438.
6. L. Brink, P. di Vecchia, and P. Howe, *Formulation of the classical and quantum dynamics of spinning particles*, Nuclear Phys. B **118** (1977), 76–94.
7. A. P. Balachandran, P. Salomonson, B. Skagerstam, and J. Winnberg, *Classical description of a particle interacting with a non-Abelian gauge field*, Phys. Rev. D **15** (1977), 2308–2317.
8. W. K. H. Panofsky and M. Phillips, *Classical electricity and magnetism*, Addison-Wesley, Reading, MA, 1962.
9. P. A. M. Dirac, *Lectures on quantum mechanics*, Yeshiva University, New York, 1964.
10. D. M. Gitman and I. V. Tyutin, *Quantization of fields with constraints*, Springer-Verlag, Berlin, 1990.
11. ———, *Canonical quantization of a relativistic particle*, Pis′ma Zh. Èksper. Teoret. Fiz. **51** (1990), no. 3, 188–190; *Classical and quantum mechanics of a relativistic particle*, Classical Quantum Gravity **7** (1990), 2131–2144.

12. S. P. Gavrilov and D. M. Gitman, *Quantization of systems with time-dependent constraints. Example of relativistic particle in plane wave*, Classical Quantum Gravity **10** (1993), 57–67.
13. L. L. Foldy and S. A. Wouthuysen, *On the Dirac theory of spin 1/2 particles and its non-relativistic limit*, Phys. Rev. **78** (1950), 29–36.
14. M. H. L. Pryce, *The mass-centre in the restricted theory of relativity and its connexion with the quantum theory of elementary particles*, Proc. Roy. Soc. A **195** (1949), 62–81.
15. G. V. Grigoryan and R. P. Grigoryan, *Canonical quantization of a D-dimensional relativistic spin particle*, Yadernaya Fiz. **53** (1991), 1062–1067; English transl., Soviet J. Nuclear Phys. **53** (1991).
16. D. M. Gitman, A. E. Gonçalves, and I. V. Tyutin, *Quantization of a pseudoclassical model of the spin 1 relativistic particle*, Internat. J. Modern Phys. A **10** (1995), 701–718.
17. E. S. Fradkin, *Green's function method in quantized field theory and quantum statistics*, Proc. FIAN, Vol. 29, "Nauka", Moscow, 1965; English transl., Consulting Bureau, New York, 1967.
18. B. M. Barbashov, *Functional integrals in quantum electrodynamics and the infrared limit of the Green's functions*, Soviet Phys. JETP **21** (1965), 402–418.
19. I. A. Batalin and E. S. Fradkin, *Quantum electrodynamics in external fields*, Teoret. Mat. Fiz. **5** (1970), 190–218; English transl. in Theoret. and Math. Phys. **5** (1970).
20. M. Henneaux and C. Teitelboim, *Relativistic quantum mechanics of supersymmetric particles*, Ann. Physics **143** (1982), 127–159.
21. N. V. Borisov and P. P. Kulish, *Path integral in superspace for a relativistic spinor particle in an external gauge field*, Teoret. Mat. Fiz. **51** (1982), 335–343; English transl. in Theoret. and Math. Phys. **51** (1982).
22. V. Ya. Fainberg and A. V. Marshakov, *Local supersymmetry and Dirac particle propagator as a path integral*, Nuclear Phys. B **306** (1988), 659–678.
23. E. S. Fradkin, D. M. Gitman, and Sh. M. Shvartsman, *Quantum electrodynamics with unstable vacuum*, Springer-Verlag, Berlin, 1991; *Path integral for relativistic particle theory*, Europhys. Lett. **15** (1991), 241–244.
24. E. S. Fradkin and D. M. Gitman, *Path integral representation for the relativistic particle propagator and BFV quantization*, Phys. Rev. D **44** (1991), 3230–3236.
25. J. Schwinger, *On gauge invariance and vacuum polarization*, Phys. Rev. **82** (1951), 664–679.
26. B. S. De Witt, *Dynamical theory in curved space*, Rev. Modern Phys. **29** (1957), 337–397.
27. F. A. Berezin, *Feynman path integrals in a phase space*, Sov. Phys. Usp. **23** (1981), 763–788; F. A. Berezin and M. A. Shubin, *Schrödinger equation*, Moscow State University, Moscow, 1983.
28. R. P. Feynman, *Space-time approach to non-relativistic quantum mechanics*, Rev. Modern Phys. **20** (1948), 367–387.
29. ———, *Mathematical formulation of the quantum theory of electromagnetic interaction*, Phys. Rev. **80** (1950), 440–457.
30. ———, *An operator calculus having applications in quantum electrodynamics*, Phys. Rev. **84** (1951), 108–128.
31. A. M. Polyakov, *Fermi–Bose transmutations induced by gauge fields*, Modern Phys. Lett. A **3** (1988), 325–328.
32. G. M. Korchemsky, *Quantum geometry of Dirac fermions*, Internat. J. Modern Phys. A **7** (1992), 339–380.
33. D. M. Gitman and Sh. M. Shvartsman, *Spinor and isospinor structure of relativistic particle propagator*, Phys. Lett. B **318** (1993), 122–126; *Errata*, Phys. Lett. B **331** (1994), 449–450.
34. D. M. Gitman and A. V. Saa, *Pseudoclassical model of spinning particle with anomalous magnetic moment*, Modern Phys. Lett. A **8** (1993), 463–468.
35. W. Pauli, *Relativistic field theories of elementary particles*, Rev. Modern Phys. **13** (1941), 203–232.
36. D. M. Gitman and A. V. Saa, *Quantization of spinning particle with anomalous magnetic moment*, Classical Quantum Gravity **10** (1993), 1447–1460.
37. P. Howe, S. Penati, M. Pernici, and P. Townsend, *A particle mechanics description of antisymmetric tensor field*, Classical Quantum Gravity **6** (1989), 1125–1140.
38. A. Barducci, R. Casalbuoni, D. Dominici, and L. Lusanna, *Pseudoclassical description of Weyl particles*, Phys. Lett. B **100** (1981), 126–130.

39. D. M. Gitman, A. E. Gonçalves, and I. V. Tyutin, *New pseudoclassical model for Weyl particles*, Phys. Rev. D **50** (1994), 5439–5442.
40. P. P. Srivastava, *Supersymmetry and pseudoclassical dynamics of particles with any spin*, Nuovo Cimento Lett. **19** (1977), 239–244.
41. V. D. Gershun and V. I. Tkach, Pis′ma Zh. Èksper. Teoret. Fiz. **29** (1979), 320–324; English transl. in JETP Lett. **29** (1979).
42. A. Barducci and L. Lusanna, *The photon in pseudoclassical mechanics*, Nuovo Cimento A **77** (1983), 39–75; *The massive photon in pseudoclassical mechanics*, J. Phys. A **16** (1983), 1993–1998.
43. J. Gomis, M. Novell, and K. Rafanelli, *Pseudoclassical model of a particle with arbitrary spin*, Phys. Rev. D **34** (1986), 1072–1075.
44. T. M. Aliev, V. Ya. Fainberg, and N. K. Pak, *Path integral for spin: A new approach*, Nuclear Phys. B **429** (1994), 321–343.
45. D. M. Gitman, S. I. Zlatev, and W. da Cruz, *Spin factor and spinor structure of Dirac propagator in constant field*, Preprint IFUSP **P-1172** (1995), 1–11.
46. A. Barducci, *Pseudoclassical description of relativistic spinning particles with anomalous magnetic moment*, Phys. Lett. B **118** (1982), 112–114.
47. A. A. Zheltukhin, *Superfield description of a particle with spin and anomalous magnetic moment*, Teoret. Mat. Fiz. **65** (1985), 151–154; English transl. in Theoret. and Math. Phys. **65** (1985).
48. G. V. Grigoryan and R. P. Grigoryan, Yadernaya Fiz. **58** (1995), 1–10; English transl. in Soviet J. Nuclear Phys..
49. D. M. Gitman and A. E. Gonçalves, *Pseudoclassical model for Weyl particle in 10 dimensions*, Preprint IFUSP **P-1158** (1995), 1–5, hep-th/9506010.
50. G. V, Grigoryan, R. P. Grigoryan, and I. V. Tyutin, *Pseudoclassical theory of Mayorana-Weyl particle*, Preprint YERPHY **1446 (16)** (1995), 1–13, hep-th/9510002.
51. D. M. Gitman, A. E. Gonçalves, and I. V. Tyutin, *Pseudoclassical supersymmetrical model for* $2+1$ *Dirac particle*, Preprint IFUSP **P-1144** (1995), 1–10, hep-th/9601065.
52. D. M. Gitman and I. V. Tyutin, *Pseudoclassical description of higher spins in* $2+1$ *dimensions*, Preprint Lebedev Inst. FIAN/TD/96-26 (1996), 1–20, hep-th/9602048.
53. D. M. Gitman and A. E. Gonçalves, *Pseudoclassical description of the massive Dirac particles in odd dimensions*, Preprint IFUSP, P-1204 (1996), 1–14.
54. M. S. Plyushchay, *The model of a free relativistic particle with fractional spin*, Internat. J. Modern Phys. A **7** (1992), 7045–7064.
55. José L. Cortés and M. S. Plyushchay, *Anyons: minimal and extended formulation*, Modern Phys. Lett. A **10** (1995), 409–418.

Instituto de Física, Universidade de São Paulo P. O. Box 66318, 05389-970-São Paulo, S. P., Brasil

Translated by THE AUTHOR

Amer. Math. Soc. Transl.
(2) Vol. **177**, 1996

Dual Waves

Renata Kallosh

ABSTRACT. We study gravitational waves in the 10-dimensional target space of superstring theory. Some of these waves have unbroken supersymmetries. They consist of the Brinkmann metric and of a 2-form field. Sigma-model duality is applied to such waves. We call the corresponding solution dual partners of gravitational waves, or dual waves. Upon Kaluza–Klein dimensional reduction to 4 dimensions, some of these dual waves become equivalent to the conformo-stationary solutions of axion-dilaton gravity. Such solutions include dilaton extreme black holes, axion-dilaton Israel–Werner–Perjes-type spacetimes and extreme charged axion-dilaton Taub–Nut solutions. The unbroken supersymmetry of the gravitational waves transfers to the unbroken supersymmetry of axion-dilaton IWP solutions.

We also describe the recently discovered duality symmetric entropy formula for the extreme black holes in string theory which is given by a unique Cartan quartic invariant of $E(7)$ group.

§1. Introduction

The fundamental reason for the existence of any supersymmetric nonrenormalization theorem is related to the fact that supersymmetric theories have a natural description in superspace, i.e., in a space with both bosonic and fermionic coordinates. The famous rules of integration over anticommuting variables were discovered by Berezin. These rules form the basis for the proof of the generic supersymmetric nonrenormalization theorems, which state that some semiclassical properties of supersymmetric theories are not changed when all quantum corrections are taken into account.

In recent years an active field of research has been the search for and also the study of nonperturbative solutions to the classical equations of motion of superstring effective field theories and the corresponding sigma-models. Many bosonic solutions of such theories have been discovered. Some special solutions turned out to have a highly nontrivial property: although bosonic, they have some unbroken supersymmetries. This means that when embedded into the right supersymmetric

1991 *Mathematics Subject Classification.* Primary 83E30; Secondary 81T30.

This work was supported by NSF grant PHY-8612280.

theory, they admit Killing spinors. Bosonic solutions with unbroken supersymmetries in theories of quantum gravity are very special because they have some kind of supersymmetric nonrenormalization property. Arguments that explain these properties were given for supersymmetric string solitons in [**1, 2**] and for extremal black holes in [**3**].

A particularly interesting kind of metric is provided by the so-called *pp-waves.*[1] Recently we have found the metrics in this class which, together with the appropriate dilaton, axion, and gauge fields, provide solutions of the lowest order superstring effective action and have unbroken supersymmetries ([**4**]). They have also been shown to be free of stringy corrections.

There exists an extensive literature on gravitational waves. We start by reviewing some results relevant for our purposes. Geometries for pp-waves are space-times admitting a covariantly constant null vector field

$$\nabla_\mu l_\nu = 0, \qquad l^\nu l_\nu = 0\,. \tag{1}$$

Spacetimes with this property were first discovered by Brinkmann in 1923 ([**5**]). In four dimensions, the metrics of these spaces can be written in the general form

$$ds^2 = 2\,du\,dv + K(u,\xi,\bar\xi)\,du^2 - d\xi\,d\bar\xi, \tag{2}$$

where u and v are light-cone coordinates defined by

$$l_\mu = \partial_\mu u, \qquad l^\mu \partial_\mu v = 1 \tag{3}$$

(thus the metric does not depend on v) and $\xi = x + iy$ and $\bar\xi = x - iy$ are complex transverse coordinates. These metrics are classified and described in detail in [**6**]. Different pp-wave spaces are characterized by different choices of the function K in equation (2). For example, when K is quadratic in ξ and $\bar\xi$,

$$K(u,\xi,\bar\xi) = f(u)\xi^2 + \bar f(u)\bar\xi^2 + g(u)\xi\bar\xi, \tag{4}$$

they are called *exact plane waves.* Plane waves (2) with K of the form

$$K(u,\xi,\bar\xi) = \delta(u)\,f(\xi,\bar\xi) \tag{5}$$

are called *shock waves.* A specific example of shock waves is given by the Aichelburg–Sexl geometry ([**7**])

$$K(u,\xi,\bar\xi) = \delta(u)\ln(\xi\bar\xi), \tag{6}$$

which describes the gravitational field of a point-like particle boosted to the speed of light.

Güven established in 1987 ([**8**]) that a solution to the lowest order superstring effective action, consisting of a generalization of the four-dimensional exact plane waves to $d = 10$ with dilaton, axion, and Yang–Mills fields, has half of the $N = 1$, $d = 10$ supersymmetries unbroken and is also a solution of the equations of motion of the superstring effective action including all the α'-corrections. Investigations on the supersymmetry of plane-fronted waves in general relativity were made even earlier. In [**9**] there is a reference to unpublished work of J. Richer who found that pp-waves are supersymmetric. Very general classes of pp-waves were found by Tod

[1]Here *pp-waves* stands for plane front waves with parallel rays.

to be supersymmetric in the context of $d = 4$, $N = 2$ ungauged supergravity in 1983 ([**10**]). Different aspects of plane wave solutions in string theory have been investigated in the last few years (see [**11, 12**]).

The existence of a covariantly constant null vector field has dramatic consequences. For instance, for the class of d-dimensional pp-waves with metrics of the form

$$ds^2 = 2\,du\,dv + K(u, x^i)\,du^2 - dx^i dx^i, \tag{7}$$

where $i, j = 1, \ldots, d-2$, the Riemann curvature is ([**12**])

$$R_{\mu\nu\rho\sigma} = -2 l_{[\mu}(\partial_{\nu]}\partial_{[\rho} K)\, l_{\sigma]}\,. \tag{8}$$

The curvature is orthogonal to l_μ in all its indices. This fact is of crucial importance in establishing that all higher order (in α') terms in the equations of motion are zero due to the vanishing of all the possible contractions of curvature tensors. Güven proved in [**8**] that the corrections to the supersymmetry transformations vanish.

In an arbitrary dimension d, the most general metrics admitting a covariantly constant null vector (1) were discovered by Brinkmann in 1925 ([**13**]):

$$ds^2 = 2\,du\,dv + A_\mu(u, x^i)\,dx^\mu du - g_{ij}(u, x^i)\,dx^i dx^j, \qquad l^\mu A_\mu = 0, \tag{9}$$

where $\mu, \nu = 0, 1, \ldots, d-1$ and $i, j = 1, \ldots, d-2$. Note that the general Brinkmann metric (9) in $d = 10$ has 8 functions $A_i(u, x^i)$ and 28 functions $g_{ij}(u, x^i)$ more than the metric (7) investigated by Güven, where only the uu-component of the metric $A_u = K/2$ was present and was quadratic in x^i.

In [**4**] we found a 10-dimensional supersymmetric generalization of Brinkmann metrics. We called this solution supersymmetric string waves (SSW). Various properties of these solutions have been studied recently in [**14–17**]. In this paper we present a review of recent results related to the dual partners of supersymmetric string waves, which we call dual waves. In addition, we describe briefly more general supersymmetric 4-dimensional configurations that may be obtained from dual waves. Duality transformations of string theory ([**18**]), which we are using here, are not the standard electromagnetic duality often used in General Relativity. This is the reason why gravitational waves are usually not related to black-hole-type solutions. From the point of view of General Relativity, those are very different geometries. However, string theory brings in a completely different concept of "equivalent" background geometries. It was understood some time ago ([**19**]) that pp-waves are dual to fundamental strings ([**1**]). The corresponding duality transformation, which is known as the sigma-model duality transformation ([**18**]), has a very particular property: it changes the value of the dilaton field $e^{2\phi}$ by the g_{xx}-component of the metric, where x is a direction on which all fields are independent. In our recent paper ([**14**]), we have established an analogous duality relation between more general solutions of the effective equations of the critical ($d = 10$) superstring theories.

A remarkable property of a class of SSW has been discovered recently ([**15**]): their dual partners are the extreme 4-dimensional dilaton black holes ([**20**]) upon Kaluza–Klein dimensional reduction. Conversely, after $d = 4$ extreme black holes are lifted up to $d = 10$, the corresponding configuration becomes a Brinkmann-type wave upon dual rotation.

In [16] we found a general class of supersymmetric solutions of a 4-dimensional axion-dilaton gravity. Such solutions include dilaton extreme black holes and axion-dilaton Israel–Werner–Perjes-type spacetimes, which also include the extreme charged axion-dilaton Taub–Nut solutions. Here we present the detailed derivation of these solutions from SSW in $d = 10$. To do this it will be necessary to investigate a more general class of SSW and to establish their relation to a very general class of solutions of 4-dimensional dilaton-axion gravity.

In this paper we limit ourselves only to the zero slope limit of the superstring effective action ($N = 1$, $d = 10$ supergravity). The coupling to the Yang–Mills multiplet will enter via α' corrections and will not be discussed here. It will be studied later along the lines of the papers [4] and [14].

The paper is organized as follows. In §2 we describe supersymmetric string waves and their dual partners. We identify some of the dual waves as the fundamental strings (see [1]) and indicate the relation between some waves and 4-dimensional black holes. In §3 we describe some details of the supersymmetric Kaluza–Klein dimensional reduction of the bosonic part of the superstring effective action. Under such a supersymmetric truncation, the unbroken supersymmetry of 10-dimensional configuration is transferred automatically to unbroken supersymmetry of the 4-dimensional configuration. In §4 we perform the supersymmetric Kaluza–Klein dimensional reduction of a special case of the dual wave and we compare it with 4-dimensional extreme black holes. In §5 we consider more general dual waves and compare them with 4-dimensional stationary solutions. In §6 we discuss this new method of generating nontrivial solutions in gravitational theories and display new possibilities of deriving 4-dimensional supersymmetric solutions of $N = 4$ supergravity interacting with $N = 4$ supersymmetric multiplets.

§2. Supersymmetric string waves and dual waves

We use the sigma-model duality of string theory and relate solutions of 4-dimensional and 10-dimensional effective actions of string theory. We limit ourselves to keeping only one scalar field, the fundamental dilaton. The pseudoscalar axion field will appear in $d = 4$ from the 3-form field strength H. The method that we develop here may give many other interesting relations for the class of solutions that will include more fields of string theory. First consider pp-waves ([5]) in some dimension d. The metric is[2]

$$ds^2 = 2\,du\,dv + 2A_u\,du\,du - \sum_{i=1}^{d-2} dx^i dx^i, \tag{10}$$

where the function A_u depends only on transverse directions x^i, $i = 1, \dots, d-2$. The equation that $A_u(x^i)$ must satisfy is

$$\triangle A_u = 0, \tag{11}$$

where the Laplacian is taken over the transverse directions only. Sigma-model duality transformations ([18]) define the changes in the metric, in the 2-form field $B_{\mu\nu}$, and in the dilaton field $e^{2\phi}$.

---

[2]The notation is that of [14, 4, and 3].

The transformation

$$(12)\quad \begin{aligned} &g'_{xx} = 1/g_{xx}, \quad g'_{x\alpha} = B_{x\alpha}/g_{xx}, \quad g'_{\alpha\beta} = g_{\alpha\beta} - (g_{x\alpha}g_{x\beta} - B_{x\alpha}B_{x\beta})/g_{xx}, \\ &B'_{x\alpha} = g_{x\alpha}/g_{xx}, \quad B'_{\alpha\beta} = B_{\alpha\beta} + 2g_{x[\alpha}B_{\beta]x}/g_{xx}, \quad \phi' = \phi - (1/2)\log|g_{xx}| \end{aligned}$$

is defined for configurations with a nonnull Killing vector in the x-direction. String theory considers such configurations as equivalent under the condition that the x-direction is compact. Leaving this issue aside, we start by presenting the upshot of the dual relation between pp-waves (10) and fundamental strings ([**1**]).

Since we have chosen the metric of our pp-waves to be u-independent and they are v-independent by the definition of pp-waves, the metric is also independent of $x = (u - v)/\sqrt{2}$ and $t = (u + v)\sqrt{2}$. A straightforward application of the sigma-model duality transformations given in (12) on pp-waves given in (10) leads to the following new solution:

$$(13)\quad ds^2 = 2e^{2\phi}\,du\,dv - \sum_{i=1}^{d-2} dx^i dx^i, \quad B = 2(1 - e^{2\phi})\,du \wedge dv, \quad e^{-2\phi} = 1 - A_u\,.$$

Let the function A_u be spherically symmetric and depend only on $r^2 = \sum_{i=1}^{d-2} x^i x^i$. The choice

$$(14)\qquad A_u = -\mu/r^{d-4}$$

solves the harmonic equation (11) at $r \neq 0$ if μ is a constant. With this choice of the function A_u, equation (13) is the solution found in [**1**] describing the field outside a fundamental string. For $d = 10$, which is the critical dimension of superstring theory, this solution looks as follows:

$$(15)\quad ds^2 = \Big(1 + \frac{\mu}{r^6}\Big)^{-1}(dt^2 - dx^2) - \sum_{i=1}^{8} dx^i dx^i, \quad B_{xt} = \frac{\mu}{r^6 + \mu}, \quad e^{-2\phi} = 1 + \frac{\mu}{r^6}\,.$$

The *dual partner* of the fundamental string is the following pp-wave:

$$(16)\qquad ds^2 = 2\,du\,dv - \frac{2\mu}{r^6}\,du\,du - \sum_{i=1}^{8} dx^i dx^i,$$

and the 2-form field and the dilaton are absent (or could be equal to some constants). Note that for $d = 10$ both the pp-waves and the fundamental strings have unbroken supersymmetries ([**8, 1**]). After this reminder of the duality between the fundamental string and pp-wave, we may proceed to display the duality between supersymmetric string waves and the "lifted" extreme black holes.

Supersymmetric string waves (SSW, [**4**]) are plane-wave-type solutions of the superstring effective action that have unbroken space-time supersymmetries. They describe dilaton, axion, and gauge fields in a stringy generalization of the Brinkmann metric. Some correspondence between the metric, the axion field, and the gauge fields is required. Here we consider only the zero slope limit of the effective string action. This limit corresponds to 10-dimensional $N = 1$ supergravity. The Yang–Mills multiplet will appear in the first order of α' string corrections. The conspiracy between the metric and the axion field is the characteristic property of

SSW and is necessary to provide unbroken supersymmetry even in the zero slope limit.

The SSW ([**4**]) in $d = 10$ are given by the Brinkmann metric ([**5**]) and the following 2-form:

$$(17)\quad ds^2 = 2\, d\tilde{u}\, d\tilde{v} + 2A_M\, d\tilde{x}^M d\tilde{u} - \sum_{i=1}^{8} d\tilde{x}^i d\tilde{x}^i, \quad B = 2A_M\, d\tilde{x}^M \wedge d\tilde{u}, \quad A_v = 0,$$

where $i = 1, \ldots, 8$, $M = 0, 1, \ldots, 9$, and we are using the following notation for the 10-dimensional coordinates: $x^M = \{\tilde{u}, \tilde{v}, \tilde{x}^i\}$. We have put a tilde over the 10-dimensional coordinates in this section since we must compare the original 10-dimensional configuration with the 4-dimensional one embedded into the 10-dimensional space. A rather nontrivial identification of coordinates describing these solutions will be required later.

Note that our SSW have a particular conspiracy between the metric and the axion field

$$(18)\qquad g_{i\tilde{u}} = A_i = B_{i\tilde{u}}\,.$$

It is interesting that the solution in which the vector function in the metric is related to the one in the axion was mentioned by Tseytlin in [**21**] as the most natural one from the point of view of the sigma-model equations. In [**22**] Duval, Horváth, and Horváthy found that such a conspiracy between the metric and the axion field leads to the absence of conformal anomaly at the two-loop level.

The equations that $A_u(\tilde{x}^i)$ and $A_i(\tilde{x}^j)$ have to satisfy are:

$$(19)\qquad \triangle A_u = 0, \qquad \triangle \partial^{[i} A^{j]} = 0,$$

where the Laplacian is taken over the transverse directions only. This solution has 8 functions A_i more than that of pp-waves (10), where only the function A_u is nonvanishing.

A straightforward application of the sigma-model duality transformations given in (12) on the SSW solution given in (17) leads to the following new supersymmetric solution of the zero slope limit equations of motion:

$$(20)\qquad \begin{aligned} ds^2 &= 2e^{2\phi}\{d\tilde{u}\, d\tilde{v} + A_i d\tilde{u}\, d\tilde{x}^i\} - \sum_{i=1}^{8} d\tilde{x}^i d\tilde{x}^i, \\ B &= -2e^{2\phi}\{A_u\, d\tilde{u} \wedge d\tilde{v} + A_i\, d\tilde{u} \wedge d\tilde{x}^i\}, \qquad e^{-2\phi} = 1 - A_u, \end{aligned}$$

where as before, the functions $A_M = \{A_u = A_u(\tilde{x}^j),\ A_v = 0,\ A_i = A_i(\tilde{x}^j)\}$ satisfy equations (19). We call these solutions *dual partners* of the waves, or for simplicity, *dual waves.*

We can make the following particular choice of the vector function A_M. First of all these functions will depend only on 3 of the transverse coordinates, $\tilde{x}^1$, $\tilde{x}^2$, $\tilde{x}^3$, corresponding to our 3-dimensional space. Secondly, we choose one of the A_i, e.g., A_4 to be related to A_u.

$$(21)\qquad A_u = -\mu/\rho, \qquad A_4 = \xi A_u, \qquad A_5 = \cdots = A_8 = 0,$$

where $\rho^2 = \sum_{i=1}^{3} \tilde{x}^i \tilde{x}^i \equiv \vec{x}^2$, and μ is a constant. We shall specify the constant ξ later. Note that equations (19) are solved outside $\rho = 0$.[3] We get

$$\begin{aligned} ds^2 &= 2e^{2\phi}\{d\tilde{u}\, d\tilde{v} + \xi(1-e^{-2\phi})\, d\tilde{x}^4 d\tilde{u}\} - \sum_{i=1}^{8} d\tilde{x}^i d\tilde{x}^i, \\ B &= -2e^{2\phi}(1-e^{-2\phi})\{d\tilde{u}\wedge d\tilde{v} + \xi\, d\tilde{u}\wedge d\tilde{x}^4\}, \qquad e^{-2\phi} = 1 + \mu/\rho\,. \end{aligned} \tag{22}$$

The solution given in (22) is different from the solution in [**1**] corresponding to the field outside a fundamental string and from our generalized FS solution, since it depends only on three coordinates and not on eight. Also we take a particular case of our solution with relations between A_4 and A_u. We perform the coordinate change

$$\hat{x} = \tilde{x}^4 + \xi\tilde{u}, \qquad \hat{v} = \tilde{v} + \xi\tilde{x}^4. \tag{23}$$

We also shift B by a constant value, since the equations of motion depend on $H = dB$ only.

The dual wave solution (22) takes the form

$$\begin{aligned} ds^2 &= 2e^{2\phi} d\tilde{u}\, d\hat{v} + \xi^2 d\tilde{u}^2 - d\hat{x}^2 - \sum_{i=1}^{3} d\tilde{x}^i d\tilde{x}^i - \sum_{i=5}^{8} d\tilde{x}^i d\tilde{x}^i, \\ B &= 2e^{2\phi} d\hat{v}\wedge d\tilde{u}, \qquad e^{-2\phi} = 1 + \mu/\rho\,. \end{aligned} \tag{24}$$

When $\xi^2 = -1$, we have

$$\begin{aligned} ds^2 &= 2e^{2\phi} d\hat{v}\, d\tilde{u} - d\tilde{u}^2 - d\hat{x}^2 - \sum_{i=1}^{3} d\tilde{x}^i d\tilde{x}^i - \sum_{i=5}^{8} d\tilde{x}^i d\tilde{x}^i, \\ B &= 2e^{2\phi} d\hat{v}\wedge d\tilde{u}, \qquad e^{-2\phi} = 1 + \mu/\rho\,. \end{aligned} \tag{25}$$

We can identify this particular dual partner of the SSW solution with the uplifted dilaton black hole if we make the following identification of coordinates

$$t = \hat{v} = \tilde{v} + \xi\tilde{x}^4, \quad x^4 = \tilde{u}, \quad x^9 = \hat{x} = \tilde{x}^4 + \xi\tilde{u}, \quad x^{1,2,3,5,\dots,8} = \tilde{x}^{1,2,3,5,\dots,8}\,. \tag{26}$$

Our dual wave becomes

$$ds^2 = 2e^{2\phi} dt\, dx^4 - \sum_{i=4}^{9} dx^i dx^i - d\vec{x}^2, \quad B = 2e^{2\phi} dt\wedge dx^4, \quad e^{-2\phi} = 1 + \mu/\rho\,. \tag{27}$$

This is an extreme electrically charged 4-dimensional black hole ([**20**]), which is embedded into 10-dimensional geometry in a stringy frame, as we are going to explain in the next section.

---

[3] In order to solve the equations (19) everywhere, it is understood that a source term at $\rho = 0$, representing an unknown object, perhaps a six-brane, must be added to these equations. We hope that it can be worked out similarly to the combined action for the macroscopic fundamental string, where the source term comes from the sigma-model action, see equations (3,1)–(3,3) in [**1**].

§3. Supersymmetric dimensional reduction

The embedding of the 4-dimensional bosonic solutions of the effective superstring action into 10-dimensional geometry is not unique in general.[4] There are different ways to identify the vector field of the charged black hole in 4 dimensions with the nondiagonal component of the metric in the extra dimensions as well as with the 2-form field. Also the identification of the 4-dimensional dilaton with the fundamental 10-dimensional dilaton and/or with some components of the metric in the extra dimension is possible.

However, the identification of the 4-dimensional solution with the 10-dimensional one becomes unique under the conditions that the supersymmetric embedding for both solutions is identified. Dimensional reduction of $N = 1$ supergravity down to $d = 4$ has been studied by Chamseddine ([**23**]) in canonical geometry. We are working in stringy metric and also in slightly different notation. In a subsequent publication, we shall present a detailed derivation of the compactification of the bosonic part of the effective action of 10-dimensional string theory which is consistent with supersymmetry ([**15**]). Here we are interested in the relation between the extreme dilaton charged black holes, which have unbroken supersymmetry ([**3**]) when imbedded into $d = 4$, $N = 4$ supergravity[5] and the corresponding 10-dimensional supersymmetric configuration. The stationary Israel–Werner–Perjes-type (IWP) solutions of 4-dimensional axion-dilaton gravity also happen to be supersymmetric when embedded into $d = 4$, $N = 4$ supergravity. For more general 4-dimensional supersymmetric configurations, which have to be embedded into $d = 4$, $N = 4$ supergravity interacting with $d = 4$, $N = 4$ supersymmetric matter multiplets, more general formulas will be required.

We start with the zero slope limit of the effective 10-dimensional superstring action. The bosonic part of the action is

$$S = \frac{1}{2}\int d^{10}x e^{-2\phi}\sqrt{-g}\Big[- R + 4(\partial\phi)^2 - \frac{3}{4}H^2\Big], \tag{28}$$

where the 10-dimensional fields are the metric, the axion and the dilaton.

We want to make a connection with the bosonic part of $N = 4$, $d = 4$ action. In this particular case we are interested in compactifying 6 space-like coordinates. All fields are assumed to be independent of six compactified dimensions. According to Chamseddine ([**23**]), dimensional reduction of $N = 1$, $d = 10$ supergravity to $d = 4$ gives $N = 4$ supergravity coupled to 6 matter multiplets. We are interested here only in dimensional reduction to $N = 4$ supergravity without matter multiplets.

Let us first reduce from $d = 10$ to $d = 5$ by trivial dimensional reduction, when we do not keep the nondiagonal components of the metric and 2-form field. We denote the 10-dimensional fields by un upper index $^{(10)}$ and the 5-dimensional fields by a hat. The 10-dimensional indices are capital letters $M, N = 0, \dots, 9$, the 5-dimensional indices will carry a hat $\hat\mu, \hat\nu = 0, \dots, 4$, and the compactified dimensions will be denoted by capital I's and J's, $I, J = 5, \dots, 9$. We take the

[4]We are grateful to E. Witten for attracting our attention to this problem.

[5]We do not know at present whether the embedding of these black holes into other theories, including the Abelian part of the Yang–Mills multiplet, will also correspond to some unbroken supersymmetries.

$d = 10$ fields to be related to the $d = 5$ ones by

$$\begin{aligned} g^{(10)}_{\hat{\mu}\hat{\nu}} &= \hat{g}_{\hat{\mu}\hat{\nu}}, \quad g^{(10)}_{I\hat{\nu}} = 0, \quad g^{(10)}_{IJ} = \eta_{IJ} = -\delta_{IJ}, \\ B^{(10)}_{\hat{\mu}\hat{\nu}} &= \hat{B}_{\hat{\mu}\hat{\nu}}, \quad B^{(10)}_{I\hat{\nu}} = 0, \quad B^{(10)}_{IJ} = 0, \quad \phi^{(10)} = \hat{\phi}. \end{aligned} \tag{29}$$

We get

$$S = \frac{1}{2}\int d^5x e^{-2\hat{\phi}}\sqrt{-\hat{g}}\Big[-\widehat{R} + 4(\partial\hat{\phi})^2 - \frac{3}{4}\widehat{H}^2\Big]. \tag{30}$$

As the second step, we reduce from $d = 5$ to $d = 4$, keeping the nondiagonal components of the metric and 2-form field. Since we are also interested in supersymmetry, we shall work with the 5-beins at this stage. The 4-dimensional indices do not carry a hat. We parameterize the 5-beins as follows:

$$(\hat{e}_{\hat{\mu}}{}^{\hat{a}}) = \begin{pmatrix} e_\mu{}^a & A_\mu \\ 0 & 1 \end{pmatrix}, \qquad (\hat{e}_{\hat{a}}{}^{\hat{\mu}}) = \begin{pmatrix} e_a{}^\mu & -A_a \\ 0 & 1 \end{pmatrix}, \tag{31}$$

where $A_a = e_a{}^\mu A_\mu$. With this parametrization, the 5-dimensional fields decompose as follows:

$$\begin{aligned} \hat{g}_{44} &= \hat{\eta}_{44} = -1, \quad \hat{g}_{4\mu} = -A_\mu, \quad \hat{g}_{\mu\nu} = g_{\mu\nu} - A_\mu A_\nu, \\ \hat{B}_{4\mu} &= B_\mu, \quad \hat{B}_{\mu\nu} = B_{\mu\nu} + A_{[\mu}B_{\nu]}, \quad \hat{\phi} = \phi, \end{aligned} \tag{32}$$

where $\{g_{\mu\nu}, B_{\mu\nu}, \phi, A_\mu, B_\mu\}$ are the 4-dimensional fields.

The 4-dimensional action for the 4-dimensional fields becomes

$$S = \frac{1}{2}\int d^4x e^{-2\phi}\sqrt{-g}\Big[-R + 4(\partial\phi)^2 - \frac{3}{4}H^2 + \frac{1}{4}F^2(A) + \frac{1}{4}F^2(B)\Big], \tag{33}$$

where

$$\begin{aligned} F_{\mu\nu}(A) &= 2\partial_{[\mu}A_{\nu]}, \qquad F_{\mu\nu}(B) = 2\partial_{[\mu}B_{\nu]}, \\ H_{\mu\nu\rho} &= \partial_{[\mu}B_{\nu\rho]} + (1/2)\{A_{[\mu}F_{\nu\rho]}(B) + B_{[\mu}F_{\nu\rho]}(A)\}. \end{aligned} \tag{34}$$

Now we study the dimensional reduction of gravitino. We are specifically interested in the supersymmetry transformation rule of gravitino in $d = 4$ supergravity without matter. This leads to identification of the matter vector fields D_μ and the supergravity vector fields V_μ,

$$D_\mu = (A_\mu - B_\mu)/2, \qquad V_\mu = (A_\mu + B_\mu)/2, \tag{35}$$

respectively. Now we want to truncate the theory, keeping only the supergravity vector field V_μ. We then have

$$V_\mu = A_\mu = B_\mu, \qquad D_\mu = 0. \tag{36}$$

The truncated action is[6]

$$S = \frac{1}{2}\int d^4x e^{-2\phi}\sqrt{-g}\Big[-R + 4(\partial\phi)^2 - \frac{3}{4}H^2 + \frac{1}{2}F^2(V)\Big], \tag{37}$$

[6]This action, which comes from the 10-dimensional theory, is slightly different from the corresponding 4-dimensional action in our previous papers, e.g., in [**3**], due to the difference in notation. The detailed explanation of this difference is given in [**15**].

where

$$F_{\mu\nu}(V) = 2\partial_{[\mu}V_{\nu]}, \qquad H_{\mu\nu\rho} = \partial_{[\mu}B_{\nu\rho]} + V_{[\mu}F_{\nu\rho]}(V)\,. \tag{38}$$

The embedding of the 4-dimensional fields in this action in $d = 10$ is as follows:

$$\begin{aligned} g^{(10)}_{\mu\nu} &= g_{\mu\nu} - V_\mu V_\nu, \quad g^{(10)}_{4\nu} = -V_\nu, \quad g^{(10)}_{44} = -1, \\ g^{(10)}_{IJ} = \eta_{IJ} &= -\delta_{IJ}, \quad B^{(10)}_{\mu\nu} = B_{\mu\nu}, \quad B^{(10)}_{4\nu} = V_\nu, \quad \phi^{(10)} = \phi\,. \end{aligned} \tag{39}$$

These formulas can be used to uplift any $U(1)$ 4-dimensional field configurations, including dilaton and axion, to a 10-dimensional field configurations in a way consistent with supersymmetry.

The conclusion of this supersymmetric dimensional reduction is the following.

i) The dilaton of the supersymmetric 4-dimensional extreme black holes is identified as a fundamental dilaton of string theory (and not one of the modulus fields).

ii) Dimensional reduction of $d = 10$ supergravity to $d = 4$ gives $N = 4$ supergravity without six matter multiplets under the condition that $g_{4\mu} = -B_{4\mu}$. Therefore the vector field of the 4-dimensional configuration is actually a nondiagonal component of the metric in the extra dimension as well as the 2-form field. This works in our case since, according to (27), we have

$$g^{(10)}_{4t} = -B^{(10)}_{4t} = -V_t = e^{2\phi}\,. \tag{40}$$

iii) The supersymmetric truncation presented above has the following important property by construction: if the 4-dimensional configuration has unbroken supersymmetries, the uplifted one also has them and vice versa, starting with the supersymmetric configuration in $d = 10$, one ends up with the supersymmetric configuration in $d = 4$. The reason for this is simple. The rules given in (39) have been derived in such a way that the equations $\Psi = 0$, $\delta\Psi{=}0$ on all fermions remain correct and still have solutions with nonvanishing Killing spinors.

§4. Uplifting the black hole

We shall use the formulas from the previous section to uplift the dilaton black hole with one vector field. The electrically charged extreme 4d black hole is given by ([**20**]):[7]

$$ds^2_{str} = e^{4\phi}dt^2 - d\vec{x}^2, \quad V = -e^{2\phi}dt, \quad B = 0, \quad e^{-2\phi} = 1 + 2M/\rho\,. \tag{41}$$

The uplifted configuration, according to (39) is

$$\begin{aligned} ds^2 &= 2e^{2\phi}dt\,dx^4 - d\vec{x}^2 - (dx^4)^2 - dx^I dx^I, \\ B^{(10)} &\equiv B^{(10)}_{MN}\,dx^M \wedge dx^N = -2e^{2\phi}dx^4 \wedge dt, \qquad \phi^{(10)} = \phi\,. \end{aligned} \tag{42}$$

Let us choose the parameter μ in the dual partner to the wave, given in (25), equal to the double mass of the black hole

$$\mu = 2M\,. \tag{43}$$

This makes the uplifted black hole (42) identical to the dual partner to the wave, given by (25).

[7] There is a difference of a $1/\sqrt{2}$ factor in the vector field compared to the one given in [**3**].

For a better understanding of black-hole-wave relation, it is useful to do the following. By adding and subtracting from the metric the term $e^{4\phi}dt^2$, we can rewrite the dual wave for $d = 10$ given in (27) as follows:

$$
\begin{aligned}
ds^2 &= e^{4\phi}dt^2 - d\vec{x}^2 - (dx^4 - e^{2\phi}dt)^2 - dx^I dx^I\,, \\
B &= -2e^{2\phi}dx^4 \wedge dt, \qquad e^{-2\phi} = 1 + 2M/\rho\,.
\end{aligned} \tag{44}
$$

Now it is easy to recognize the 4-dimensional metric in the first two terms in the metric and, in the third term, the nondiagonal component of the 10-dimensional metric, which, together with the nondiagonal component of the 2-form, plays the role of the vector field in the 4-dimensional geometry.

To show that our 10-dimensional solution is the embedding of the extreme 4d black hole, we may also present the 10-dimensional metric in Kaluza–Klein parametrization, where we have a 5d metric $g_{\hat{\mu}\hat{\nu}} \times x^I$-flat space ($\hat{\mu} = 0, 1, 2, 3, 4$, $I = 5, \ldots, 9$).

(45)

$$
g_{\hat{\mu}\hat{\nu}} = \begin{pmatrix} 0 & 0 & 0 & 0 & e^{2\phi} \\ 0 & -1 & 0 & 0 & 0 \\ 0 & 0 & -1 & 0 & 0 \\ 0 & 0 & 0 & -1 & 0 \\ e^{2\phi} & 0 & 0 & 0 & -1 \end{pmatrix} = \begin{pmatrix} g_{00} + A_0 A_0 g_{44} & 0 & 0 & 0 & A_0 g_{44} \\ 0 & -1 & 0 & 0 & 0 \\ 0 & 0 & -1 & 0 & 0 \\ 0 & 0 & 0 & -1 & 0 \\ A_0 g_{44} & 0 & 0 & 0 & g_{44} \end{pmatrix},
$$

where

$$
g_{00} = e^{4\phi} = (A_0)^2, \quad g_{11} = -1, \quad g_{22} = -1, \quad g_{33} = -1, \quad g_{44} = -1\,.
$$

Thus the 10-dimensional nondiagonal component of the metric $g_{04} = A_0 g_{44}$ together with the part of B_{04} forms the 4-dimensional vector field, as is usual in Kaluza–Klein theory. The relation is (see equation (35))

$$
A_0 = -V_t = e^{2\phi} = (g_{04}^{(10)} + B_{04}^{(10)})/2\,. \tag{46}
$$

The left-hand side of this equation gives a nice simple form of the dual partner of the wave metric

$$
ds^2 = 2e^{2\phi}dt\, dx^4 - d\vec{x}^2 - (dx^4)^2 - dx^I dx^I\,, \tag{47}
$$

the right-hand side shows that if this metric is rewritten in a more complicated form by replacing the zero in the upper right corner of the matrix by $g_{00} + A_0 A_0 g_{44}$, we can recognize the g_{00}-piece of the 4-dimensional black hole.

The case $\xi^2 = -1$, which gives the 4-dimensional black hole in Minkowski space with the signature $(1,3)$ times the compact 6-dimensional space with the signature $(0,6)$, corresponds to a complex 10-dimensional wave in the space with the signature $(1,9)$.

$$
ds^2 = 2\, d\tilde{u}\, d\tilde{v} - \frac{4M}{\rho}\, d\tilde{u}\,(d\tilde{u} - i\, d\tilde{x}^4) - \sum_{i=1}^{8} d\tilde{x}^i d\tilde{x}^i, \qquad B = -i\,\frac{4M}{\rho}\, d\tilde{u} \wedge d\tilde{x}^4\,. \tag{48}
$$

By performing a rotation $i\tilde{x}^4 = \tilde{\tau}$ one can get

$$
ds^2 = 2\, d\tilde{u}\, d\tilde{v} - \frac{4M}{\rho}\, d\tilde{u}\,(d\tilde{u} - d\tilde{\tau}) + d\tilde{\tau}^2 - \sum_{i=1}^{7} d\tilde{x}^i d\tilde{x}^i, \qquad B = -\frac{4M}{\rho}\, d\tilde{u} \wedge d\tilde{\tau}\,. \tag{49}
$$

This makes the wave real but with the signature of the space $(2,8)$.

Thus we may conclude that string theory considers as dual partners the extreme 4d electrically charged dilaton black hole embedded into 10-dimensional geometry, as given in (44) or (42), and the Brinkmann-type 10-dimensional wave (48), (49).

If we choose $\xi^2 = 1$, we get the stringy equivalence between Brinkmann-type 10-dimensional wave

$$ds^2 = 2\,d\tilde{u}\,d\tilde{v} - \frac{4M}{\rho}\,d\tilde{u}\,(d\tilde{u} - d\tilde{x}^4) - \sum_{i=1}^{8} d\tilde{x}^i d\tilde{x}^i, \qquad B = -\frac{4M}{\rho}\,d\tilde{u}\wedge d\tilde{x}^4, \tag{50}$$

and lifted Euclidean 4-dimensional electrically charged dilaton black hole with the signature $(0,4)$ and the 6-dimensional space has the signature $(1,5)$,

$$\begin{aligned} ds^2 &= -e^{4\phi}dt^2 - d\vec{x}^2 + (dx^4 + e^{2\phi}dt)^2 - dx^I dx^I, \\ B &= -2e^{2\phi}dx^4\wedge dt, \qquad e^{-2\phi} = 1 + 2M/\rho\,. \end{aligned} \tag{51}$$

With this choice of the signature, the gravitational wave does not have imaginary components. However, the fact that the metric, as well as the 2-form field of the gravitational wave in $d = 10$ having an imaginary component, is dual to the lifted black hole in Minkowski space is strange. Note that this is necessary only if one insists that the $d = 10$ space as well as the $d = 4$ space are both Minkowski spaces. One can avoid imaginary components by allowing changes in the signature of the space-time when performing duality and dimensional reduction as explained above. Still this remains a puzzle.

§5. From waves to IWP and vice versa

We consider again only the zero slop limit of SSW ([**4**]) in $d = 10$. To generate a dual wave that gives the 4-dimensional IWP solutions after dimensional reduction, we make the following choices for the vector A_M in the 10-dimensional SSW (17). The components A_M are taken to depend only on three of the transverse coordinates, $\tilde{x}^1$, $\tilde{x}^2$, $\tilde{x}^3$, which will ultimately correspond to our 3-dimensional space. We choose one component, e.g., A_4 to be related to A_u according to $A_4 = \xi A_u$, where $\xi^2 = \pm 1$ depending on the signature of spacetime. Recall that from (20) A_u is related to the dilaton by $A_u = 1 - e^{-2\phi}$. For the remaining components, we take only A_1, A_2, A_3 to be nonvanishing, and we relabel these as ω_1, ω_2, ω_3 for obvious reasons. To summarize, we have

$$A_4 = \xi A_u, \quad A_1 \equiv \omega_1, \quad A_2 \equiv \omega_2, \quad A_3 \equiv \omega_3, \quad A_5 = \cdots = A_8 = 0\,. \tag{52}$$

We get the following wave:

$$\begin{aligned} ds^2 &= 2e^{2\phi}\{d\tilde{u}\,d\tilde{v} + \xi\,(1 - e^{-2\phi})\,d\tilde{x}^4 d\tilde{u} + \omega_i\,d\tilde{u}\,d\tilde{x}^i\} - \sum_{i=1}^{8} d\tilde{x}^i d\tilde{x}^i, \\ B &= -2e^{2\phi}(1 - e^{-2\phi})\{d\tilde{u}\wedge d\tilde{v} + \xi\,d\tilde{u}\wedge d\tilde{x}^4 + \omega_i\,d\tilde{u}\wedge d\tilde{x}^i\}, \quad e^{-2\phi} = 1 + \mu/\rho\,. \end{aligned} \tag{53}$$

The equations that the functions $e^{-2\phi}$ and ω_i have to satisfy for the gravitational wave to be a solution of equations of motion following from the action (28), are:

$$\triangle e^{-2\phi} = 0, \qquad \triangle\partial^{[i}\omega^{j]} = 0, \tag{54}$$

where again the Laplacian is taken over the transverse directions only. These equations follow from (19).

Below we repeat exactly the same steps that allowed us to show that in the absence of ω_i the dual partner of a wave is an extreme black hole after dimensional reduction. If we redefine $\tilde{v}$,

$$\hat{v} = \tilde{v} + \xi \tilde{x}^4, \tag{55}$$

the solution takes the form

$$\begin{aligned} ds^2 &= e^{2\phi} 2\, d\tilde{u}\,(d\hat{v} + \omega_i\, dx^i) - 2\xi\, d\tilde{u}\, d\tilde{x}^4 - \sum_{i=1}^{8} (d\tilde{x}^i)^2, \\ B &= 2\,(1 - e^{2\phi})\, d\tilde{u} \wedge d\hat{v} - 2e^{2\phi}\omega_i\, d\tilde{u} \wedge d\tilde{x}^i. \end{aligned} \tag{56}$$

The dilaton-axion field λ satisfies the harmonic equation (71). We may bring the metric to the form

$$ds^2 = 2e^{2\phi} d\tilde{u}\,(d\hat{v} + \omega_i\, d\tilde{x}^i) + \xi^2 d\tilde{u}^2 - \sum_{i=1}^{3} d\tilde{x}^i d\tilde{x}^i - (d\hat{x})^2 - \sum_{i=5}^{8} d\tilde{x}^i d\tilde{x}^i, \tag{57}$$

where

$$\hat{x} = \tilde{x}^4 + \xi \tilde{u}\,. \tag{58}$$

One can shift B by a constant value, since the equations of motion depend on $H = dB$ only. We get

$$B = 2e^{2\phi}(d\hat{v} + \omega_i\, d\tilde{x}^i) \wedge d\tilde{u}\,. \tag{59}$$

When $\xi^2 = -1$, we have

$$\begin{aligned} ds^2 &= 2e^{2\phi} d\tilde{u}\,(d\hat{v} + \omega_i\, d\tilde{x}^i) - d\tilde{u}^2 - \sum_{i=1}^{3} d\tilde{x}^i d\tilde{x}^i - (d\hat{x})^2 - \sum_{i=5}^{8} d\tilde{x}^i d\tilde{x}^i, \\ B &= 2e^{2\phi}(d\hat{v} + \omega_i\, d\tilde{x}^i) \wedge d\tilde{u}, \quad \triangle e^{-2\phi} = 0, \quad \triangle \partial^{[i}\omega^{j]} = 0\,. \end{aligned} \tag{60}$$

In particular, the identification of the coordinates in the dual wave solution with those in the uplifted IWP solution can be performed,

$$t = \hat{v} = \tilde{v} + \xi \tilde{x}^4, \quad x^4 = \tilde{u}, \quad x^9 = \hat{x} = \tilde{x}^4 + \xi \tilde{u}, \quad x^{1,2,3,5,\ldots,8} = \tilde{x}^{1,2,3,5,\ldots,8}\,. \tag{61}$$

After all these steps, our 10-dimensional dual wave becomes

$$\begin{aligned} ds^2 &= 2e^{2\phi} dx^4 (dt + \vec{\omega} \cdot d\vec{x}) - \sum_{i=4}^{9} dx^i dx^i - d\vec{x}^2, \\ B &= -2e^{2\phi} dx^4 \wedge (dt + \vec{\omega} \cdot d\vec{x})\,. \end{aligned} \tag{62}$$

This allows us to identify the stationary supersymmetric axion-dilaton IWP solution ([**16**]), embedded into 10d geometry in the stringy metric. All arguments given in [**15**] about the uniqueness of the embedding into higher dimension are valid in this case also, once the supersymmetry is identified.

To recognize this as the lifted IWP solution, we add and subtract from the metric the term $e^{4\phi}(dt + \vec{\omega} \cdot d\vec{x})^2$. We can then rewrite the dual wave metric (61) as

$$ds^2 = e^{4\phi}(dt + \vec{\omega} \cdot d\vec{x})^2 - d\vec{x}^2 - (dx^4 - e^{2\phi}(dt + \vec{\omega} \cdot d\vec{x}))^2 - \sum_{i=5}^{9} dx^i dx^i . \tag{63}$$

The first two terms now give the string metric for the 4-dimensional IWP solutions. The nondiagonal components $g_{\mu 4}$ in the third term are interpreted as the 4-dimensional gauge field components, showing the Kaluza–Klein origin of the gauge field in this construction. Note that the 4-dimensional vector field components are also equal to the off-diagonal components of the 2-form gauge field, giving the overall identifications

$$g_{t4} = B_{t4} = e^{2\phi} = -V_t, \qquad g_{i4} = B_{i4} = e^{2\phi}\omega_i = -V_i . \tag{64}$$

The dilaton of the IWP solution is identified with the fundamental dilaton of string theory, rather than with one of the modulus fields. The *axion* is identified with the 4-dimensional part of the 3-form field strength H given in (38). Note that these components of H come totally from the second term in (38), since the 4-dimensional $B_{\mu\nu}$ vanishes.

To explain this relation better, let us recall that the stationary supersymmetric axion-dilaton IWP solution ([**16**]) is

$$\begin{aligned} ds^2_{str} &= e^{4\phi}(dt + \omega_i\, dx^i)^2 - d\vec{x}^2, \quad \partial_{[i}\omega_{j]} = -\frac{1}{2}\,\epsilon_{ijk}\partial_k a, \\ A_\mu &= \frac{1}{\sqrt{2}}\, e^{2\phi}(1, \omega_i), \quad \triangle e^{-2\phi} = 0, \quad \triangle \partial^{[i}\omega^{j]} = 0 . \end{aligned} \tag{65}$$

The Kaluza–Klein parametrization of the metric means the following relation between the higher-dimensional metric g_{MN} and the reduced one $g_{\mu\nu}$:

$$g_{MN} = \begin{pmatrix} g_{\mu\nu} + A_\mu{}^k g_{kl} A_\nu{}^l & A_\mu{}^k g_{ik} \\ A_\nu{}^k g_{kj} & g_{ij} \end{pmatrix} . \tag{66}$$

To show that our 10d solution is the embedding of the axion-dilaton 4d IWP metric we may present the 10d metric in Kaluza–Klein parametrization, where we have a 5d metric $(g_{\hat{\mu}\hat{\nu}} \times x^I)$-flat space ($\hat{\mu}$, $\hat{\nu} = 0, 1, 2, 3, 4$), $I, J = 5, \ldots, 9$:

$$\begin{aligned} g_{\hat{\mu}\hat{\nu}} &= \begin{pmatrix} 0 & 0 & 0 & 0 & e^{2\phi} \\ 0 & -1 & 0 & 0 & e^{2\phi}\omega_1 \\ 0 & 0 & -1 & 0 & e^{2\phi}\omega_2 \\ 0 & 0 & 0 & -1 & e^{2\phi}\omega_3 \\ e^{2\phi} & e^{2\phi}\omega_1 & e^{2\phi}\omega_2 & e^{2\phi}\omega_3 & -1 \end{pmatrix} \\ &= \begin{pmatrix} g_{00}+A_0{}^4 A_0{}^4 g_{44} & g_{10}+A_1{}^4 A_0{}^4 g_{44} & g_{20}+A_2{}^4 A_0{}^4 g_{44} & g_{01}+A_3{}^4 A_0{}^4 g_{44} & A_0{}^4 g_{44} \\ g_{01}+A_0{}^4 A_1{}^4 g_{44} & -1 & 0 & 0 & A_1{}^4 g_{44} \\ g_{02}+A_0{}^4 A_2{}^4 g_{44} & 0 & -1 & 0 & A_2{}^4 g_{44} \\ g_{03}+A_0{}^4 A_3{}^4 g_{44} & 0 & 0 & -1 & A_3{}^4 g_{44} \\ A_0{}^4 g_{44} & A_1{}^4 g_{44} & A_2{}^4 g_{44} & A_3{}^4 g_{44} & g_{44} \end{pmatrix} , \end{aligned} \tag{67}$$

where $g_{00} = e^{4\phi} = (A_0^4)^2$, $g_{ii} = -1, g_{0i} = e^{4\phi}\omega_i$. The 10-dimensional nondiagonal components of the metric $g_{04} = A_0^4 g_{44}$, $g_{0i} = A_i{}^4 g_{44}$, together with the parts of

B_{04}, B_{0i}, form the 4d vector field, as usual in Kaluza–Klein theory. The relation is (see (65))

$$A_0 = -V_t = e^{2\phi} = (g_{04} + B_{04})/2, \qquad A_i = -V_i = e^{2\phi}\omega_i = (g_{i4} + B_{i4})/2\,. \tag{68}$$

The left-hand side of equation (67) yields a nice simple form of the dual partner of the wave metric

$$ds^2 = 2e^{2\phi}(dt + \omega_i\, dx^i)\, dx^4 - dx^i dx^i - (dx^4)^2 - dx^I dx^I\,. \tag{69}$$

The right-hand side shows that if this metric is rewritten in a much more complicated form by replacing many zeros in the upper right corner of the matrix by terms like $g_{00} + A_0^4 A_0^4 g_{44}$, we can recognize the g_{00}-piece and other pieces of the 4-dimensional axion-dilaton IWP metrics.

We may use the 3-dimensional antisymmetric tensor ϵ_{ijk} to express $\partial_{[i}\omega_{j]}$ through the axion field $a(x^i)$,

$$\partial_{[i}\omega_{j]} = -\epsilon_{ijk}\partial_k a/2\,. \tag{70}$$

By introducing the standard dilaton-axion complex field $\lambda = a + ie^{-2\phi}$, we may rewrite equations (11) as follows:

$$\triangle\lambda = 0\,. \tag{71}$$

This accomplishes the derivation of the 4-dimensional IWP axion-dilaton solutions from the 10-dimensional gravitational waves.

§6. Conclusion

Can we actually consider the dual relation between waves and lifted black holes as something more than pure algebraic curiosity? We believe that the answer to this question is positive. The dual relation displayed above was established at the zero slope limit of the effective action of the superstring theory. The issue of α'-corrections in string theory has been studied extensively for the waves in [**4**, **14**]. The pp-waves have the best known properties of absence of such quantum corrections ([**8**]). The SSW are known to have the special property of the absence of α'-corrections under the condition that the non-Abelian Yang–Mills fields are added to the configuration which at the zero slope limit $\alpha' = 0$ consists only of the metric and 2-form ([**4**]).

It was explained in [**14**] that the importance of sigma-model duality between supersymmetric configurations is in the fact that the structure of α' corrections is under control for the dual solution if it was under control for the original solution. In this way we have found that the nice properties of the pp-waves ([**8**]) are carried over to the fundamental string solutions. The present investigation shows that the electrically charged extreme black hole embedded into 10-dimensional geometry may need to be supplemented by some non-Abelian Yang–Mills configuration to avoid possible α'-corrections. In this respect we would like to emphasize that the study of the properties of quantum corrections established via duality may become a powerful mechanism in quantum theory, despite the strange imaginary factors in the waves, which are the dual partners of the uplifted black holes.

At the very minimal level, one can regard the method developed above, which consists of stringy duality combined with Kaluza–Klein dimensional reduction, as a solution generating method. This method has the advantage of generating new supersymmetric solutions from the original ones. If we did not know that extreme 4-dimensional black holes are supersymmetric, we would discover this via the supersymmetric properties of 10-dimensional waves. If we had not found the IWP axion-dilaton spaces in the 4-dimensional world, we could have found them from SSW via dual rotation and dimensional reduction. Simultaneously, we would establish the existence of unbroken supersymmetry of IWP geometries.

The SSW solutions found in [**4**] are not the most general supersymmetric wave solutions; more general ones may exist. However, if we look only at those SSW solutions that we know are supersymmetric, could we still use them to produce more general 4-dimensional configurations with unbroken supersymmetries? The answer to this question is positive. Indeed, we may relax some of the conditions used in equation (52): we may consider A_4 unrelated to A_u and choose $A_5, \dots, A_9$ all nonvanishing. The net result of this is the following. The fields which become part of the six vector matter supermultiplet upon dimensional reduction of dual waves are nonvanishing for new solutions. Indeed, by choosing A_4 proportional to A_u, we have achieved that the metric in the x^4 direction is flat. Otherwise, g_{44} would not be equal to a constant. The choice $A_5 \neq 0$ will make the 6-dimensional space nonflat, $g_{45} = -B_{45} = e^{2\phi} A_5$, etc. The combination of vector fields which enter the matter multiplet will be nonvanishing for these solutions, since the definition of vector fields will include the above-mentioned g_{4I}-component of the metric and the B_{4I}-component of the 2-form. The corresponding 4-dimensional theory upon dimensional reduction will contain additional vectors and scalars as well as pseudoscalars. By construction, we shall have a bosonic action and configurations which solve its equation of motion. These configurations will have half of the unbroken supersymmetries when embedded into $N = 4$ supergravity with six abelian vector multiplets. We can also take into account the Yang–Mills part of the 10-dimensional gravitational wave solution SSW. This will lead to 4-dimensional solutions of the equations of motion that have unbroken supersymmetry when embedded into $N = 4$ supergravity with six abelian vector supermultiplets and non-Abelian Yang–Mills supermultiplet. The action and the corresponding solutions will be presented in explicit form in a future publication.

Acknowledgments. I am very grateful to my collaborators E. Bergshoeff, D. Kastor, and T. Ortín who made it possible to establish some deep properties of gravitational waves in string theory and their relationship to the most interesting 4-dimensional geometries, including extreme black holes.

Note added in proof. After the paper above was written there were various new developments in the field of supersymmetric gravity. The one which we would like to describe here concerns the new striking improvement in our understanding of extreme black holes in string theory.

It had been realized some time ago ([**3**]) that the entropy of supersymmetric black holes in absence of supersymmetry anomalies is not corrected by quantum corrections on the basis of Berezin's rules of integration in supersymmetric theories. At that time we had studied a relatively small subset of extreme black holes, which have been called $U(1) \times U(1)$ stringy black holes. The corresponding bosonic solutions were embedded into $N = 4$ supergravity and we have found a very nice area formula for those black holes which have 1/4 of unbroken supersymmetry. The Bekenstein-Hawking entropy was shown to be proportional to the 1/4 of the area of the horizon

$$S = \frac{A}{4} = 2\pi|pq|, \tag{72}$$

which in turn was given by a product of conserved electric q and magnetic p charges in each of the $U(1)$ groups. Due to Dirac quantization these charges are quantized and have to be integer numbers. More recently in [**25**], Strominger and Vafa came up with the idea of how to explain the entropy formulas of the type shown above from the point of view of microphysics, by counting the states of the string theory.

It has been also realized some time ago that the extreme black holes carry duality properties of string theory and supergravity via a sophisticated dependence of the area of the black holes from the charges. For example, consider black holes in $N = 8$ supergravity, which is a low energy effective action of type II string theory compactified on $\mathbf{T}^6$. The $N = 8$ supergravity has a hidden symmetry of equations of motion under the group $E_{7(7)}$ as was discovered by Cremmer and Julia in [**26**] back in 1979. More recently in [**27**], Hull and Townsend provided some arguments that the discrete subgroup of E_7 may be an exact symmetry of the string theory. The corresponding discrete subgroup is called $E_7(\mathbf{Z})$ and the symmetry is called U duality. Recently in [**28**], we conjectured that the area of the black hole horizon is given by the square root of the quartic invariant J of the E(7) group, as build by Cartan:

$$S = \pi\sqrt{J} \tag{73}$$

$$\begin{aligned} J = {} & q^{ij}p_{jk}q^{kl}p_{li} - \tfrac{1}{4}q^{ij}p_{ij}q^{kl}p_{kl} \\ & + \tfrac{1}{96}\left(\epsilon^{ijklmnop}\, p_{ij}p_{kl}p_{mn}p_{op} + \epsilon_{ijklmnop}\, q^{ij}q^{kl}q^{mn}q^{op}\right), \end{aligned} \tag{74}$$

and the electric charges q^{ij} and magnetic ones p_{ij}, $i, j = 1, \ldots, 8$ span the 56-dimensional space. The theory has 28-gauge groups, each may have electric and magnetic charge. The formula above in particular reduces to (72) when only two charges are present.

This $E(7)$ formula (74) went through various checks and seems to be consistent with all supersymmetric black hole solutions known at present. The formula also has a considerable predictive power in terms of being able to accommodate those black hole solutions which have not been found yet. It also serves the purpose of continuous effort to explain the entropy of classical black hole solutions via quantum mechanical counting of states and to solve the long-standing problem of quantum gravity.

The fact that the generic supersymmetric nonrenormalization theorem for the extreme black holes in case of maximally extended $N = 8$ supersymmetry protects from quantum corrections a unique quartic invariant of the exceptional group $E(7)$ is particularly satisfying: no other structure exists with $E(7)$ symmetry which can be constructed from one fundamental 56-dimensional representation. Thus, the properties of the exceptional groups have found a new way into theories with anticommuting variables, which had been constructed by Berezin. He would love this development...

References

1. A. Dabholkar, G. Gibbons, J. A. Harvey, and F. Ruiz, Nuclear Phys. B **340** (1990), 33.
2. C. G. Callan Jr., J. A. Harvey, and A. Strominger, Nuclear Phys. B **359** (1991), 611; **367** (1991), 60.
3. R. E. Kallosh, A. D. Linde, T. M. Ortí n, A. W. Peet, and A. Van Proeyen, Phys. Rev. **46** (1992), 5236.
4. E. Bergshoeff, R. Kallosh, and T. Ortín, Phys. Rev. D **47** (1993), 5444.
5. H. W. Brinkmann, Proc. Nat. Acad. Sci. U.S.A. **9** (1923), 1.
6. D. Kramer, H. Stephani, M. MacCallum, and E. Herlt, *Exact solutions of Einstein's field equations*, Cambridge U.P., 1980.
7. P. Aichelburg and R. Sexl, Gen. Relativity Gravitatiton **2** (1971), 303.
8. R. Güven, Phys. Lett. B **191** (1987), 265.
9. G. W. Gibbons and C. M. Hull, Phys. Lett. B **109** (1982), 190.
10. K. P. Tod, Phys. Lett. B **121** (1983), 241.
11. D. Amati and C. Klimčik, Phys. Lett B **219** (1988), 443; D. Amati, M. Ciafaloni, and Veneziano and references therein, Nuclear Phys. B **347** (1990), 550; H. J. de Vega, M. Ramón Medrano, and N. Sánchez and references therein, Nuclear Phys. B **374** (1992), 425.
12. G. T. Horowitz and A. R. Steif, Phys. Rev. Lett. **64** (1990), 260; T. Horowitz and references therein, in: Proceedings of Strings '90, College Station, Texas, March 1990, World Scientific, 1991.
13. H. W. Brinkmann, Math. Ann. **94** (1925), 119.
14. E. Bergshoeff, I. Entrop, and R. Kallosh, *Exact duality in string effective action*, Phys. Rev. D. **49** (1994), 6663–6673.
15. E. Bergshoeff, R. Kallosh, and T. Ortín, *Black-hole-wave duality in string theory*, Phys. Rev. D **50** (1994), 5188.
16. R. Kallosh, D. Kastor, T. Ortín, and T. Torma, Phys.Rev. D **50** (1994), 6374;C. Johnson and R. C. Meyers, Phys. Rev. D **50** (1994), 6512; D. V. Galt′sov and O. V. Kechkin,, Phys. Rev. D **50** (1994), 7394.
17. E. Bergshoeff, R. Kallosh, and T. Ortín, *Duality versus supersymmetry and compactification*, Phys. Rev. D **51** (1995), 3009 (to appear).
18. T. Buscher, Phys. Lett. B **159** (1985), 127; **194** (1987), 59; **201** (1988), 466.
19. J. H. Horne, G. T. Horowitz, and A. R. Steif, Phys. Rev. Lett. **68** (1992), 568.
20. G. W. Gibbons, Nuclear Phys. B **207** (1982), 337; G. W. Gibbons and K. Maeda, Nuclear Phys. B **298** (1988), 741; D. Garfinkle, G. T. Horowitz, and A. Strominger, Phys. Rev. D **43** (1991), 3140.
21. A. A. Tseytlin, Nuclear Phys. B **390** (1993), 153.
22. C. Duval, Z. Horváth, and P. Horváthy, Phys. Lett. B **313** (1993), 10.
23. A. Chamseddine, Nuclear Phys. B **185** (1981), 403.
24. G. T. Horowitz and A. A. Tseytlin, Phys. Rev. Lett. **73** (1994), 3351; Phys. Rev. D **51** (1995), 2896.
25. A. Strominger and C. Vafa, *Microscopic Origin of the Bekenstein-Hawking Entropy*, Harvard University preprint HUTP-96-A002 (1996), hep-th/9601029.
26. E. Cremmer and B. Julia, Nuclear Phys. B **B159** (1979), 141.
27. C. M. Hull and P. K. Townsend, Nuclear Phys. B **B438** (1995), 109.
28. R. Kallosh and B. Kol, *E(7) Symmetric area of the black hole horizon*, hep-th/9602014.

DEPARTMENT OF PHYSICS, STANFORD UNIVERSITY, STANFORD, CALIFORNIA 94305, USA
E-mail address: kallosh@physics.stanford.edu

Amer. Math. Soc. Transl.
(2) Vol. **177**, 1996

Geometric Quantization in the Fock Space

V. P. Maslov and O. Yu. Shvedov

ABSTRACT. We investigate an infinite-dimensional analog of the theory of Lagrangian manifolds with complex germs. To such a manifold, we assign a canonical operator that depends on creation and annihilation operators. This operator is by definition the geometric quantization for isotropic manifolds with complex germs. We prove that for secondary quantized equations this quantization is the asymptotics for the Cauchy problem. Results of Berezin are used thoroughly in the construction of the canonical operator and in proofs of the theorems.

§1. Introduction

The construction of asymptotic solutions for multiparticle Schrödinger and Liouville equations as the number of particles tends to infinity was investigated in [**15, 16**]. The construction is as follows. Both Schrödinger and Liouville multiparticle equations may be presented in a unified form through creation and annihilation operators:

$$i\frac{\partial\widehat{\Phi}}{\partial t} = \left(T_{mn}\hat{\psi}_m^+\hat{\psi}_n^- + \frac{\varepsilon}{2}V_{klrs}\hat{\psi}_k^+\hat{\psi}_l^+\hat{\psi}_r^-\hat{\psi}_s^-\right)\widehat{\Phi}. \tag{1}$$

Here $\widehat{\Phi}$ is an element of the Fock space, $\hat{\psi}_j^{\pm}$ are creation and annihilation operators in this space ([**2, 12**]), and we sum over the repeated indices, $m, n, k, l, r, s = 1, 2, \dots$.

The coefficients T_{mn} and V_{klrs} are as follows.

1. For the Schrödinger equation

$$\begin{aligned} T_{mn} &= \int dx\, f_m^*(x)(-\Delta/2 + U(x))\, f_n(x), \\ V_{klrs} &= \int dx\, dy\, f_k^*(x)\, f_l^*(y)\, V(x,y)\, f_r(y)\, f_s(x), \end{aligned} \qquad x, y \in \mathbb{R}^\nu,\ \nu \in \mathbb{N},$$

1991 *Mathematics Subject Classification.* Primary 81Q20.

This work was supported by the International Science Foundation, grant MFO000.

where $\{f_1, f_2, \dots\}$ is an orthonormal basis in $L^2(\mathbb{R}^\nu)$, Δ is the Laplace operator in $\mathbb{R}^\nu$, U is an external potential, V is the potential of the interparticle interaction. The form (1) of the multiparticle Schrödinger equation originates in the papers [**4–6, 9**].

2. For the Liouville equation

$$T_{mn} = i \int dp\, dq\, f_m^*(p,q) \left(\frac{\partial U}{\partial q}(q) \frac{\partial}{\partial p} - p \frac{\partial}{\partial q} \right) f_n(p,q),$$

$$V_{klrs} = i \int dp_1\, dp_2\, dq_1\, dq_2\, f_k^*(p_1,q_1) f_l^*(p_2,q_2)$$
$$\times \left(\frac{\partial V(q_1,q_2)}{\partial q_1} \frac{\partial}{\partial p_1} + \frac{\partial V(q_1,q_2)}{\partial q_2} \frac{\partial}{\partial p_2} \right) f_r(p_1,q_1) f_s(p_2,q_2),$$
$$p, q, p_1, q_1, p_2, q_2 \in \mathbb{R}^\nu, \quad \nu \in \mathbb{N},$$

where $\{f_1, f_2, \dots\}$ is an orthonormal basis in $L^2(\mathbb{R}^{2\nu})$, U and V are, as in the previous case, an external potential and the interparticle interaction potential, respectively. Schönberg [**19, 20**] was the first to suggest the form (1) for the multiparticle Liouville equation (see also [**18**]).

The asymptotics of the solution of equation (1) was constructed as follows. After the substitution $\sqrt{\varepsilon}\hat{\psi}_j^\pm = \hat{\phi}_j^\pm$, equation (1) becomes

$$i\varepsilon \frac{\partial \widehat{\Phi}}{\partial t} = \mathcal{H}(\hat{\phi}_j^+, \hat{\phi}_j^-) \widehat{\Phi}, \tag{2}$$

for

$$\mathcal{H}(\hat{\phi}_j^+, \hat{\phi}_j^-) = T_{mn} \hat{\phi}_m^+ \hat{\phi}_n^- + \tfrac{1}{2} V_{klrs} \hat{\phi}_k^+ \hat{\phi}_l^+ \hat{\phi}_r^- \hat{\phi}_s^-,$$

where

$$[\hat{\phi}_j^-, \hat{\phi}_l^+] = \varepsilon \delta_{jl}. \tag{3}$$

Property (3) allows us to apply quasi-classical methods to equation (2), ε being the parameter of the quasi-classical expansion.

Since the mean number of particles in the state $\widehat{\Phi}$ is equal to (see [**2, 12**])

$$N = \frac{(\widehat{\Phi}, \hat{\psi}_l^+ \hat{\psi}_l^- \widehat{\Phi})}{(\widehat{\Phi}, \widehat{\Phi})} = \frac{1}{\varepsilon} \frac{(\widehat{\Phi}, \hat{\phi}_l^+ \hat{\phi}_l^- \widehat{\Phi})}{(\widehat{\Phi}, \widehat{\Phi})}, \qquad l = 1, 2, \dots,$$

and the quantity $(\widehat{\Phi}, \hat{\phi}_l^+ \hat{\phi}_l^- \widehat{\Phi})/(\widehat{\Phi}, \widehat{\Phi})$ is of order ε^0, quasi-classical methods yield approximate solutions of equation (1) as $\varepsilon \to 0$, $N \to \infty$, $\varepsilon N \to \alpha = \text{const}$.

To equation (2) there corresponds a classical infinite-dimensional Hamiltonian system with the Hamiltonian $\mathcal{H}((Q_j - iP_j)/\sqrt{2}, (Q_j + iP_j)/\sqrt{2})$.

The results relating to statistical physics may be rigorously justified. For some of them such justification was given in [**16**].

An analogous parameter arises in quantum field theory. However, the consideration of formal asymptotic expansions over this small parameter does not allow us to justify this asymptotics, since quantum field theory itself does not have a rigorous mathematical meaning. Some approaches to this problem have been developed only in simple special cases (see [**22, 7**]), and without its solution a justification of heuristic asymptotics is of course impossible. Up to now, only the postulated perturbation theory series makes sense in quantum field theory. One can hope that a geometric quantization related with the "classical" equations for the case in which all the commutators tend to zero will be postulated as well.

The concept of geometric quantization has been essentially introduced in the works of Bohr, Sommerfeld, and de Broglie, even before the works of Schrödinger, Heisenberg, and Dirac. It is also a rapidly developing concept in modern mathematics ([**8, 10**]).

This paper is organized as follows. Section 2 contains the definition of the canonical operator in the Fock space. We define objects of geometric quantization; namely, Lagrangian manifolds with complex germs. We assign elements of the Fock space to each such manifold, and we also justify the relationship between the traditional definition of canonical operator and ours. These definitions differ because of divergencies in the infinite-dimensional case. The traditional definition is not applicable in this case. In §3 we define the canonical transformation of a Lagrangian manifold with complex germ and justify the germ axioms. We show that the canonical operator approximately satisfies the corresponding secondary quantized equations. The main theorem is formulated in §4. Section 5 contains a construction of another asymptotics with the help of complex germ creation and annihilation operators. Many examples are mentioned in sections 4 and 5. We prove the main theorem in section 6.

§2. The canonical operator

In the present paper we consider finite-dimensional isotropic manifolds with infinite-dimensional complex germ in an infinite-dimensional phase space (see the definition below). We assign a canonical operator to such a manifold. For simplicity, we give a definition of the canonical operator that is not absolutely invariant. This definition, however, is sufficient for constructing asymptotic solutions to the Cauchy problem.

In the case of quantum field theory, this canonical operator satisfies formally the corresponding secondary quantized equations as $\varepsilon \to 0$, provided that the initial conditions for these equations correspond to isotropic finite-dimensional manifolds (or, in the stationary case, if there exist stable isotropic manifolds). In specific examples such isotropic manifolds usually have dimension 1 or 0.

2.1. **Objects of geometric quantization.** Let us define the objects of geometric quantization.

By $\mathcal{H}^n$ we denote the space of complex square summable symmetric functions of n variables $i_1, \dots, i_n \in \mathbb{N}$. In $\mathcal{H}^n$ we introduce an inner product by the formula

$$(f, g) = \sum_{i_1,\dots,i_n=1}^{\infty} f^*_{i_1\dots i_n} g_{i_1\dots i_n}, \qquad f, g \in \mathcal{H}^n.$$

Denote by $\mathcal{H}$ the Fock space $\bigoplus_{n=0}^{\infty} \mathcal{H}^n$, and by $\widehat{\Phi}^{(k)} \in \mathcal{H}^k$ the kth component of $\widehat{\Phi} \in \mathcal{H}$.

Consider creation and annihilation operators acting in $\mathcal{H}$ in the following way:

$$(\hat\psi_j^-\widehat\Phi)^{(k-1)}_{j_1\dots j_{k-1}} = k^{1/2}\widehat\Phi^{(k)}_{j_1\dots j_{k-1}j},$$
$$(\hat\psi_j^+\widehat\Phi)^{(k)}_{j_1\dots j_k} = k^{-1/2}\sum_{i=1}^{k}\widehat\Phi^{(k-1)}_{j_1\dots j_{i-1}j_{i+1}\dots j_k}\delta_{jj_i}, \qquad j, j_l \in \mathbb{N}.$$

By $\widehat\Phi_0$ denote the following element of the space $\mathcal{H}$: $\widehat\Phi_0^{(0)} = 1$, $\widehat\Phi_0^{(i)} = 0$, $i \geqslant 1$. As usual, elements of the space $\mathcal{H}$ are called *state vectors in the Fock presentation.*

In the present paper we shall also use the presentation of elements of $\mathcal{H}$ as Berezin generating functionals ([**2**]). To each element $\widehat\Phi \in \mathcal{H}$ we assign the generating functional

$$\Phi(a_1^*, a_2^*, \dots) = \sum_{n=0}^{\infty}\frac{1}{\sqrt{n!}}\sum_{i_1,\dots,i_n=1}^{\infty}\widehat\Phi^{(n)}_{i_1\dots i_n}a^*_{i_1}\dots a^*_{i_n}, \qquad a_j^* \in \mathbb{C},\ \sum_{j=1}^{\infty}|a_j^*|^2 < \infty.$$

In this presentation, the creation operators $\hat\psi_j^+$ are multiplications by a_j^*, and the annihilation operators $\hat\psi_j^-$ are the derivation operators $\partial/\partial a_j^*$ ([**2**]).

We shall also make use of the Schrödinger coordinate presentation (or the Q-presentation). In this presentation we assign to each element $\widehat\Phi \in \mathcal{H}$ the functional

$$\Phi_Q(q_1, q_2, \dots) = \sum_{n=0}^{\infty}\frac{1}{\sqrt{n!}}\sum_{i_1,\dots,i_n=1}^{\infty}\widehat\Phi^{(n)}_{i_1\dots i_n}\prod_{l=1}^{n}\left(q_{i_l} - \varepsilon\frac{\partial}{\partial q_{i_l}}\right)\frac{1}{\sqrt{2\varepsilon}}\exp\left(-\sum_{i=1}^{\infty}\frac{q_i^2}{2\varepsilon}\right).$$

Then the creation and annihilation operators $\hat\psi_j^\pm$ can be written in the form $\hat\psi_j^\pm = (q_j \mp \varepsilon\partial/\partial q_j)/\sqrt{2\varepsilon}$.

By $\mathcal{L}$ denote the set

$$\mathcal{L} = \left\{(P_1, Q_1; P_2, Q_2; \dots) : P_i \in \mathbb{R},\ Q_i \in \mathbb{R},\ \sum_{i=1}^{\infty}(P_i^2 + Q_i^2) < \infty\right\}.$$

Now let us define finite-dimensional isotropic manifolds and the corresponding Lagrangian manifolds with complex germs. Let Λ^k be a k-dimensional surface in $\mathcal{L}$, and $\tau_1, \dots, \tau_k$ local coordinates on Λ^k. For the sake of brevity, denote the sequence $(P_1, P_2, \dots)$ by P, the sequence $(Q_1, Q_2, \dots)$ by Q, and the sequence $(\tau_1, \dots, \tau_k)$ by τ.

DEFINITION 1. A manifold $\Lambda^k = \{P = P(\tau),\ Q = Q(\tau)\}$ is called *isotropic* if the following axioms hold.

(m1) For any $\lambda_1, \dots, \lambda_k$, $\lambda_i \in \{0, 1, 2, \dots\}$, $i = 1, \dots, k$, the derivatives

$$\frac{\partial^{\lambda_1+\dots+\lambda_k}}{\partial\tau_1^{\lambda_1}\dots\partial\tau_k^{\lambda_k}}P_j, \qquad \frac{\partial^{\lambda_1+\dots+\lambda_k}}{\partial\tau_1^{\lambda_1}\dots\partial\tau_k^{\lambda_k}}Q_j, \qquad j = 1, 2, \dots,$$

exist and the following series converges:

$$\sum_{j=1}^{\infty}\left[\left(\frac{\partial^{\lambda_1+\dots+\lambda_k}}{\partial\tau_1^{\lambda_1}\dots\partial\tau_k^{\lambda_k}}P_j\right)^2 + \left(\frac{\partial^{\lambda_1+\dots+\lambda_k}}{\partial\tau_1^{\lambda_1}\dots\partial\tau_k^{\lambda_k}}Q_j\right)^2\right].$$

(m2) If

$$\sum_{j=1}^{\infty}\left[\left(\sum_{m=1}^{k}\frac{\partial P_j}{\partial \tau_m}\,\xi_m\right)^2+\left(\sum_{m=1}^{k}\frac{\partial Q_j}{\partial \tau_m}\,\xi_m\right)^2\right]=0,$$

for some $\xi_m \in \mathbb{R}$, then $\xi_m = 0$, $m = 1, \dots, k$.

(m3) We have

$$\sum_{j=1}^{\infty}\left(\frac{\partial P_j}{\partial \tau_m}\frac{\partial Q_j}{\partial \tau_n}-\frac{\partial P_j}{\partial \tau_n}\frac{\partial Q_j}{\partial \tau_m}\right)=0, \qquad m, n = 1, \dots, k.$$

Denote by M^k the universal covering space of the isotropic manifold Λ^k.

DEFINITION 2. A *complex germ* is a set of planes $r(\tau)$, $\tau \in M^k$, in the abstract space $\mathbb{R}^\infty \times \mathbb{R}^\infty$. Each plane consists of the vectors

$$\begin{pmatrix} w_1(\alpha,\tau)\\ w_2(\alpha,\tau)\\ \vdots \\ z_1(\alpha,\tau)\\ z_2(\alpha,\tau)\\ \vdots \end{pmatrix}, \qquad \begin{aligned} w_i(\alpha,\tau) &= \sum_{j=1}^{\infty} B_{ij}(\tau)\,\alpha_j, \qquad i = 1, 2, \dots,\\ z_i(\alpha,\tau) &= \sum_{j=1}^{\infty} C_{ij}(\tau)\,\alpha_j, \qquad i = 1, 2, \dots, \end{aligned}$$

where α ranges over infinite sequences $\alpha = (\alpha_1, \alpha_2, \dots)$ ($\alpha_i \in \mathbb{C}$, $i = 1, 2, \dots$, $\sum_{i=1}^{\infty} |\alpha_i|^2 < \infty$) and the following axioms hold.

(r1) Let τ', τ'' be two points of the universal covering space M^k such that they project into one and the same point of Λ^k. Then there exists a unitary operator $A(\tau', \tau'')$ such that

$$B(\tau'') = B(\tau')A(\tau',\tau''), \qquad C(\tau'') = C(\tau')A(\tau',\tau'').$$

(r2) For $a = 1, \dots, k$, $i = 1, 2, \dots$

$$B_{ia}(\tau) = \frac{\partial P_i(\tau)}{\partial \tau_a}, \qquad C_{ia}(\tau) = \frac{\partial Q_i(\tau)}{\partial \tau_a}.$$

(r3) We have

$$B^T(\tau)C(\tau) - C^T(\tau)B(\tau) = 0.$$

(Here B^T and C^T are the transpose matrices to B and C, respectively.)

(r4) We have

$$C^+(\tau)B(\tau) - B^+(\tau)C(\tau) = iL,$$

where L is the diagonal matrix with first k diagonal elements equal to 0 and all others to 1.

(r5) For any $\lambda_1, \dots, \lambda_k$, $\lambda_i \in \{0, 1, 2, \dots\}$, $i = 1, \dots, k$, the derivatives

$$F_{mn}^{(\lambda_1,\dots,\lambda_k)} = \frac{\partial^{\lambda_1+\dots+\lambda_k}}{\partial \tau_1^{\lambda_1}\cdots\partial \tau_k^{\lambda_k}}\left(\frac{C_{mn}(\tau)+iB_{mn}(\tau)}{\sqrt{2}}\right),$$

$$G_{mn}^{(\lambda_1,\dots,\lambda_k)} = \frac{\partial^{\lambda_1+\dots+\lambda_k}}{\partial \tau_1^{\lambda_1}\cdots\partial \tau_k^{\lambda_k}}\left(\frac{C_{mn}(\tau)-iB_{mn}(\tau)}{\sqrt{2}}\right),$$

exist, the operators $G^{(\lambda_1,\dots,\lambda_k)}$ are bounded and the operators $F^{(\lambda_1,\dots,\lambda_k)}$ are Hilbert–Schmidt operators.

(r6) The operator $(C(\tau) - iB(\tau))/\sqrt{2}$ has a bounded inverse operator.

A pair consisting of an isotropic manifold and a complex germ will be called a *Lagrangian manifold with complex germ*, since together they form a germ of an infinite-dimensional complex Lagrangian manifold.

REMARK 1. Let $r(\tau')$, $r(\tau'')$ be planes at two points τ', τ'' of the universal covering space M^k that project into one and the same point of the isotropic manifold Λ^k. It then follows from the axiom (r1) that these two planes differ from each other only by a parametrization.

REMARK 2. In the finite-dimensional case, axioms (r2)–(r4) are equivalent to the ordinary axioms of complex germ theory ([**13, 14, 1**]).

REMARK 3. For a zero-dimensional isotropic manifold, the axioms for the complex germ mean that the matrix

$$\begin{pmatrix} G & \bar{F} \\ F & \bar{G} \end{pmatrix}$$

is the matrix of a Berezin canonical transformation (see [**2**]).

To each isotropic manifold $\Lambda^k = \{P = P(\tau),\ Q = Q(\tau)\}$ we shall assign the function $\phi_l(\tau) = (Q_l(\tau) + iP_l(\tau))/\sqrt{2}$.

2.2. **Heuristic motivation for the definition of canonical operator.** Here we describe a heuristic method for deriving state vectors, which are approximate solutions to secondary-quantized equations (2) and correspond to Lagrangian manifolds with complex germs.

Equation (2) in the Q-presentation has the form of the infinite-dimensional Schrödinger equation,

$$i\varepsilon\frac{\partial\Phi_Q}{\partial t} = H\left(\frac{\widehat{Q}_j - i\widehat{P}_j}{\sqrt{2}}, \frac{\widehat{Q}_j + i\widehat{P}_j}{\sqrt{2}}\right)\Phi_Q, \tag{4}$$

where the operators $\widehat{Q}_j$ are multiplications by q_j, operators $\widehat{P}_j$ are the derivation operators $-i\varepsilon\partial/\partial q_j$. Thus, $\widehat{Q}_j$ and $\widehat{P}_j$ are infinite-dimensional analogs of coordinate and momentum operators, while ε is an analog of the Planck constant. As we mentioned above, one can apply semiclassical techniques to equation (4) as ε tends to zero.

Semiclassical approximate solutions of the type

$$\phi(q)\exp\left(\frac{i}{\varepsilon}S(q)\right), \tag{5}$$

where the sequence $(q_1, q_2, \dots)$ is denoted by q and S is a real functional, are widely used in physics (see, for example, [**12**]). Functionals (5) are of order $O(1)$ as $\varepsilon \to 0$ for all q.

In this paper we consider yet another type of asymptotic solution to quantized equations. These solutions are not small and oscillate rapidly only if the distance between the point q and the surface

$$\{Q(\tau),\ \tau \in \Lambda^k\} \tag{6}$$

is of order $O(\varepsilon^{1/2})$ as ε tends to zero, i.e., the expression

$$\min_{\tau\in\Lambda^k} \sum_{i=1}^{\infty} (q_i - Q_i(\tau))^2 \tag{7}$$

is of order $O(\varepsilon)$. If the expression (7) is of order $O(\varepsilon^{1-\delta})$, $\delta > 0$, then these solutions are exponentially small.

First of all, consider the case of a zero-dimensional Lagrangian manifold, i.e., a point $(P_1, Q_1, P_2, Q_2, \dots) \in \mathcal{L}$. The complex germ method leads to asymptotics of the type

$$\Phi_Q(q) = \mathcal{F}\Big(\frac{q-Q}{\sqrt{\varepsilon}}\Big) \exp\Big(\frac{i}{\varepsilon}\sum_{i=1}^{\infty} P_i(q_i - Q_i)\Big). \tag{8}$$

The function $\mathcal{F}$ in formula (8) descreases rapidly as its argument tends to infinity and can be expressed via the complex germ (see subsection 2.5 for more details).

Let us describe the Fock presentation for the functional (8). For simplicity, we begin with considering the case $P = Q = 0$. If

$$\mathcal{F}\Big(\frac{q}{\sqrt{\varepsilon}}\Big) = \mathcal{F}_0\Big(\frac{q}{\sqrt{\varepsilon}}\Big) = \exp\Big(-\sum_{i=1}^{\infty}\frac{q_i^2}{2\varepsilon}\Big), \tag{9}$$

then the functional (8) corresponds to the vacuum state vector $\widehat{\Phi}_0$ in the Fock presentation. If functional $\mathcal{F}$ is arbitrary, then we can extract the factor (9) from the functional (8) and represent the corresponding state vector in the form

$$\mathcal{F}_1\Big(\frac{\widehat{Q}}{\sqrt{\varepsilon}}\Big)\widehat{\Phi}_0, \tag{10}$$

where $\mathcal{F}_1 = \mathcal{F}/\mathcal{F}_0$. The operator $\widehat{Q}/\sqrt{\varepsilon}$ may be expressed as a function of creation and annihilation operators

$$\frac{\widehat{Q}_i}{\sqrt{\varepsilon}} = \frac{\hat{\psi}_i^+ + \hat{\psi}_i^-}{\sqrt{2}}, \tag{11}$$

which does not depend on ε. Any function of the operator (11) transforms the vacuum vector into an ε-independent element of the Fock space, which corresponds to the functional (8) if $P = Q = 0$.

The general case may be reduced to the case considered above. Namely, the functional (8) may be presented in the form

$$\exp\Big(\frac{i}{\varepsilon}P_l q_l\Big)\exp\Big(-Q_l\frac{\delta}{\delta q_l}\Big)\mathcal{F}_0\Big(\frac{q}{\sqrt{\varepsilon}}\Big), \tag{12}$$

where $l = 1, 2, \dots$. Expression (12) may be simplified using the Baker–Hausdorff formula

$$e^{A_1+A_2} = e^{A_1}e^{A_2}e^{-[A_1,A_2]/2} \tag{13}$$

for the operators

$$A_1 = iP_l\frac{q_l}{\varepsilon}, \quad A_2 = -Q_l\frac{\delta}{\delta q_l}, \qquad l = 1, 2, \dots .$$

From

$$A_1 + A_2 = \frac{1}{\sqrt{\varepsilon}}\sum_{l=1}^{\infty}(\hat{\psi}_l^+\phi_l - \hat{\psi}_l^-\phi_l^*)$$

we obtain the following state vector in the Fock presentation corresponding to the functional (12):

$$U_\phi \exp\Big(-\frac{i}{2\varepsilon}\sum_{l=1}^{\infty}P_lQ_l\Big)\widehat{Y}, \tag{14}$$

where

$$U_\phi = \exp\Big(\frac{1}{\sqrt{\varepsilon}}\sum_{l=1}^{\infty}(\hat{\psi}_l^+\phi_l - \hat{\psi}_l^-\phi_l^*)\Big),$$

and $\widehat{Y}$ is an ε-independent state vector, which can be expressed through creation operators ([**2**]):

$$\widehat{Y} = Y(\hat{\psi}^+)\widehat{\Phi}_0, \qquad Y(\hat{\psi}^+) = \sum_{n=0}^{\infty}\frac{1}{\sqrt{n!}}Y^{(n)}_{i_1,\dots,i_n}\hat{\psi}^+_{i_1}\cdots\hat{\psi}^+_{i_n}.$$

From the Baker–Hausdorff formula and the commutation relations ([**2**])

$$U_\phi^{-1}\hat{\psi}_l^+U_\phi = \hat{\psi}_l^+ + \phi_l^*/\sqrt{\varepsilon}, \qquad U_\phi^{-1}\hat{\psi}_l^-U_\phi = \hat{\psi}_l^- + \phi_l/\sqrt{\varepsilon}, \tag{15}$$

which can be verified by using the commutation relations between creation and annihilation operators, we see that the vector (14) may be expressed in the form

$$\Psi_{\phi,Y} = Y\Big(\hat{\psi}^+ - \frac{\phi^*}{\sqrt{\varepsilon}}\Big)\exp\Big(\frac{1}{\varepsilon}\Big[g + \sum_{l=1}^{\infty}\phi_l(\sqrt{\varepsilon}\hat{\psi}_l^+ - \phi_l^*)\Big]\Big)\widehat{\Phi}_0, \tag{16}$$

where

$$g = \phi_l^*\phi_l/2 + (\phi_l^*\phi_l^* - \phi_l\phi_l)/4, \qquad l = 1, 2, \dots . \tag{17}$$

We shall use the Gaussian functionals Y

$$Y(\hat{\psi}^+) = c\exp(\hat{\psi}_i^+M_{ij}\hat{\psi}_j^+/2), \qquad i, j = 1, 2, \dots, \tag{18}$$

where c is a constant, M is a Hilbert–Schmidt operator, $\|M\| < 1$.

Thus we obtained state vector (16) corresponding to a zero-dimensional isotropic manifold. In 2.3 we assign a matrix M to each complex germ. The relationship between complex germ theories in the Q-presentation and in the Fock space is discussed in more detail in subsection 2.5.

Now consider the asymptotics in the Q-presentation concentrated in a neighborhood of the surface (6) if $k > 0$. In this subsection we discuss the particular case of a manifold Λ^k satisfying the conditions

$$Q_{k+1} = Q_{k+2} = \dots = P_{k+1} = P_{k+2} = \dots = 0. \tag{19}$$

The general case will be considered in subsection 2.5.

Complex germ asymptotics in the case in which we have (19) has the form

$$\mathcal{F}\Big(q_1,\dots,q_k;\frac{q_{k+1}}{\sqrt{\varepsilon}},\frac{q_{k+2}}{\sqrt{\varepsilon}},\dots\Big)\exp\Big(\frac{i}{\varepsilon}S(q_1,\dots,q_k)\Big), \tag{20}$$

where $\mathcal{F}$ decreases rapidly at infinity and S is a real function. The isotropic manifold corresponding to the asymptotics (20) may be parametrized by the coordinates $\tau_a = Q_a$, $a = 1,\dots,k$,

$$Q_a(\tau) = \tau_a, \qquad P_a(\tau) = \partial S/\partial \tau_a\,.$$

In order to write the Fock presentation of the functional (20) it is convenient to reduce it to the functionals (8). Namely, consider the following superposition of expressions (8)

$$\begin{aligned}&\int \frac{d\tau}{\varepsilon^{k/2}} f\Big(\tau_1,\dots,\tau_k;\frac{q_1-\tau_1}{\sqrt{\varepsilon}},\dots,\frac{q_k-\tau_k}{\sqrt{\varepsilon}};\frac{q_{k+1}}{\sqrt{\varepsilon}},\frac{q_{k+2}}{\sqrt{\varepsilon}},\dots\Big)\\ &\qquad\times\exp\Big(\frac{i}{\varepsilon}\Big[S(\tau)+\sum_{a=1}^{k}\frac{\partial S}{\partial\tau_a}(\tau)(q_a-\tau_a)\Big]\Big)\end{aligned} \tag{21}$$

and choose a function f rapidly decreasing at infinity in order to make the functional (21) approximately equal to (20). The substitution

$$\xi_a = (q_a-\tau_a)/\sqrt{\varepsilon}, \qquad a = 1,\dots,k,$$

transforms integral (21) into:

$$\begin{aligned}&\int d\xi_1\cdots d\xi_k\, f\Big(q_1-\xi_1\sqrt{\varepsilon},\dots,q_k-\xi_k\sqrt{\varepsilon};\xi_1,\dots,\xi_k;\frac{q_{k+1}}{\sqrt{\varepsilon}},\frac{q_{k+2}}{\sqrt{\varepsilon}},\dots\Big)\\ &\qquad\times\exp\Big(\frac{i}{\varepsilon}\Big[S(q-\xi\sqrt{\varepsilon})+\sum_{a=1}^{k}\frac{\partial S}{\partial q_a}(q-\xi\sqrt{\varepsilon})\xi_a\sqrt{\varepsilon}\Big]\Big).\end{aligned} \tag{22}$$

As $\varepsilon\to 0$, $\xi = \text{const}$, the exponent in formula (22) has the form

$$\exp\Big[-\frac{i}{2}\xi_a\frac{\partial^2 S(q)}{\partial q_a\partial q_b}\xi_b+\frac{i}{\varepsilon}S(q)\Big].$$

Taking into account that f decreases rapidly we obtain that integrals (21) and (20) are approximately equal to

$$\begin{aligned}&\mathcal{F}\Big(q_1,\dots,q_k;\frac{q_{k+1}}{\sqrt{\varepsilon}},\frac{q_{k+2}}{\sqrt{\varepsilon}},\dots\Big)\\ &\quad=\int d^k\xi\, f\Big(q_1,\dots,q_k;\xi_1,\dots,\xi_k;\frac{q_{k+1}}{\sqrt{\varepsilon}},\frac{q_{k+2}}{\sqrt{\varepsilon}},\dots\Big)\exp\Big[-\frac{i}{2}\xi_a\frac{\partial^2 S(q)}{\partial q_a\partial q_b}\xi_b\Big].\end{aligned} \tag{23}$$

Of course, there are many such functions f.

The functional (21) has the following form in the Fock presentation

$$\int\frac{d\tau}{\varepsilon^{k/2}}Y\Big(\hat\psi^+-\frac{\phi^*}{\sqrt{\varepsilon}}\Big)\exp\Big(\frac{1}{\varepsilon}\Big[g(\tau)+iS(\tau)+\sum_{l=1}^{\infty}\phi_l(\tau)(\sqrt{\varepsilon}\,\hat\psi_l^+-\phi_l^*(\tau))\Big]\Big)\widehat\Phi_0\,. \tag{24}$$

In formula (24), $g(\tau)$ is of the form (17). We shall usually use functionals Y of the form (18). The vector (24) will be multiplied by $\varepsilon^{k/4}$ so as to ensure that its norm is of order $O(1)$. Notice also that

$$g(\tau)+iS(\tau)=\int_{\tau^{(0)}}^{\tau}\phi_l\,d\phi_l^*+g(\tau^{(0)})+iS(\tau^{(0)})\,.$$

We shall define a canonical operator by a formula analogous to (24) in subsection 2.4; some auxiliary lemmas will be proved in 2.3. The relationship between the expression (24) and the traditional canonical operator corresponding to an arbitrary Lagrangian manifold with complex germ will be discussed in 2.5.

2.3. Some auxiliary lemmas.

LEMMA 1. *Let L be an operator in l^2 that maps the sequence $(\xi_1,\xi_2,\dots)$ into the sequence $(0,\dots,0,\xi_{k+1},\xi_{k+2},\dots)$, and let Y be a bounded operator in l^2 with the following properties*:

1. $(\xi,Y\xi)\geqslant 0$, $\xi\in l^2$;
2. *if* $L\xi=0$, $(\xi,Y\xi)=0$, *then* $\xi=0$.

Then there exists a positive κ such that $(\xi,(L+Y)\xi)\geqslant\kappa(\xi,\xi)$ for any ξ.

PROOF. It follows from the assumptions of the lemma that

$$(\xi,(L+Y)\xi)\geqslant(L\xi,L\xi)\,. \tag{25}$$

On the other hand, property 2 implies that for $L\xi=0$

$$(\xi,Y\xi)\geqslant\sigma(\xi,\xi),\qquad\sigma>0\,.$$

It follows that

$$\begin{aligned}(\xi,(L+Y)\xi)&=((E-L)\xi,Y(E-L)\xi)+(L\xi,Y(E-L)\xi)\\&\quad+((E-L)\xi,YL\xi)+(L\xi,(YL+L)\xi)\\&\geqslant\sigma(\xi,\xi)-2\|Y\|\sqrt{(L\xi,L\xi)(\xi,\xi)}-(\|Y\|+\sigma)(L\xi,L\xi),\end{aligned}\tag{26}$$

where $\|Y\|=\sup_{\|\xi\|=1}\|Y\xi\|$.

Suppose that for any $\delta>0$ there exists a vector $\xi\in l^2$ such that $(\xi,(L+Y)\xi)<\delta(\xi,\xi)$. It follows from (25) that $(L\xi,L\xi)<\delta(\xi,\xi)$ and (26) implies

$$(\xi,(L+Y)\xi)\geqslant(\sigma-(\|Y\|+\sigma)\delta-2\|Y\|\sqrt{\delta})(\xi,\xi)\,.$$

This inequality for δ small enough contradicts the condition $(\xi,(L+Y)\xi)<\delta(\xi,\xi)$. The contradiction obtained proves Lemma 1. □

Denote by $W_{ab}(\tau)$, $\tau\in\Lambda^k$, $a,b=1,\dots,k$, the inverse matrix to the matrix

$$\sum_{i=1}^{\infty}\frac{\partial\phi_i^*}{\partial\tau_a}\frac{\partial\phi_i}{\partial\tau_b},$$

which is invertible by axiom (m2). We set

$$M_{ij}(\tau)=((C+iB)(\tau)(C-iB)^{-1}(\tau))_{ij}-\sum_{a,b=1}^{k}\frac{\partial\phi_i}{\partial\tau_a}W_{ab}(\tau)\frac{\partial\phi_j}{\partial\tau_b}\,. \tag{27}$$

LEMMA 2. *The operator M has the following properties:*
1. *M is a Hilbert–Schmidt operator;*
2. $\|M\| < 1$;
3. $\sum_{j=1}^{\infty} M_{ij} \dfrac{\partial \phi_j^*}{\partial \tau_a} = 0, \quad a = 1, \dots, k.$

PROOF. Since the operators

$$C + iB \quad \text{and} \quad \sum_{a,b=1}^{k} \frac{\partial \phi_i}{\partial \tau_a} W_{ab} \frac{\partial \phi_j}{\partial \tau_b}$$

are Hilbert–Schmidt operators, and the operator $(C - iB)^{-1}$ is bounded, we see that M is a Hilbert–Schmidt operator.

Axiom (r2) and the definition (27) of M imply

$$\sum_{j=1}^{\infty} M_{ij} \frac{\partial \phi_j^*}{\partial \tau_a} = 0, \qquad a = 1, \dots, k.$$

Further, the expression (27) implies that

$$\begin{aligned} E - M^+ M = &((C - iB)^+)^{-1} \\ &\times \Big((C + iB)^+ \sum_{a,b=1}^{k} \sum_{l=1}^{\infty} \frac{\partial \phi_l}{\partial \tau_a} \times W_{ab} \frac{\partial \phi_l}{\partial \tau_b} (C + iB) + 2L \Big) (C - iB)^{-1}. \end{aligned}$$

It follows from Lemma 1 that for some $\rho > 0$ we have $(\xi, (E - M^+ M)\xi) > \rho(\xi, \xi)$. This implies $\|M\| < 1$. Lemma 2 is proved.

2.4. **Definition of the canonical operator.** Consider an aggregate consisting of a number $\varepsilon > 0$, a sequence ϕ_j, $\sum_{j=1}^{\infty} |\phi_j|^2 < \infty$, and a Hilbert–Schmidt operator $M\colon l^2 \to l^2$, $\|M\| < 1$. Assign the following element of $\mathcal{H}$ to such an aggregate:

$$\widehat{\Phi}_{\phi,M} = \exp\Big\{ \frac{1}{\varepsilon} \phi_j (\sqrt{\varepsilon} \hat{\psi}_j^+ - \phi_j^*) + \frac{1}{2\varepsilon} (\hat{\psi}_j^+ \sqrt{\varepsilon} - \phi_j^*) M_{ij} (\hat{\psi}_j^+ \sqrt{\varepsilon} - \phi_j^*) \Big\} \widehat{\Phi}_0, \tag{28}$$

where we assume summation over repeating indices, $j = 1, 2, \dots$.

REMARK 4. Expanding the exponent in formula (28) into a power series in the variable ε, we obtain the lth component of $\widehat{\Phi}_{\phi,M} \in \mathcal{H}$ in the following form:

$$(\widehat{\Phi}_{\phi,M})^{(l)}_{i_1 \dots i_l} = \sum_{k=0}^{[l/2]} \frac{c}{\sqrt{l!}\, 2^k k!} a_{i_1} \cdots a_{i_l} \sum_{1 \leqslant j_1 \neq \dots \neq j_{2k} \leqslant l} \frac{M_{i_{j_1} i_{j_2}} \cdots M_{i_{j_{2k-1}} i_{j_{2k}}}}{a_{i_{j_1}} \cdots a_{i_{j_{2k}}}}, \tag{29}$$

where

$$\begin{aligned} a_m &= (\phi_m - M_{mn} \phi_n^*)/\sqrt{\varepsilon}, & m, n &= 1, 2, \dots, \\ c &= \exp\Big\{ -\frac{1}{\varepsilon} \phi_j^* \phi_j - \frac{1}{2\varepsilon} \phi_i^* M_{ij} \phi_j^* \Big\}, & i, j &= 1, 2, \dots. \end{aligned}$$

Consider a Lagrangian manifold with complex germ satisfying the quantization conditions

(30) $$\frac{1}{2\pi\varepsilon}\oint_l P_i\,dQ_i = n, \qquad n \in \mathbb{Z},$$

where l is an arbitrary closed path on the isotropic manifold Λ^k.

Let f be an infinitely differentiable finite function on the isotropic manifold Λ^k, and let $\tau^{(0)}$ be an arbitrary point in Λ^k.

Introduce notation

$$g = \frac{\phi_l^*(\tau^{(0)})\,\phi_l(\tau^{(0)})}{2} + \frac{\phi_l^*(\tau^{(0)})\,\phi_l^*(\tau^{(0)}) - \phi_l(\tau^{(0)})\,\phi_l(\tau^{(0)})}{4}, \qquad l = 1, 2, \dots,$$

$$F(\tau) = \frac{C(\tau) + iB(\tau)}{\sqrt{2}}, \quad G(\tau) = \frac{C(\tau) - iB(\tau)}{\sqrt{2}}, \qquad \tau \in \Lambda^k.$$

To the Lagrangian manifold with complex germ $[\Lambda^k, r]$, the function f, the point $\tau^{(0)} \in \Lambda^k$, and the number $\varepsilon > 0$ we assign the following element of $\mathcal{H}$:

(31)
$$\widehat{\mathcal{K}}^{\varepsilon}_{[\Lambda^k,r],\tau^{(0)}} f = \int_{\Lambda^k} \frac{d\sigma\, f(\tau)}{(2\pi)^{k/2}\varepsilon^{k/4}} \frac{\exp\left(\frac{i}{\varepsilon}[g + \int_{\tau^{(0)}}^{\tau} \phi_i(\tau')\, d\phi_i^*(\tau')]\right)}{\sqrt[4]{\det G^+(\tau) G(\tau)}} \widehat{\Phi}^{\varepsilon}_{\phi(\tau), M(\tau)},$$
$$i = 1, 2, \dots,$$

where M is defined by formula (27), $d\sigma$ is the following measure on Λ^k

$$d\sigma = \sqrt{\det \frac{\partial \phi_i^*}{\partial \tau_a}(\tau) \frac{\partial \phi_i}{\partial \tau_b}(\tau)}\, d\tau,$$

$d\tau \equiv d\tau_1 \cdots d\tau_k$, $a, b = 1, \dots, k$, $i = 1, 2, \dots$. This measure does not depend on the choice of local coordinates $\tau_1, \dots, \tau_k$ on Λ^k.

REMARK 5. Since the manifold Λ^k is isotropic, the following integral does not depend locally on the path:

$$\int_{\tau^{(0)}}^{\tau} \phi_i(\tau')\, d\phi_i^*(\tau').$$

REMARK 6. The integrand in (31) is a single-valued function due to the quantization conditions (30).

REMARK 7. It follows from axiom (r4) that

(32) $$(C - iB)^+(C - iB) = 2L + (C + iB)^+(C + iB).$$

Formula (32) implies that the operator

$$\left(\frac{C(\tau) - iB(\tau)}{\sqrt{2}}\right)^+ \left(\frac{C(\tau) - iB(\tau)}{\sqrt{2}}\right) - E$$

is an operator of trace class and therefore the Fredholm determinant for the operator

$$\left(\frac{C(\tau) - iB(\tau)}{\sqrt{2}}\right)^+ \left(\frac{C(\tau) - iB(\tau)}{\sqrt{2}}\right)$$

is defined.

REMARK 8. By the axiom (r6), we have

$$\det\left[\left(\frac{C(\tau)-iB(\tau)}{\sqrt{2}}\right)^{+}\left(\frac{C(\tau)-iB(\tau)}{\sqrt{2}}\right)\right]>0.$$

2.5. **Relationship with the usual definition of canonical operator.** We now compare definition (31) of the canonical operator and the traditional definition of a canonical operator ([**13, 1**]). Let us write out the vectors (28) and (31) in the Q-presentation. Consider the auxiliary presentation with the wave function depending on l momenta and an infinite number of coordinates. This presentation will allow us to include the case of focal points into our considerations.

Let I be a finite set of positive integers, $I=\{i_1,\dots,i_l\}$. Denote by Λ the diagonal matrix with $\Lambda_{jj}=1$ for $j\notin I$, $\Lambda_{jj}=-j$ for $j\in I$. To each set I we assign the following presentation (I-presentation) of the vectors of $\mathcal{H}$. To each element of $\mathcal{H}$ we assign the functional

$$\Phi_I(q_1^I,q_2^I,\dots)=\int\frac{dq_{i_1}\cdots dq_{i_l}}{(2\pi\varepsilon)^{l/2}}\exp\Big(-\frac{i}{\varepsilon}\sum_{j\in I}q_j^Iq_j\Big)$$
$$\times\,\Phi_Q(q_1^I,\dots,q_{i_1-1}^I,q_{i_1},q_{i_1+1}^I,\dots,q_{i_l-1}^I,q_{i_l},q_{i_l+1}^I,\dots).$$

The Q-presentation is a special case of the I-presentation for $I=\varnothing$.

In order to write out the I-presentations of the vectors (28) and (31), we use the formula (see [**2**])

$$\Phi(a^*)=\int K_z(a^*)\,\Phi(z^*)\,e^{-z^*z}\prod dz^*dz.$$

Here $K_z(a^*)$ is the generating functional corresponding to the vector

$$K_z=\exp\Big(\sum_{j=1}^{\infty}z_j\hat\psi_j^+\Big)\widehat{\Phi}_0,$$

the measure in the functional integral is defined in [**2**], and $\widehat{\Phi}$ is an arbitrary element of $\mathcal{H}$.

Introduce the following notation:

$$P_j^I(\tau)=\begin{cases}-Q_j(\tau), & j\in I,\\ P_j(\tau), & j\notin I,\end{cases}\qquad Q_j^I(\tau)=\begin{cases}P_j(\tau), & j\in I,\\ Q_j(\tau), & j\notin I,\end{cases}$$
$$\phi_j^I(\tau)=(Q_j^I(\tau)+iP_j^I(\tau))/\sqrt{2}=\Lambda_{jm}\phi_m(\tau),\qquad j,m=1,2,\dots,$$
$$z_j^I=\Lambda_{jm}z_m,\quad z_j^{*I}=\Lambda_{jm}^*z_m^*,\qquad j,m=1,2,\dots.$$

The functional

$$(K_z)_I(q_1^I,q_2^I,\dots)=\exp\Big\{\frac{1}{\sqrt{2\varepsilon}}z_m^I\Big(q_m^I-\varepsilon\frac{\partial}{\partial q_m^I}\Big)\Big\}\exp\Big\{-\frac{1}{2\varepsilon}q_m^Iq_m^I\Big\}$$
$$=\exp\Big\{-\frac{1}{2}z_m^Iz_m^I+\sqrt{\frac{2}{\varepsilon}}z_m^Iq_m^I-\frac{1}{2\varepsilon}q_m^Iq_m^I\Big\},\qquad m=1,2,\dots,$$

corresponds to the vector $\widehat{K}_z$ in the I-presentation.

This expression follows from the Baker–Hausdorff formula

$$e^{A_1+A_2} = e^{A_1}e^{A_2}e^{-\frac{1}{2}[A_1,A_2]}, \qquad A_1 = \frac{1}{\sqrt{2\varepsilon}} z^I_m q^I_m,\ A_2 = -\sqrt{\frac{\varepsilon}{2}}\, z^I_m \frac{\partial}{\partial q^I_m}$$

and the equality

$$\exp\left\{a_i \frac{\partial}{\partial q^I_i}\right\} f(q^I_1, q^I_2, \dots) = f(q^I_1 + a_1, q^I_2 + a_2, \dots).$$

It now follows that the functional corresponding to the vector $\widehat{\Phi}_{\phi,M}$ has the following form in the I-presentation:

$$\begin{aligned}
&(\Phi_{\phi,M})_I(q^I_1, q^I_2, \dots) \\
&= \int \prod dz^* dz \exp\bigg(- z^{*I}_l z^I_l - \frac{1}{2} z^I_l z^I_l + \sqrt{\frac{2}{\varepsilon}}\, z^I_l q^I_l - \frac{1}{2\varepsilon} q^I_l q^I_l \\
&\qquad + \frac{1}{2\varepsilon}(z^{*I}_l\sqrt{\varepsilon} - \phi^{*I}_l) M^I_{lm} (z^{*I}_m\sqrt{\varepsilon} - \phi^{*I}_m) \\
&\qquad + \frac{1}{\varepsilon}\phi^I_l(\sqrt{\varepsilon} z^{I*}_l - \phi^{I*}_l)\bigg), \\
&l, m = 1, 2, \dots,
\end{aligned}$$

where $M^I = \Lambda M \Lambda$.

By [2] this integral is equal to

$$\begin{aligned}
&(\Phi_{\phi,M})_I(q^I_1, q^I_2, \dots) \\
&= \exp\bigg\{ \frac{1}{\varepsilon}\frac{\phi^I_l - \phi^{*I}_l}{\sqrt{2}}\left(q^I_l - \frac{\phi^I_l + \phi^{*I}_l}{\sqrt{2}}\right) \\
&\qquad + \frac{i}{2\varepsilon}\left(q^I_m - \frac{\phi^I_m + \phi^{I*}_m}{\sqrt{2}}\right) A^I_{mn} \left(q^I_n - \frac{\phi^I_n + \phi^{I*}_n}{\sqrt{2}}\right) \\
&\qquad - \frac{\phi^{*I}_l \phi^I_l}{2\varepsilon} + \frac{\phi^I_l\phi^I_l - \phi^{I*}_l\phi^{I*}_l}{4\varepsilon}\bigg\} \qquad (33) \\
&\quad \times \frac{1}{\sqrt{\det(E + M^I)}}, \qquad l, m, n = 1, 2, \dots, \\
&A^I = i(E - M^I)(E + M^I)^{-1}.
\end{aligned}$$

REMARK 9. Not all elements of $\mathcal{H}$ of the form (28) can be presented in the form (33). Indeed, $\det(E + M^I)$ exists only if the operator M^I is an operator of trace class, while (28) may be defined in other cases as well. For example,

$$\exp\left\{\frac{1}{4}\sum_{n=1}^{\infty}\frac{1}{n}(\hat{\psi}^+_n)^2\right\}\widehat{\Phi}_0 \in \mathcal{H},$$

but this vector cannot be written in the Q-presentation. This means that the definition of a canonical operator in the Fock presentation (31) is more general than the corresponding definition in the Q-presentation.

Now consider the vector (31) in the I-presentation. By (31) and (33) we have

(34)
$$
\begin{aligned}
&(\mathcal{K}^{\varepsilon}_{[\Lambda^k,r],\tau^{(0)}}f)_I(q_1^I,q_2^I,\dots)\\
&=\int_{\Lambda^k}\frac{d\tau\, f(\tau)\sqrt{\det\frac{\partial\phi_l^*}{\partial\tau_a}(\tau)\,\frac{\partial\phi_l}{\partial\tau_b}(\tau)}}{(2\pi)^{k/2}\varepsilon^{k/4}}\\
&\times\frac{\exp\left(\frac{i}{\varepsilon}\int_{\tau^{(0)}}^{\tau}P_l^I(\tau')\,dQ_l^I(\tau')+\frac{i}{\varepsilon}P_l^I(\tau)(q_l^I-Q_l^I(\tau))+\frac{i}{2\varepsilon}(q_l^I-Q_l^I(\tau))A_{lm}^I(\tau)(q_m^I-Q_m^I(\tau))\right)}{\sqrt[4]{\det\left[\left(\frac{C(\tau)-iB(\tau)}{\sqrt{2}}\right)^+\left(\frac{C(\tau)-iB(\tau)}{\sqrt{2}}\right)\right]}\sqrt{\det(E+M^I(\tau))}}\\
&\times\exp\Big(-\frac{i}{\varepsilon}\sum_{j\in I}P_j(\tau^{(0)})Q_j(\tau^{(0)})\Big),\qquad l,m=1,2,\dots,\ a,b=1,\dots,k,
\end{aligned}
$$

where A_{lm}^I is the matrix of the operator $A^I(\tau)=i(E+\Lambda M(\tau)\Lambda)^{-1}(E-\Lambda M(\tau)\Lambda)$, and $M(\tau)$ is determined by formula (27).

In particular, in the Q-presentation we have

(35)
$$
\begin{aligned}
&(\mathcal{K}^{\varepsilon}_{[\Lambda^k,r],\tau^{(0)}}f)_Q(q_1,q_2,\dots)\\
&=\int_{\Lambda^k}\frac{d\tau\, f(\tau)\sqrt{\det\frac{\partial\phi_l^*}{\partial\tau_a}(\tau)\,\frac{\partial\phi_l}{\partial\tau_b}(\tau)}}{(2\pi)^{k/2}\varepsilon^{k/4}}\\
&\times\frac{\exp\left(\frac{i}{\varepsilon}\int_{\tau^{(0)}}^{\tau}P_l(\tau')\,dQ_l(\tau')+\frac{i}{\varepsilon}P_l(\tau)(q_l-Q_l(\tau))+\frac{i}{2\varepsilon}(q_l-Q_l(\tau))A_{lm}(\tau)(q_m-Q_m(\tau))\right)}{\sqrt[4]{\det\left[\left(\frac{C(\tau)-iB(\tau)}{\sqrt{2}}\right)^+\left(\frac{C(\tau)-iB(\tau)}{\sqrt{2}}\right)\right]}\sqrt{\det(E+M(\tau))}},\\
&\qquad l,m=1,2,\dots,\ a,b=1,\dots,k,\qquad A(\tau)=i(E+M(\tau))^{-1}(E-M(\tau)).
\end{aligned}
$$

We now present a heuristic justification of the relationship between the definition of a canonical operator in the Q-presentation (35) and the traditional definition of the canonical operator ([**13, 1**]).

Cover the support of the function f on the isotropic manifold Λ^k by a finite number of domains Ω_α, $\alpha=1,\dots,M$, such that each domain has a one-to-one projection onto one of the coordinate planes of the form

$$
P_j=0,\quad j\notin I_\alpha,\qquad Q_j=0,\quad j\in I_\alpha,\qquad I_\alpha=\{i_1^\alpha,\dots,i_{l_\alpha}^\alpha\}\subset\mathbb{N}.
$$

Consider a partition of unity $1=\sum_{\alpha=1}^M e_\alpha(\tau)$, where the functions $e_\alpha(\tau)$ are infinitely differentiable with the support in Ω_α. Introduce the notation $f_\alpha(\tau)=f(\tau)e_\alpha(\tau)$.

The canonical operator in the Q-presentation has the form

(36)
$$
\begin{aligned}
&(\mathcal{K}^{\varepsilon}_{[\Lambda^k,r],\tau^{(0)}}f)_Q(q_1,q_2,\dots)\\
&=\sum_{\alpha=1}^M\int\frac{dq_{i_1}^\alpha\cdots dq_{i_{l_\alpha}}^\alpha}{(2\pi\varepsilon)^{l_\alpha/2}}\cdot\exp\Big(-\frac{i}{\varepsilon}\sum_{j\in I_\alpha}q_j^Iq_j\Big)\\
&\times(\mathcal{K}^{\varepsilon}_{[\Lambda^k,r]\tau^{(0)}}f_\alpha)_{I_\alpha}(q_1^I,\dots,q_{i_1^\alpha-1}^I,q_{i_1^\alpha},q_{i_1^\alpha+1}^I,\dots,q_{i_l^\alpha-1}^I,q_{i_l^\alpha},q_{i_l^\alpha+1}^I,\dots).
\end{aligned}
$$

Now we calculate the integral in formula (34) that gives $(\mathcal{K}^{\varepsilon}_{[\Lambda^k,r],\tau^{(0)}}f_\alpha)_{I_\alpha}$. Introduce following notation:

$$
\begin{aligned}
B_{mn}^I&=-C_{mn},\qquad C_{mn}^I=B_{mn},\qquad m\in I,\\
B_{mn}^I&=B_{mn},\qquad C_{mn}^I=C_{mn},\qquad m\notin I,
\end{aligned}
$$

and let S_{ab} be the inverse matrix to the matrix

$$\frac{\partial Q_i^I}{\partial \tau_a} A_{ij}^I \frac{\partial Q_j^I}{\partial \tau_b} - \frac{\partial P_i^I}{\partial \tau_a} \frac{\partial Q_i^I}{\partial \tau_b}, \qquad a, b, = 1, \dots, k, \ i, j = 1, 2, \dots .$$

STATEMENT. *The following relations are valid:*

(37)
$$(B^I (C^I)^{-1})_{mn} = A_{mn}^I - \left(A_{mr}^I \frac{\partial Q_r^I}{\partial \tau_a} - \frac{\partial P_r^I}{\partial \tau_a} \right) S_{ab} \left(\frac{\partial Q_j^I}{\partial \tau_b} A_{jn}^I - \frac{\partial P_n^I}{\partial_b} \right), \quad j, m, n, r = 1, 2, \dots, \ a, b = 1, \dots, k,$$

(38)
$$\frac{\det \left(\frac{\partial \phi_m^*}{\partial \tau_a} \frac{\partial \phi_m}{\partial \tau_b} \right)}{\det (E + M^I)} = \det \left[\frac{1}{2} (E - i B^I (C^I)^{-1}) \right] \times \det \left[\frac{1}{i} \left(\frac{\partial Q_m^I}{\partial \tau_a} A_{mn}^I \frac{\partial Q_n^I}{\partial \tau_b} - \frac{\partial P_m^I}{\partial \tau_a} \frac{\partial Q_m^I}{\partial \tau_b} \right) \right], \quad m, n = 1, 2, \dots, \ a, b = 1, \dots, k.$$

PROOF. Property (37) follows from the following expressions for $B^I (C^I)^{-1}$ and A^I:

$$A^I = i (E + M^I)^{-1} (E - M^I),$$
$$B^I (C^I)^{-1} = i (E + N^I)^{-1} (E - N^I),$$

where

$$N_{mn}^I = M_{mn}^I + \frac{\partial \phi_m^I}{\partial \tau_a} W_{ab} \frac{\partial \phi_n^I}{\partial \tau_b}, \qquad a, b = 1, \dots, k, \ m, n = 1, 2, \dots,$$

W_{ab} is the inverse matrix to the matrix

$$\frac{\partial \phi_m^*}{\partial \tau_a} \frac{\partial \phi_m}{\partial \tau_b}, \qquad a, b = 1, \dots, k, \ m = 1, 2, \dots,$$

and from the property $M_{mn} \partial \phi_m^{I*} / \partial \tau_a = 0$.

The proof of formula (38) is based on the following lemma.

LEMMA 3. *Let $y^a, z^a \in l^2$, $a = 1, \dots, k$, and let R be the operator in l^2 of the form*

$$R\kappa = \kappa - \sum_{c=1}^{k} y^c (z^c, \kappa), \qquad \kappa \in l^2.$$

Then $\det R = \det (\delta_{ab} - (y^a, z^b))$, $a, b = 1, \dots, k$.

PROOF. Choose an orthonormal basis $\{e_s, \ s = 1, 2, \dots\}$ in l^2 such that only the first k components of the vectors y^a, $a = 1, \dots, k$, differ from zero. Then

$$\det R = \det \left(\delta_{ij} - \sum_{c=1}^{\infty} (e_i, y^c)(z^c, e_j) \right), \qquad i, j = 1, \dots, k.$$

Then, since the functions

$$\det(\delta_{ab} - \alpha(y^a, z^b)) \quad \text{and} \quad \det\left(\delta_{ij} - \alpha\sum_{c=1}^{\infty}(e_i, y^c)(z^c, e_j)\right)$$

are polynomials in α, and the Taylor series expansions of their logarithms as $\alpha \to 0$ coincide, these functions coincide as well.

This completes the proof of Lemma 3.

Lemma 3 implies that

$$\begin{aligned}
&\det\left(\delta_{mn} + (E+M)^{-1}_{ms}\frac{\partial\phi_s}{\partial\tau_a}W_{ab}\frac{\partial\phi_n}{\partial\tau_b}\right)\\
&\quad = \det\left(\delta_{ab} + W_{ac}\frac{\partial\phi_m}{\partial\tau_c}(E+M)^{-1}_{mn}\frac{\partial\phi_n}{\partial\tau_b}\right),\\
&\quad a, b, c = 1, \dots, k, \ \ s, m, n = 1, 2, \dots .
\end{aligned}$$

Formula (38) now follows. The statement is proved.

Now we can write

$$\begin{aligned}
&(\mathcal{K}^{\varepsilon}_{[\Lambda^k, r], \tau^{(0)}} f_\alpha)_{I_\alpha}(q_1^{I_\alpha}, q_2^{I_\alpha}, \dots)\\
&\quad = \int_{\Lambda^k} \frac{d\tau\, f(\tau)\, e^{\pi i l_\alpha/4}}{(2\pi)^{k/2}\varepsilon^{k/4}} \frac{\exp\left(\frac{i}{2}\operatorname{Arg}\det\frac{C-iB}{\sqrt{2}}\right)}{\sqrt{\det(C^{I_\alpha}\sqrt{2})}}\\
&\qquad \times \sqrt{\det\frac{1}{i}\left(\frac{\partial Q_m^{I_\alpha}}{\partial\tau_a}(\tau)\, A^{I_\alpha}_{mn}(\tau)\frac{\partial Q_n^{I_\alpha}}{\partial\tau_b}(\tau) - \frac{\partial P_m^{I_\alpha}}{\partial\tau_a}(\tau)\frac{\partial Q_m^{I_\alpha}}{\partial\tau_b}(\tau)\right)}\\
&\qquad \times \exp\Bigg(\frac{i}{\varepsilon}\int_{\tau^{(0)}}^{\tau} P_m^{I_\alpha}(\tau')\, dQ_m^{I_\alpha}(\tau') + \frac{i}{\varepsilon}P_m^{I_\alpha}(\tau)(q_m^{I_\alpha} - Q_m^{I_\alpha}(\tau))\\
&\qquad\qquad + \frac{i}{2\varepsilon}(q_m^{I_\alpha} - Q_m^{I_\alpha}(\tau))\, A^{I_\alpha}_{mn}(\tau)(q_n^{I_\alpha} - Q_n^{I_\alpha}(\tau))\\
&\qquad\qquad - \frac{i}{\varepsilon}\sum_{j\in I} P_j(\tau^{(0)})\, Q_j(\tau^{(0)})\Bigg),\\
&\qquad m, n = 1, 2, \dots, \ \ a, b = 1, \dots, k.
\end{aligned} \tag{39}$$

We choose the sign of $\sqrt{\det(C^{I_\alpha}\sqrt{2})}$ so that

$$\frac{\operatorname{Re}\sqrt{\det(C^{I_\alpha}\sqrt{2})}}{e^{\pi i l_\alpha/4}\sqrt{\det((C^{I_\alpha} - iB^{I_\alpha})/\sqrt{2})}} > 0$$

(this real part cannot vanish since $\operatorname{Re}\det(C^{I_\alpha}\sqrt{2})/(i^{l_\alpha}\det((C^{I_\alpha} - iB^{I_\alpha})/\sqrt{2})) > 0$).

Let us find the asymptotics of the integral (39). Since $\operatorname{Im} A^{I_\alpha} > 0$ only the domain in $\mathbb{R}^k$ where $(q_m - Q_m(\tau))(q_m - Q_m(\tau)) \sim \varepsilon$ provides a nonexponentially small contribution. The minimum point of $(q_m - Q_m(\tau))(q_m - Q_m(\tau))$ is denoted by $\tilde{\tau}(q)$.

Expanding the exponent in a neighborhood of the point $\tilde{\tau}$ we obtain

$$\int_{\tau^{(0)}}^{\tau} P_m^{I_\alpha}(\tau')\,dQ_m^{I_\alpha}(\tau') + P_m^{I_\alpha}(\tau)(q_m^{I_\alpha} - Q_m^{I_\alpha}(\tau)) + \frac{1}{2}(q_m^{I_\alpha} - Q_m^{I_\alpha}(\tau))\,A_{mn}^{I_\alpha}(\tau)(q_n^{I_\alpha} - Q_n^{I_\alpha}(\tau))$$

$$= \int_{\tau^{(0)}}^{\tilde{\tau}} P_m^{I_\alpha}(\tau')\,dQ_m^{I_\alpha}(\tau') + P_m^{I_\alpha}(\tilde{\tau})\xi_m\sqrt{\varepsilon} + \frac{\varepsilon}{2}\xi_m A_{mn}^{I_\alpha}(\tilde{\tau})\xi_n - \frac{\varepsilon}{2}\frac{\partial P_m^{I_\alpha}}{\partial\tilde{\tau}_a}(\tilde{\tau})\frac{\partial Q_m^{I_\alpha}}{\partial\tilde{\tau}_b}(\tilde{\tau})t_a t_b + \varepsilon\left(\frac{\partial P_m^{I_\alpha}}{\partial\tilde{\tau}_a}(\tilde{\tau}) - \frac{\partial Q_n^{I_\alpha}}{\partial\tilde{\tau}_a}(\tilde{\tau})A_{mn}^{I_\alpha}(\tilde{\tau})\right)t_a\xi_m$$

$$+ \frac{\varepsilon}{2}\frac{\partial Q_m^{I_\alpha}}{\partial\tilde{\tau}_a}(\tilde{\tau})A_{mn}^{I_\alpha}(\tilde{\tau})\frac{\partial Q_n^{I_\alpha}}{\partial\tilde{\tau}_b}(\tilde{\tau})t_a t_b + \dots, \qquad m, n = 1, 2, \dots,$$

$$a, b = 1, \dots, k, \quad t_a = (\tau_a - \tilde{\tau}_a)/\sqrt{\varepsilon}, \quad \xi_m = (q_m - Q_m(\tilde{\tau}))/\sqrt{\varepsilon}.$$

Integrating over t_a, ξ_m we obtain the following asymptotics

(40)
$$(\mathcal{K}^{\varepsilon}_{[\Lambda^k, r], \tau^{(0)}} f_\alpha)_{I_\alpha}(q_1^{I_\alpha}, q_2^{I_\alpha}, \dots) = \frac{\exp\left(i\pi l_\alpha/4 + \frac{i}{2}\operatorname{Arg}\det\left(\frac{C^{I_\alpha} - iB^{I_\alpha}}{\sqrt{2}}(\tilde{\tau})\right)\right)}{\sqrt{\det(C^{I_\alpha}(\tilde{\tau})\sqrt{2})}}$$

$$\times f(\tilde{\tau})\varepsilon^{k/4}\exp\left\{\frac{i}{\varepsilon}\int_{\tau^{(0)}}^{\tilde{\tau}} P_m^{I_\alpha}(\tau')\,dQ_m^{I_\alpha}(\tau') + P_m^{I_\alpha}(\tilde{\tau})(q_m^{I_\alpha} - Q_m^{I_\alpha}(\tilde{\tau})) + \frac{1}{2}(q_m^{I_\alpha} - Q_m^{I_\alpha}(\tilde{\tau}))(B^{I_\alpha}(C^{I_\alpha})^{-1})_{mn}(q_n - Q_n^{I_\alpha}(\tilde{\tau})) - \frac{i}{\varepsilon}\sum_{j\in I} P_j(\tau^{(0)})\,Q_j(\tau^{(0)})\right\},$$

$m, n = 1, 2, \dots$.

REMARK 10. It follows that the definition of a canonical operator in the present paper differs from the classical definition only by the factor

$$\operatorname{const}\exp\left(\frac{i}{2}\operatorname{Arg}\det\frac{C - iB}{\sqrt{2}}\right).$$

This factor changes the quantization conditions and the transport equation.

However, his heuristic consideration will not be used in the proof of the theorem.

§3. Time evolution of a Lagrangian manifold with complex germ

Now consider the canonical transformation of a Lagrangian manifold with complex germ.

Let

$$H(\phi^*, \phi) = \sum_{m=1}^{s}\sum_{n=1}^{s} H^{(m,n)}_{i_1\dots i_m j_1\dots j_n}\phi^*_{i_1}\cdots\phi^*_{i_m}\phi_{j_1}\cdots\phi_{j_m}, \tag{41}$$

$$\bar{H}^{(m,n)}_{i_1\dots i_m j_1\dots j_n} = H^{(n,m)}_{j_1\dots j_n i_1\dots i_m},$$

where $H^{(m,n)}_{i_1\dots i_m j_1\dots j_n}$ is symmetric separately over $i_1,\dots,i_m$ and over $j_1,\dots,j_n$.

DEFINITION 3. We say that a canonical transformation $\mathcal{D}^t_H$, $t\in[0,T]$, of a Lagrangian manifold with complex germ $[\Lambda^k_t, r_t]$ *corresponds to the Hamiltonian* H, if the following conditions hold.

1. On the segment $[0,T]$ there exists a solution $\phi_j(\tau,t)$ to the Cauchy problem

$$
\begin{aligned}
&i\dot\phi_j = \frac{\partial H}{\partial \phi_j^*}(\phi_j^*,\phi), && \\
&\phi_j(\tau,t)|_{t=0} = \phi_j^0(\tau) = \frac{Q_j^0(\tau)-iP_j^0(\tau)}{\sqrt2}, && (P^0(\tau),Q^0(\tau))\in\Lambda_0^k,
\end{aligned}
\tag{42}
$$

such that for any $\lambda_1,\dots,\lambda_k$, $\lambda_i\in\{0,1,2,\dots\}$, $i=1,\dots,k$, the derivatives

$$
\frac{\partial^{\lambda_1+\dots+\lambda_k}}{\partial\tau_1^{\lambda_1}\dots\partial\tau_k^{\lambda_k}}\,\phi_j(\tau,t)
$$

exist, and the series

$$
\sum_{j=1}^{\infty}\left|\frac{\partial^{\lambda_1+\dots+\lambda_k}}{\partial\tau_1^{\lambda_1}\dots\partial\tau_k^{\lambda_k}}\,\phi_j(\tau,t)\right|^2
$$

converges.

2. On the segment $[0,T]$ there exists a solution to the following Cauchy problem

$$
\begin{aligned}
&i\dot\Pi_{mn}(\tau,t) = -\frac{\partial^2 H}{\partial\phi_m\partial\phi_l^*}\,\Pi_{ln}(\tau,t) - \frac{\partial^2 H}{\partial\phi_m\partial\phi_l}\,\Omega_{ln}(\tau,t),\\
&i\dot\Omega_{mn}(\tau,t) = \frac{\partial^2 H}{\partial\phi_m^*\partial\phi_l^*}\,\Pi_{ln}(\tau,t) + \frac{\partial^2 H}{\partial\phi_m^*\partial\phi_l}\,\Omega_{ln}(\tau,t),\\
&\Pi_{mn}(\tau,0)=\delta_{mn},\quad \Omega_{mn}(\tau,0)=0 \qquad m,n,k=1,2,\dots
\end{aligned}
\tag{43}
$$

(the arguments $\phi^*(\tau,t)$, $\phi(\tau,t)$ at the derivatives of H are omitted).

Matrices Π and Ω satisfy the following conditions: for any $\lambda_1,\dots,\lambda_k$, $\lambda_i\in\{0,1,2,\dots\}$, $i=1,\dots,k$, the derivatives

$$
\Pi^{(\lambda_1,\dots,\lambda_k)}_{mn} = \frac{\partial^{\lambda_1+\dots+\lambda_k}}{\partial\tau_1^{\lambda_1}\dots\partial\tau_k^{\lambda_k}}\,\Pi_{mn}(\tau,t),\quad \Omega^{(\lambda_1,\dots,\lambda_k)}_{mn} = \frac{\partial^{\lambda_1+\dots+\lambda_k}}{\partial\tau_1^{\lambda_1}\dots\partial\tau_k^{\lambda_k}}\,\Omega_{mn}(\tau,t)
$$

exist, the operators $\Pi^{(\lambda_1,\dots,\lambda_k)}$ are bounded, and the operators $\Omega^{(\lambda_1,\dots,\lambda_k)}$ are Hilbert–Schmidt operators.

A canonical transformation $\mathcal{D}^t_H$ *of a Lagrangian manifold with complex germ* is a set of transformations mapping the manifold Λ_0^k into the manifold

$$
g^t_H\Lambda^k_0 = \Lambda^k_t = \left\{P_j = \frac{\phi_j(\tau,t)-\phi_j^*(\tau,t)}{\sqrt2 i},\ Q_j = \frac{\phi_j(\tau,t)+\phi_j^*(\tau,t)}{\sqrt2}\right\},
$$

where $\phi_j(\tau,t)$ is a solution of the system (42), and mapping the matrices $B(\tau,0)$, $C(\tau,0)$ into the matrices $B(\tau,t)$, $C(\tau,t)$ such that

$$
\begin{pmatrix} C-iB\\ C+iB\end{pmatrix}(\tau,t) = \begin{pmatrix}\Pi & \bar\Omega\\ \Omega & \bar\Pi\end{pmatrix}(\tau,t)\begin{pmatrix} C-iB\\ C+iB\end{pmatrix}(\tau,0).
\tag{44}
$$

REMARK 11. The substitution

$$\phi_j = (Q_j + iP_j)/\sqrt{2}, \qquad \phi_j^* = (Q_j - iP_j)/\sqrt{2}$$

makes equations (42) Hamiltonian with the Hamiltonian

$$\mathcal{H}((Q_j - iP_j)/\sqrt{2}, (Q_j + iP_j)/\sqrt{2}).$$

REMARK 12. The equations for the matrices B and C are equations in variations for this Hamiltonian system (see [**14, 1**]).

LEMMA 4. *The pair consisting of the manifold Λ_t^k and the complex germ corresponding to the matrices $B(\tau,t)$, $C(\tau,t)$ is a Lagrangian manifold with complex germ.*

PROOF. First of all, let us check that the matrix

$$\begin{pmatrix} \Pi & \bar{\Omega} \\ \Omega & \bar{\Pi} \end{pmatrix} \tag{45}$$

is the matrix of a proper canonical transformation, i.e.,

$$\Pi^+\Pi - \Omega^+\Omega = E, \quad \Omega^T\Pi = \Pi^T\Omega, \quad \Pi\Pi^+ - \bar{\Omega}\Omega^T = E, \quad \Omega\Pi^+ = \bar{\Pi}\Omega^T. \tag{46}$$

At the initial moment these properties obviously hold. Equations (43) imply that

$$\begin{aligned} (\Pi^+\Pi - \Omega^+\Omega)^\cdot &= 0, & (\Omega^T\Pi - \Pi^T\Omega)^\cdot &= 0, \\ (\Pi\Pi^+ - \bar{\Omega}\Omega^T)^\cdot &= 0, & (\Omega\Pi^+ - \bar{\Pi}\Omega^T)^\cdot &= 0. \end{aligned}$$

Therefore, properties (46) hold at any moment of time.

Now let us check the axioms of a Lagrangian manifold with complex germ. Properties (46) imply the germ axioms (r3), (r4) and axiom (m3). Axiom (m2) follows from the invertibility of the matrix (45). Axioms (m1) and (r5) follow immediately from Definition 3. Axiom (r1) follows from the linearity of system (43). Axiom (r2) follows from equation (42).

Berezin showed in [**2**] that for the matrix of a proper canonical transformation (45) we have $\|\Pi^{-1}\bar{\Omega}\| < 1$ and the operator Π^{-1} is bounded. Formula (32) implies $\|(C+iB)(0,\tau)(C-iB)^{-1}(0,\tau)\| \leqslant 1$, so

$$\|\Pi^{-1}\bar{\Omega}(C+iB)(0,\tau)(C-iB)^{-1}(0,\tau)\| < 1.$$

Thus, the operator

$$\begin{aligned} &(C+iB)^{-1}(0,\tau)(E+\Pi^{-1}\bar{\Omega}(C+iB)(0,\tau)(C-iB)^{-1}(0,\tau))^{-1}\Pi^{-1}(t,\tau) \\ &\quad = (C-iB)^{-1}(t,\tau) \end{aligned}$$

is bounded, so axiom (r6) is also satisfied. Lemma 4 is proved.

It is also easy to check that the quantization condition (30) holds at any moment of time provided it holds at the initial moment.

REMARK 13. In the case of a zero-dimensional isotropic manifold, the preservation of the germ axioms has the following meaning: the product of two matrices of proper canonical transformations ([**2**])

$$\begin{pmatrix} \Pi & \bar\Omega \\ \Omega & \bar\Pi \end{pmatrix} \quad \text{and} \quad \begin{pmatrix} G & \bar F \\ F & \bar G \end{pmatrix}$$

is a matrix of a proper canonical transformation.

§4. Relationship between geometric and canonical quantization and some examples

Suppose $H(\overset{2}{\sqrt{\varepsilon}\hat\psi^+}, \overset{1}{\sqrt{\varepsilon}\hat\psi^-})$ is a selfadjoint operator in $\mathcal{H}$ of the form

$$H(\overset{2}{\sqrt{\varepsilon}\hat\psi^+}, \overset{1}{\sqrt{\varepsilon}\hat\psi^-}) = \sum_{m,n=1}^{s} H^{(m,n)}_{i_1\dots i_m j_1\dots j_n} \varepsilon^{(m+n)/2} \hat\psi^+_{i_1}\cdots\hat\psi^+_{i_m}\hat\psi^-_{j_1}\cdots\hat\psi^-_{j_n}. \tag{47}$$

Consider the Cauchy problem for the equation

$$\begin{aligned} & i\,\frac{\partial\widehat\Phi}{\partial t} = \frac{1}{\varepsilon}\,H(\sqrt{\varepsilon}\hat\psi^+, \sqrt{\varepsilon}\hat\psi^-)\,\widehat\Phi, \qquad \widehat\Phi(t)\in\mathcal{H}, \\ & \widehat\Phi|_{t=0} = \widehat{\mathcal{K}}^{\varepsilon}_{[\Lambda^k_0, r_0], \tau^{(0)}} f_0\,. \end{aligned} \tag{48}$$

Introduce the notation

(49)

$$\begin{aligned} \widehat\Psi^{\varepsilon}(t) = \exp\Bigg\{ & \frac{\phi^*_l(\tau^{(0)},0)\,\phi_l(\tau^{(0)},0)}{2\varepsilon} \\ & + \frac{\phi^*_l(\tau^{(0)},0)\,\phi^*_l(\tau^{(0)},0) - \phi_l(\tau^{(0)},0)\,\phi_l(\tau^{(0)},0)}{4\varepsilon}\Bigg\} \\ \times \int & \frac{d\tau\, f(\tau,t)}{(2\pi)^{k/2}\varepsilon^{k/4}} \, \frac{\sqrt{\det \frac{\partial\phi^*_l}{\partial\tau_a}(\tau,t)\,\frac{\partial\phi_l}{\partial\tau_b}(\tau,t)}}{\sqrt[4]{\det\left[\left(\frac{C(\tau,t)-iB(\tau,t)}{\sqrt2}\right)^+\left(\frac{C(\tau,t)-iB(\tau,t)}{\sqrt2}\right)\right]}} \\ \times \exp\Bigg\{ & \frac{1}{\varepsilon}\int_{(\tau^{(0)},0)}^{(\tau,t)} (\phi_l(\tau',t')\,d\phi^*_l(\tau',t') \\ & - iH(\phi^*(\tau',t'),\phi(\tau',t'))\,dt')\Bigg\}\widehat\Phi_{\phi(\tau,t),M(\tau,t)} \\ & l = 1,2,\dots,\quad a,b = 1,\dots,k\,. \end{aligned}$$

where

$$\begin{aligned} f(\tau,t) = \exp\Bigg\{ & -\frac{i}{4}\int_0^t dt' \left(\frac{\partial^2 H}{\partial\phi_m\partial\phi_n}\,(C+iB)(C-iB)^{-1}\right)_{mn} \\ & + \left(\frac{\partial^2 H}{\partial\phi^*_m\partial\phi^*_n}\,\overline{(C+iB)(C-iB)^{-1}}\right)_{mn}\Bigg\} f_0(\tau). \end{aligned} \tag{50}$$

THEOREM 1. *Suppose $H^{(m,n)}_{i_1\dots i_m j_1\dots j_n}$ are set of numbers such that*

$$\|H(\overset{2}{\sqrt{\varepsilon}\hat\psi^+}, \overset{1}{\sqrt{\varepsilon}\hat\psi^-})\widehat{\Phi}_{\phi(\tau,t),M(\tau,t)}\| \leqslant C_1, \qquad \varepsilon\in(0,\varepsilon_0),$$

where $M(\tau,t)$ is defined by formula (27). Let the canonical transformation $\mathcal{D}^t_H$ correspond to the Hamiltonian H, and let the series

$$\sum_{n,m=1}^{\infty}\left|\frac{\partial^2 H}{\partial\phi_m\partial\phi_n}\right|^2$$

converge. Then the solution of Cauchy problem (48) can be presented in the form

$$\widehat{\Phi}(t)=\widehat{\Psi}^{\varepsilon}(t)+\hat\delta^{(1)}_{\varepsilon}(t), \qquad (\hat\delta^{(1)}_{\varepsilon}(t),\hat\delta^{(1)}_{\varepsilon}(t)) \underset{\varepsilon\to 0}{\longrightarrow} 0,$$
$$(\widehat{\Phi}(t),\widehat{\Phi}(t))=O(1) \quad \text{as } \varepsilon\to 0\,.$$

A proof of this theorem will be given below as a corollary of a more general statement. Hence it will be proved that the canonical operator does indeed give the asymptotics of the Cauchy problem solution, i.e., that geometric quantization is compatible with the canonical one.

Now we consider some examples.

EXAMPLE 1. The following approximate solution of equation (48) coinciding up to a constant factor with the vector $\widehat{\Phi}_{\phi(t),M(t)}$ corresponds to a zero-dimensional isotropic manifold:

$$\widehat{\Psi}^{\varepsilon}(t)=\exp\left\{\frac{\phi^*_l(0)\,\phi_l(0)}{2\varepsilon}+\frac{\phi^*_l(0)\,\phi^*_l(0)-\phi_l(0)\,\phi_l(0)}{4\varepsilon}\right\}$$
$$\times\frac{f_0\exp\left(-\frac{i}{4}\int_0^t\left(\frac{\partial^2 H}{\partial\phi_m\partial\phi_n}M_{mn}(t')+\frac{\partial^2 H}{\partial\phi^*_m\partial\phi^*_n}M^*_{mn}(t')\right)dt'\right)}{\sqrt[4]{\det\left[\left(\frac{C(t)-iB(t)}{\sqrt2}\right)^+\left(\frac{C(t)-iB(t)}{\sqrt2}\right)\right]}}$$
$$\times\exp\left(\frac{1}{\varepsilon}\int_0^t(\phi_l(t')\,\dot\phi^*_l(t')-iH)\,dt'\right)\widehat{\Phi}_{\phi(t),M(t)}, \qquad m,n,l=1,2,\dots,$$

where we omit arguments $\phi^*(t')$, $\phi(t')$ at H.

The matrix M satisfies the equation

$$i\dot M_{mn}=\frac{\partial^2 H}{\partial\phi^*_m\partial\phi^*_n}+\frac{\partial^2 H}{\partial\phi^*_m\partial\phi_s}M_{sn}+M_{ms}\frac{\partial^2 H}{\partial\phi_s\partial\phi^*_n}+M_{ms}\frac{\partial^2 H}{\partial\phi_s\partial\phi_r}M_{rn},$$
$$m,s,n,r,=1,2,\dots\,.$$

In the Fock presentation, this vector coincides with the vector (29) up to a constant factor. Note that all the components of the vector corresponding to a zero-dimensional isotropic manifold differ from zero.

EXAMPLE 2 [**16, 17**]. Let H be of the form

$$H(\phi^*, \phi) = \sum_{n=1}^{s} H^{(n)}_{i_1\dots i_n j_1 \dots j_n} \phi^*_{i_1} \cdots \phi^*_{i_n} \phi_{j_1} \cdots \phi_{j_n}\,.$$

The family g^t_H maps a one-dimensional isotropic manifold of the form $\phi_j(\tau) = \tilde{\phi}^0_j \exp(i\tau)$, $\tau \in [0, 2\pi)$, into an isotropic manifold of the form

$$\phi_j(\tau, t) = \tilde{\phi}_j(t) \exp(i\tau), \qquad \tau \in [0, 2\pi),$$

where ϕ_j is a solution of the equations (42).

As usual, set $F = (C + iB)/\sqrt{2}$, $G = (C - iB)/\sqrt{2}$.

The equations in variations for F and G

$$\begin{aligned} i\dot{F}_{mn} &= \frac{\partial^2 H}{\partial\phi^*_m \partial\phi^*_r} G_{rn} + \frac{\partial^2 H}{\partial\phi^*_m \partial\phi_r} F_{rn}, \\ i\dot{G}_{mn} &= -\frac{\partial^2 H}{\partial\phi_m \partial\phi^*_r} G_{rn} - \frac{\partial^2 H}{\partial\phi_m \partial\phi_r} F_{rn}, \end{aligned} \qquad m, r, n = 1, 2, \dots,$$

have a solution of the form

$$F(\tau, t) = \widetilde{F}(t)\, e^{i\tau}, \qquad G(\tau, t) = \widetilde{G}(t)\, e^{-i\tau}.$$

The expressions

$$\det G^+ G, \qquad \det \frac{\partial\phi^*_l}{\partial\tau_a} \frac{\partial\phi_l}{\partial\tau_b} = \tilde{\phi}^*_l \tilde{\phi}_l, \quad l = 1, 2, \dots,$$

do not depend on τ. If f_0 does not depend on τ the function $f(\tau, t)$ does not depend on τ as well.

In this case, therefore, vector (49) has the following form:

$$\begin{aligned} \widehat{\Psi}^\varepsilon(t) &= \exp\left\{ \frac{\tilde{\phi}^*_l(0)\, \tilde{\phi}_l(0)}{2\varepsilon} + \frac{\tilde{\phi}^*_l(0)\, \tilde{\phi}^*_l(0) - \tilde{\phi}_l(0)\, \tilde{\phi}_l(0)}{4\varepsilon} \right\} \\ &\times \frac{f(t)\sqrt{\tilde{\phi}^*_l \tilde{\phi}_l}}{\sqrt[4]{\det \widetilde{G}^+(t)\, \widetilde{G}(t)}} \exp\left(\frac{1}{\varepsilon} \int_0^t (\tilde{\phi}_l(t')\, \dot{\tilde{\phi}}_l^{\,*}(t') - iH)\, dt' \right) \\ &\times \int \frac{d\tau}{\sqrt{2\pi}\, \varepsilon^{1/4}} \exp\left(-\frac{i}{\varepsilon} \tau \tilde{\phi}^*_l \tilde{\phi}_l + \frac{1}{\varepsilon} \tilde{\phi}_j(t) (\sqrt{\varepsilon} \hat{\psi}^+_j e^{i\tau} - \tilde{\phi}^*_j(t)) \right) \\ &\times \exp\left(\frac{1}{2\varepsilon} (\hat{\psi}^+_j \sqrt{\varepsilon}\, e^{i\tau} - \phi^*_j(t))\, \widetilde{M}_{jl} (\hat{\psi}^+_l \sqrt{\varepsilon}\, e^{i\tau} - \phi^*_l(t)) \right) \widehat{\Phi}_0, \\ &\quad j, l = 1, 2, \dots\,. \end{aligned}$$

The quantization condition can be written as

$$\tilde{\phi}^*_l \tilde{\phi}_l = \varepsilon N, \qquad N \in \mathbb{Z}\,.$$

It is easy to see that integrating over τ maps all the components of $\widehat{\Psi}^\varepsilon$ but the Nth one to zero. This component has the following form (up to a normalizing factor):

$$(51)\quad \begin{aligned} (\Psi^\varepsilon)^{(N)}_{i_1\dots i_N} &= \tilde{\phi}_{i_1}\cdots\tilde{\phi}_{i_N}\sum_{k=0}^{[N/2]}\frac{1}{2^k k!}\sum_{1\leqslant j_1\neq\dots\neq j_{2k}\leqslant N}\frac{\widetilde{M}_{i_{j_1}i_{j_2}}\cdots\widetilde{M}_{i_{j_{2k-1}}i_{j_{2k}}}}{\tilde{\phi}_{i_{j_1}}\cdots\tilde{\phi}_{i_{j_{2k}}}}\,c(t),\\ c(t) &= \frac{f(t)}{\sqrt[4]{\det G^+(t)\,G(t)}}\exp\left\{\frac{1}{\varepsilon}\int_0^t(\tilde{\phi}_l(t')\dot{\tilde{\phi}}_l^*(t')-iH)\,dt'\right\},\\ \widetilde{M}_{lj} &= (FG^{-1})_{lj}-\tilde{\phi}_l\tilde{\phi}_j/(\tilde{\phi}^*_m\tilde{\phi}_m),\qquad l,j,m=1,2,\dots. \end{aligned}$$

Formula (51) coincides with the one obtained in [**15**] for the stationary case. It may be also obtained by taking the Nth component of the state vector corresponding to a germ in a point (see [**16, 17**]).

The matrix M that vanishes at the initial moment may differ from zero at other moments of time. It now follows that an N-particle wave function cannot be decomposed into a product of one-particle wave functions (i.e., in the form (51) with $M=0$) as $\varepsilon\to 0$, $N\to\infty$, $\varepsilon N\to$ const, even for the case in which it is decomposed into such a product at the initial moment. For the case of classical state mechanics, this statement was stated and proved in [**16**].

EXAMPLE 3. The approach developed here may be also used formally in the case of quantum field theory, for scalar quantum electrodynamics, for example. The formulas, however, look simpler for the theory of a real scalar field with self-interaction. This theory describes an approximation of the interaction of π-mesons ([**3, 21**]). The Lagrangian for the theory is of the form

$$(52)\qquad \mathcal{L}=\frac{1}{2}(\partial_\mu\phi)(\partial^\mu\phi)-\frac{m^2}{2}\phi^2-\frac{g}{4}\phi^4,\qquad \mu=0,1,\dots,d-1,$$

where d is the space-time dimension.

The Hamiltonian

$$(53)\quad \mathcal{H}(\hat{p}(\cdot),\hat{q}(\cdot))=\int d^{d-1}\mathbf{x}\left(\frac{1}{2}\hat{p}^2(\mathbf{x})+\frac{1}{2}(\vec{\nabla}\hat{q}(\mathbf{x}))^2+\frac{m^2}{2}\hat{q}^2(\mathbf{x})+\frac{g}{4}\hat{q}^4(\mathbf{x})\right),$$
$$d^{d-1}\mathbf{x}=dx_1\cdots dx_{d-1},$$

corresponds to the Lagrangian (52).

After the substitution $\sqrt{g}\hat{p}=\widehat{P}$, $\sqrt{g}\hat{q}=\widehat{Q}$, the Hamiltonian (53) becomes

$$\mathcal{H}(\widehat{P}(\cdot),\widehat{Q}(\cdot))=\frac{1}{g}\int d\mathbf{x}\left(\frac{1}{2}\widehat{P}^2(\mathbf{x})+\frac{1}{2}(\vec{\nabla}\widehat{Q}(\mathbf{x}))^2+\frac{m^2}{2}\widehat{Q}^2(\mathbf{x})+\frac{1}{4}\widehat{Q}^4(\mathbf{x})\right),$$
$$[\widehat{Q}(\mathbf{x}),\widehat{P}(\mathbf{y})]=ig\delta(\mathbf{x}-\mathbf{y}).$$

The theory may be interpreted in terms of particles with the help of the following creation and annihilation operators:

$$\hat{q}(\mathbf{x})=\frac{1}{(2\pi)^{(d-1)/2}}\int\frac{d\mathbf{l}}{\sqrt{2\sqrt{\mathbf{l}^2+m^2}}}(\hat{\psi}^+(\mathbf{l})\,e^{-i\mathbf{l}\mathbf{x}}+\hat{\psi}^-(\mathbf{l})\,e^{i\mathbf{l}\mathbf{x}}),$$

$$\hat{p}(\mathbf{x})=\frac{i}{(2\pi)^{(d-1)/2}}\int d\mathbf{l}\sqrt{\frac{\sqrt{\mathbf{l}^2+m^2}}{2}}(\hat{\psi}^+(\mathbf{l})\,e^{-i\mathbf{l}\mathbf{x}}-\hat{\psi}^-(\mathbf{l})\,e^{i\mathbf{l}\mathbf{x}}).$$

According to the outlined scheme, one may assign solutions of secondary quantized equations of type (49) to each Lagrangian manifold with complex germ. To construct such an asymptotics, one must solve a Hamiltonian system and a system of equations in variations. The Hamiltonian system has the form

$$\begin{aligned}\dot{Q}(\mathbf{x},t) &= P(\mathbf{x},t),\\ \dot{P}(\mathbf{x},t) &= \Delta Q(\mathbf{x},t) - m^2 Q(\mathbf{x},t) - Q^3(\mathbf{x},t)\end{aligned}$$

and the equations in variations over the variables δP, δQ are

$$\begin{aligned}\dot{B}(\mathbf{x},t,\tau) &= \Delta C(\mathbf{x},t,\tau) - m^2 C(\mathbf{x},t,\tau) - 3C(\mathbf{x},t,\tau)Q^2(\mathbf{x},t,\tau),\\ \dot{C}(\mathbf{x},t,\tau) &= B(\mathbf{x},t,\tau)\,.\end{aligned}$$

The outlined approach may be applied without preliminary regularization if

$$\int d\mathbf{x}\Big(Q(\mathbf{x},t,\tau)\sqrt{m^2-\Delta}\,Q(\mathbf{x},t,\tau) + P(\mathbf{x},t,\tau)\,\frac{1}{\sqrt{m^2-\Delta}}\,P(\mathbf{x},t,\tau)\Big) < \infty,$$

and the operator $BC^{-1} - i\sqrt{m^2-\Delta}$ is a Hilbert–Schmidt operator.

If these conditions are not satisfied, regularization must be carefully performed.

The concept of geometric quantization may also be applied to scalar quantum electrodynamics in the α-gauge ([**3**]) with the Lagrangian

$$\begin{aligned}\mathcal{L} = &-\frac{1}{4}\,(\partial_\mu A_\nu - \partial_\nu A_\mu)(\partial^\mu A^\nu - \partial^\nu A^\mu) - \frac{1}{2\alpha}\,(\partial_\mu A^\mu)^2\\ &+ (\partial_\mu + ieA_\mu)\,\Phi^*(\partial_\mu - ieA_\mu)\,\Phi - m^2\Phi^*\Phi - \frac{g}{4}\,(\Phi^*\Phi)^2.\end{aligned}$$

§5. Complex germ creation and annihilation operators

The theory of complex germs allows one to construct not only asymptotic solutions of type (49) for secondary quantized equations. Other asymptotic solutions of equation (48) may be obtained with the help of the so-called germ creation and annihilation operators.

Suppose a basis on the complex germ is chosen so that the matrix A from axiom (r1) for complex germs is diagonal. Consider the following creation and annihilation operators on a Lagrangian manifold with complex germ:

$$\begin{aligned}\bar{\mathcal{A}}_\alpha(\tau,t) &= \bar{G}_{m\alpha}(\tau,t)\Big(\hat{\psi}_m^+ - \frac{\phi_m^*(\tau,t)}{\sqrt{\varepsilon}}\Big) - \bar{F}_{m\alpha}(\tau,t)\Big(\hat{\psi}_m^- - \frac{\phi_m(\tau,t)}{\sqrt{\varepsilon}}\Big),\\ \mathcal{A}_\alpha(\tau,t) &= G_{m\alpha}(\tau,t)\Big(\hat{\psi}_m^- - \frac{\phi_m(\tau,t)}{\sqrt{\varepsilon}}\Big) - F_{m\alpha}(\tau,t)\Big(\hat{\psi}_m^+ - \frac{\phi_m^*(\tau,t)}{\sqrt{\varepsilon}}\Big),\end{aligned} \tag{54}$$

$$F = (C+iB)/\sqrt{2},\quad G = (C-iB)/\sqrt{2},\quad \alpha = k+1, k+2, \dots,\ m = 1,2,\dots.$$

Let

$$\nu_\alpha \in \{0,1,\dots\},\quad \alpha = k+1,k+2,\dots,\quad \sum_{\alpha=k+1}^{\infty} \nu_\alpha < \infty\,.$$

Further, let $l(\tau',\tau'')$ be a closed path on Λ^k such that it is covered by a path on M^k issuing from the point τ' and ending at τ''. Suppose that the following quantization condition holds for any such path:

$$\frac{1}{2\pi\varepsilon}\oint_{l(\tau',\tau'')} P_m\,dQ_m = \sum_{\alpha=k+1}^{\infty} \gamma_\alpha(\tau',\tau'')\,\nu_\alpha + n, \qquad n\in\mathbb{Z}, \tag{55}$$

where $\gamma_\alpha(\tau',\tau'')$ are determined by the condition $A_{\alpha,\alpha}(\tau',\tau'') = e^{i\gamma_\alpha(\tau',\tau'')}$.

Consider the following element of $\mathcal{H}$:

$$\begin{aligned}\widehat{\tilde{\Psi}}^\varepsilon(t) = \int \frac{d\tau\, f(\tau,t)}{(2\pi)^{k/2}\varepsilon^{k/4}} & \frac{\sqrt{\det \frac{\partial\phi_l}{\partial\tau_a}(\tau,t)\,\frac{\partial\phi_l^*}{\partial\tau_b}(\tau,t)}}{\sqrt[4]{\det G^+(\tau,t)\,G(\tau,t)}}\\ &\times \exp\Big(\frac{i}{\varepsilon}\,S(\tau,t)\Big)\bar{\mathcal{A}}_{k+1}^{\nu_{k+1}}\bar{\mathcal{A}}_{k+2}^{\nu_{k+2}}\cdots\widehat{\Phi}_{\phi(\tau,t),M(\tau,t)},\\ &l=1,2,\dots,\quad a,b=1,\dots,k,\end{aligned} \tag{56}$$

where

$$\begin{aligned}S(\tau,t) &= \frac{1}{2i}\,\phi_l^*(\tau^{(0)},0)\,\phi_l(\tau^{(0)},0)\\ &+\frac{1}{4i}\,(\phi_l^*(\tau^{(0)},0)\,\phi_l^*(\tau^{(0)},0) - \phi_l(\tau^{(0)},0)\,\phi_l(\tau^{(0)},0))\\ &+\int_{(\tau^{(0)},0)}^{(\tau,t)}\Big(\frac{1}{i}\,\phi_l(\tau',t')\,d\phi_l^*(\tau',t') - H(\phi^*(\tau',t'),\phi(\tau',t'))\,dt'\Big),\end{aligned}$$

and $f(\tau,t)$ and $M(\tau,t)$ are determined from formulas (50) and (27), respectively. Let $\widehat{\Phi}(t)$ be the solution of (48) satisfying the initial condition $\widehat{\Phi}(0)=\widehat{\tilde{\Psi}}^\varepsilon(0)$.

THEOREM 2. *Suppose*

$$\|H(\sqrt{\varepsilon}\psi^+,\sqrt{\varepsilon}\psi^-)\bar{\mathcal{A}}_{k+1}^{\nu_{k+1}}\bar{\mathcal{A}}_{k+2}^{\nu_{k+2}}\cdots\widehat{\Phi}_{\phi(\tau,t),M(\tau,t)}\| \leqslant C_2, \qquad \varepsilon\in(0,\varepsilon_0)\,.$$

Let the canonical transformation $\mathcal{D}_H^t$ correspond to the Hamiltonian H and the series

$$\sum_{m,n=1}^{\infty}\left|\frac{\partial^2 H}{\partial\phi_m\partial\phi_n}\right|^2$$

converge. Then

$$\widehat{\Phi}(t) = \widehat{\tilde{\Psi}}^\varepsilon(t) + \hat{\delta}_\varepsilon(t), \qquad (\hat{\delta}_\varepsilon(t),\hat{\delta}_\varepsilon(t)) \xrightarrow[\varepsilon\to 0]{} 0,$$
$$(\widehat{\Phi}(t),\widehat{\Phi}(t)) = O(1), \quad \varepsilon\to 0\,.$$

REMARK 14. The canonical operator may be modified for the case of a C-Lagrangian manifold ([**13**]), when $\gamma_\alpha=0$,

Let $f^{(n)}_{\alpha_1,\dots,\alpha_n}(\tau)$, where $\alpha_i = k+1,k+2,\dots$, $n=0,1,\dots$, be a set of infinitely differentiable functions with compact support on a Lagrangian manifold with complex germ $[\Lambda^k, r]$ such that the series

$$\sum_{n=0}^{\infty} \sum_{\alpha_1,\dots,\alpha_n=k+1}^{\infty} |f^{(n)}_{\alpha_1,\dots,\alpha_n}(\tau)|^2$$

converges for all τ.

The following element of $\mathcal{H}$ is assigned to the Lagrangian manifold with complex germ $[\Lambda^k, r]$, the set of functions $f^{(n)}_{\alpha_1,\dots,\alpha_n}(\tau)$, the point $\tau^{(0)} \in \Lambda^k$, and the number $\varepsilon > 0$:

$$\begin{aligned} \widehat{\mathcal{K}}^{\varepsilon}_{[\Lambda^k,r]\tau^{(0)}} f = \int \frac{d\tau}{(2\pi)^{k/2}\varepsilon^{k/4}} \frac{\sqrt{\det \frac{\partial \phi_l}{\partial \tau_a}(\tau) \frac{\partial \phi_l^*}{\partial \tau_b}(\tau)}}{\sqrt[4]{\det G^+(\tau) G(\tau)}} \exp\left(\frac{i}{\varepsilon} S(\tau)\right) \\ \times \sum_{n=0}^{\infty} \sum_{\alpha_1,\dots,\alpha_n=k+1}^{\infty} \frac{1}{\sqrt{n!}} f^{(n)}_{\alpha_1,\dots,\alpha_n}(\tau) \bar{\mathcal{A}}_{\alpha_1} \cdots \bar{\mathcal{A}}_{\alpha_n} \widehat{\Phi}_{\phi(\tau),M(\tau)}, \\ l = 1, 2, \dots, \quad a, b = 1, \dots, k. \end{aligned} \tag{57}$$

If $f^{(0)} = f$ and the other functions $f^{(n)}$ vanish, then the canonical operator (57) is equal to the canonical operator (31).

EXAMPLE 4. Consider the Hamiltonian from Example 2 and suppose $\phi_l(t)$ is a solution to equation (42) of the form $\phi_l(t) = \tilde{\phi}_l \exp(-i\Omega t)$.

Suppose that there exist matrices $\widetilde{F}$ and $\widetilde{G}$ such that

$$\begin{gathered} (\beta_l + \Omega) \widetilde{G}_{ml} = \sum_{r=1}^{\infty} \frac{\partial^2 H}{\partial \phi_m \partial \phi_r^*} \widetilde{G}_{rl} + \sum_{r=1}^{\infty} \frac{\partial^2 H}{\partial \phi_m \partial \phi_r} \widetilde{F}_{rl}, \\ -(\beta_l - \Omega) \widetilde{F}_{ml} = \sum_{r=1}^{\infty} \frac{\partial^2 H}{\partial \phi_m^* \partial \phi_r^*} \widetilde{G}_{rl} + \sum_{r=1}^{\infty} \frac{\partial^2 H}{\partial \phi_m^* \partial \phi_r} \widetilde{F}_{rl}, \\ \beta_l \in \mathbb{R}, \quad \widetilde{G}_{m1} = \tilde{\phi}_m^*, \quad \widetilde{F}_{m1} = -\tilde{\phi}_m, \quad \widetilde{F}^T \widetilde{G} = \widetilde{G}^T \widetilde{F}, \\ \widetilde{G}^*_{m\alpha} \widetilde{G}_{m\beta} - \widetilde{F}^*_{m\alpha} \widetilde{F}_{m\beta} = \delta_{\alpha\beta}, \quad m = 1, 2, \dots, \quad \alpha, \beta = 2, 3, \dots, \end{gathered}$$

where $\widetilde{F}$ is a Hilbert–Schmidt operator and $\widetilde{G}$ has a bounded inverse operator.

Consider a Lagrangian manifold with complex germ corresponding to the following function ϕ_l:

$$\phi_l(\tau, t) = \tilde{\phi}_l \exp(i(\tau - \Omega t))$$

with matrices F and G given by

$$\begin{aligned} F_{ml}(\tau, t) &= \widetilde{F}_{ml} \exp(-i(\beta_l/\Omega - 1)(\tau - \Omega t)), \\ G_{ml}(\tau, t) &= \widetilde{G}_{ml} \exp(-i(\beta_l/\Omega + 1)(\tau - \Omega t)). \end{aligned}$$

The matrix $A(0, 2\pi)$ is

$$A(0, 2\pi) = \operatorname{diag}\{1, \exp(2\pi i \beta_2/\Omega), \exp(2\pi i \beta_3/\Omega), \dots\},$$

and the quantization condition is

$$\tilde{\phi}_l^* \tilde{\phi}_l = \varepsilon N + \varepsilon \sum_{\lambda=2}^{\infty} \frac{\beta_\lambda \nu_\lambda}{\Omega}.$$

The vector (56) depends on time as $\exp(-i\mathcal{E}t)$, where

$$\mathcal{E}=\frac{1}{\varepsilon}H(\phi^*,\phi)+\frac{1}{4}\sum_{m,n=1}^{\infty}\left(\frac{\partial^2 H}{\partial\phi_m\partial\phi_n}(FG^{-1})_{mn}+\frac{\partial^2 H}{\partial\phi^*_m\partial\phi^*_n}(FG^{-1})^*_{mn}\right).$$

Hence this is an approximate solution of the stationary equation

$$\mathcal{E}\widehat{\Phi}=\frac{1}{\varepsilon}H(\hat{\psi}^+\sqrt{\varepsilon},\hat{\psi}^-\sqrt{\varepsilon})\widehat{\Phi}. \tag{58}$$

EXAMPLE 5. Now consider the complex germ given by the following matrices:

$$F_{ml}(\tau,t)=F_{ml}\exp(-i\Omega t+i\tau)\exp(i\beta_l t),$$
$$G_{ml}(\tau,t)=G_{ml}\exp(i\Omega t-i\tau)\exp(i\beta_l t).$$

The matrix $A(0,2\pi)$ is the identity matrix, and the quantization condition has the form

$$\tilde{\phi}^*_l\tilde{\phi}_l=\varepsilon N.$$

In this example the vector (56) is also an approximate solution of equation (58) for

$$\mathcal{E}=\frac{1}{\varepsilon}H(\phi^*,\phi)+\frac{1}{4}\left(\frac{\partial^2 H}{\partial\phi_m\partial\phi_n}(FG^{-1})_{mn}+\frac{\partial^2 H}{\partial\phi^*_m\partial\phi^*_n}(FG^{-1})^*_{mn}\right)+\sum_{\lambda=2}^{\infty}\beta_\lambda\nu_\lambda,$$
$$m,n=1,2\dots.$$

This solution coincides with that given in [**15**].

Now let us find a relation between the results of Examples 4 and 5. Let $\phi_{(1)_m}$, $\phi_{(2)_m}$ be solutions of the equations

$$\Omega_1\phi_{(1)_m}=\frac{\partial\mathcal{H}}{\partial\phi^*_{(1)_m}},\qquad \Omega_2\phi_{(2)_m}=\frac{\partial\mathcal{H}}{\partial\phi^*_{(2)_m}}$$

such that

$$\phi^*_{(1)_m}\phi_{(1)_m}=\varepsilon N+\varepsilon\sum_{\lambda=1}^{\infty}\frac{\beta_\lambda\nu_\lambda}{\Omega_1},\qquad \phi^*_{(2)_m}\phi_{(2)_m}=\varepsilon N,$$
$$\phi_{(1)_m}=\phi_{(2)_m}+\chi_m,\qquad \|\chi\|=O(\varepsilon),\qquad \Omega_2-\Omega_1=O(\varepsilon).$$

We have

$$H(\phi^*_{(1)},\phi_{(1)})-H(\phi^*_{(2)},\phi_{(2)})=\Omega_2(\phi_{(2)_m}\chi^*_m+\phi^*_{(2)_m}\chi_m)+O(\varepsilon^2),$$
$$(\phi_{(1)},\phi_{(1)})-(\phi_{(2)},\phi_{(2)})=\phi_{(2)_m}\chi^*_m+\phi^*_{(2)_m}\chi_m+O(\varepsilon^2),\qquad m=1,2,\dots.$$

It now follows that the values of $\mathcal{E}$ in the two examples coincide up to $O(\varepsilon^1)$.

REMARK 15. We have shown for a specific example how different quantization conditions may be used with a simultaneous change of the transport equation. It is only convenience that determines the choice of the quantization conditions. In the finite-dimensional case, a quantization condition of the form

$$\frac{1}{2\pi\varepsilon}\oint_{l(\tau',\tau'')} P_m\,dQ_m = \sum_{\alpha=k+1}^{D} \gamma_\alpha(\tau',\tau'')\left(\nu_\alpha+\frac{1}{2}\right)+n, \qquad n\in\mathbb{Z},$$

is often used ([**1**]).

In the infinite-dimensional case, this condition does not make sense even for the simplest Hamiltonian

$$H(\phi^*,\phi)=\sum_{i=1}^{\infty}\phi_i^*\phi_i$$

and for the isotropic manifold with complex germ as in Example 4.

EXAMPLE 6. We describe one more (heuristic) way to deduce a formula for $\mathcal{E}$ (see [**11**]).

Note that if a family $\mathcal{D}_H^t$ acts on a two-dimensional isotropic manifold with complex germ in the following way

$$\begin{aligned}(g^t\phi)(\tau_1,\tau_2)&=\phi(\tau_1+\Omega_1 t,\tau_2+\Omega_2 t),\\ F(\tau_1,\tau_2,t)&=F(\tau_1+\Omega_1 t,\tau_2+\Omega_2 t),\\ G(\tau_1,\tau_2,t)&=G(\tau_1+\Omega_1 t,\tau_2+\Omega_2 t),\end{aligned}$$

then multiplication of the function $f(\tau_1,\tau_2,0)$ by $\exp(i(n_1\tau_1+n_2\tau_2))$ leads to the multiplication of the vector (56) by $\exp(-i(n_1\Omega_1+n_2\Omega_2)t)2$, $n_1,n_2\in\mathbb{Z}$

Now consider "almost invariant" two-dimensional isotropic manifolds close to the one-dimensional manifolds of Examples 4 and 5:

$$\phi_l(t,\tau_1,\tau_2)=e^{-i(\tau_1-\Omega t)}(\phi_l+\delta(\widetilde{F}_{lm}e^{i(\tau_2+\beta^m t)}+\widetilde{G}_{lm}^*e^{-i(\tau_2+\beta^m t)})), \qquad \delta\to 0.$$

The values of $\mathcal{E}$ corresponding to these almost invariant manifolds are

$$\mathcal{E}_m^{(\mu)}=\mathcal{E}_m^{(0)}+\beta_m\mu_m, \qquad \mu_m\in\mathbb{Z}.$$

Since these values must agree with the value for a neighborhood of a one-dimensional manifold, the last value is of the form

$$\mathcal{E}^{(0)}+\sum_{m=1}^{\infty}\beta_m\nu_m, \qquad \nu_m\in\{0,1,2,\dots\}.$$

§6. Proof of Theorem 2

Now let us prove Theorem 2 stated in the previous section.

Introduce the notation

$$\begin{aligned}S(\tau)=\frac{1}{2i}\phi_j^*(\tau^{(0)})\phi_j(\tau^{(0)})&+\frac{1}{4i}(\phi_j^*(\tau^{(0)})\phi_j^*(\tau^{(0)})-\phi_j(\tau^{(0)})\phi_j(\tau^{(0)}))\\ &+\frac{1}{i}\int_{\tau^{(0)}}^{\tau}\phi_j(\tau')\,d\phi_j^*(\tau'), \qquad j=1,2,\dots.\end{aligned}$$

To the set of numbers $\mathcal{D}_{ij}(\tau)$, $i,j=1,2,\dots$ and $(Y_a^n)_{i_1\dots i_n}(\tau)$, $a=1,2$, $n=1,\dots,s$, $i_p=1,2,\dots$, assign the following operators in $\mathcal{H}$:

$$Y_a\left(\tau,\hat\psi^+-\frac{\phi^*(\tau)}{\sqrt\varepsilon}\right)=\sum_{n=0}^{s}(Y_a^n)_{i_1\dots i_n}(\tau)\left(\hat\psi^+_{i_1}-\frac{\phi^*_{i_1}(\tau)}{\sqrt\varepsilon}\right)\cdots\left(\hat\psi^+_{i_n}-\frac{\phi^*_{i_n}(\tau)}{\sqrt\varepsilon}\right)$$
$$\times\exp\left\{\frac12\left(\hat\psi^+_i-\frac{\phi^*_i(\tau)}{\sqrt\varepsilon}\right)\mathcal{D}_{ij}(\tau)\left(\hat\psi^+_i-\frac{\phi^*_i(\tau)}{\sqrt\varepsilon}\right)\right\},$$
$$i,j,i_1,\dots,i_n=1,2,\dots,\quad a=1,2.$$

Further, denote

$$\widehat\Phi^\varepsilon_a=\int_{\Lambda^k}\frac{d\tau\exp\left(\frac{i}{\varepsilon}S(\tau)\right)}{\varepsilon^{k/4}}Y_a\left(\tau,\hat\psi^+-\frac{\phi^*(\tau)}{\sqrt\varepsilon}\right)\exp\left(\frac1\varepsilon\,\phi_j(\tau)(\hat\psi^+_j\sqrt\varepsilon-\phi^*_j(\tau))\right)\widehat\Phi_0,$$
$$a=1,2,\ j=1,2,\dots.$$

The commutation relations for the operators $\psi^\pm_j$ imply

$$\widehat\Phi^\varepsilon_a=\int\frac{d\tau\exp\left(\frac{i}{\varepsilon}S(\tau)-\frac{1}{2\varepsilon}\phi_j(\tau)\phi^*_j(\tau)\right)}{\varepsilon^{k/4}}$$
$$\times\exp\left(\frac{\phi_j(\tau)\hat\psi^+_j-\phi^*_j(\tau)\hat\psi^-_j}{\sqrt\varepsilon}\right)Y_a(\tau,\hat\psi^+)\widehat\Phi_0,\qquad j=1,2,\dots,\ a=1,2.$$

LEMMA 5. *Let the matrix $\mathcal{D}_{ij}(\tau)$ correspond to a Hilbert–Schmidt operator $\mathcal{D}$ in l^2, $\|\mathcal{D}\|<1$, with $\mathcal{D}_{ij}(\tau)$ and $(Y_a^n)_{i_1\dots i_n}(\tau)$ being smooth functions of τ, the functions $(Y_a^n)_{i_1\dots i_n}(\tau)$ having compact supports, and*

$$\sum_{i_1\dots i_n=1}^{\infty}|(Y_a^n)_{i_1\dots i_n}(\tau)|^2<c,\qquad n=1,\dots,s,\ a=1,2.$$

Then

$$(\widehat\Phi^\varepsilon_1,\widehat\Phi^\varepsilon_2)\xrightarrow[\varepsilon\to0]{}\int_{\Lambda^k}d\tau\int_{\mathbb{R}^k}d\xi\Big(Y_1(\tau,\hat\psi^+)\widehat\Phi_0,$$
$$\exp\left(\xi_b\left(\frac{\partial\phi_j}{\partial\tau_b}(\tau)\hat\psi^+_j-\frac{\partial\phi^*_j}{\partial\tau_b}(\tau)\hat\psi^-_j\right)\right)Y_2(\tau,\hat\psi^+)\widehat\Phi_0\Big),\tag{59}$$
$$b=1,\dots,k,\quad\xi\in\mathbb{R}^k,\quad j=1,2,\dots.$$

PROOF. We have

$$(\widehat\Phi^\varepsilon_1,\widehat\Phi^\varepsilon_2)=\int\frac{d\tau\,d\tau'}{\varepsilon^{k/2}}\exp\left(\frac{i}{\varepsilon}(S(\tau')-S^*(\tau))-\frac{1}{2\varepsilon}(\phi_j(\tau)\phi^*_j(\tau)+\phi_j(\tau')\phi^*_j(\tau'))\right)$$
$$\times\exp\left(\frac{1}{2\varepsilon}(\phi^*_j(\tau)\phi_j(\tau')-\phi^*_j(\tau')\phi_j(\tau))\right)$$
$$\times\Big(\widehat\Phi_0,Y_1^*(\tau,\hat\psi^-)\exp\Big(\frac{1}{\sqrt\varepsilon}(\hat\psi^+_j(\phi_j(\tau')-\phi_j(\tau))$$
$$-\hat\psi^-_j(\phi^*_j(\tau')-\phi^*_j(\tau)))\Big)Y_2(\tau',\hat\psi^+)\widehat\Phi_0\Big),$$

$j=1,2,\dots.$

It is easy to show that the contribution of the integration domain

$$(\tau_b - \tau_b')(\tau_b - \tau_b') > \delta_1, \quad \delta_1 > 0,\ b = 1, \dots, k,$$

to the integral (59) is exponentially small. For $(\tau_b - \tau_b')(\tau_b - \tau_b') \leqslant \delta_1 \varepsilon^{1/2-\lambda}$ the integrand may be presented as

$$\frac{1}{\varepsilon^{k/2}} \exp(\sqrt{\varepsilon}\xi_b\xi_c\xi_d R_{bcd}(\tau,\tau')) \\ \times \biggl(\widehat{\Phi}_0, Y_1^*(\tau, \hat{\psi}^-) \exp\biggl(\xi_b\Bigl(\hat{\psi}_j^+ \frac{\partial\phi_j}{\partial\tau_b} - \hat{\psi}_j^- \frac{\partial\phi_j^*}{\partial\tau_b}\Bigr) \\ + \sqrt{\varepsilon}(\hat{\psi}_j^+ J_j(\tau,\tau') - \hat{\psi}_j^- J_j(\tau,\tau'))\biggr) Y_2(\tau', \hat{\psi}^+)\widehat{\Phi}_0\biggr),$$

$$j = 1, 2, \dots, \qquad b, c, d = 1, \dots, k,$$

$$\xi_b = \frac{\tau_b' - \tau_b}{\sqrt{\varepsilon}}, \quad |R_{bcd}(\tau,\tau')| < \text{const}, \quad \sum_{j=1}^{\infty} |J_j(\tau,\tau')|^2 < \text{const}\,.$$

Integrating over τ and ξ as $\varepsilon \to 0$, we obtain (59). Lemma 5 is proved.

COROLLARY. $\|\widehat{\tilde{\Psi}}^\varepsilon(t)\| = O(1)$ *as* $\varepsilon \to 0$.

PROOF. It is sufficient to prove that only the integral

$$\int d^k\xi \biggl(\widehat{\Phi}_0, \exp(\hat{\psi}_m^- M_{mn}^* \hat{\psi}_n^-/2)\, \hat{a}_{k+1}^{\nu_{k+1}} \hat{a}_{k+2}^{\nu_{k+2}} \cdots \exp\Bigl\{\xi_b\Bigl(\frac{\partial\phi_m}{\partial\tau_b}\hat{\psi}_m^+ - \frac{\partial\phi_m^*}{\partial\tau_b}\hat{\psi}_m^-\Bigr)\Bigr\} \\ \times \hat{a}_{k+1}^{+\nu_{k+1}} \hat{a}_{k+2}^{+\nu_{k+2}} \cdots \exp(\hat{\psi}_m^+ M_{mn} \hat{\psi}_n^+)\widehat{\Phi}_0\biggr), \tag{60}$$

$$\hat{a}_\alpha^+ = \bar{G}_{m\alpha}\hat{\psi}_m^+ - \bar{F}_{m\alpha}\hat{\psi}_m^-, \qquad m, n = 1, 2, \dots,\ b = 1, \dots, k,$$
$$\hat{a}_\alpha = G_{m\alpha}\hat{\psi}_m^- - F_{m\alpha}\hat{\psi}_m^+,$$

differs from zero. First we show that

$$\int d^k\xi \biggl(\widehat{X}, \exp\Bigl(\xi_b\Bigl(\frac{\partial\phi_m}{\partial\tau_b}\hat{\psi}_m^+ - \frac{\partial\phi_m^*}{\partial\tau_b}\hat{\psi}_m^-\Bigr)\Bigr)\hat{a}_\beta \exp\Bigl(\frac{\hat{\psi}_m^+ M_{mn}\hat{\psi}_n^+}{2}\Bigr)\widehat{\Phi}_0\biggr) = 0, \tag{61}$$

$$\beta \in \{k+1, k+2, \dots\}, \quad b = 1, \dots, k,\ m, n = 1, 2, \dots,$$
$$\widehat{X} = \hat{a}_{\alpha_1}^+ \cdots \hat{a}_{\alpha_r}^+ \hat{a}_{\beta_1} \cdots \hat{a}_{\beta_s} \exp(\hat{\psi}_m^+ M_{mn} \hat{\psi}_n^+/2)\widehat{\Phi}_0,$$
$$\alpha_1, \dots, \alpha_r, \beta_1, \dots, \beta_s \in \{k+1, k+2, \dots\}.$$

Indeed, using the presentation of the inner product in the left-hand side of (61) as a functional integral (see [**2**])

$$\int d^k\xi \int \mathcal{D}z^*\mathcal{D}z X(z^*)\Bigl(G_{m\beta}\frac{\partial}{\partial z_m} - F_{m\beta}z_m\Bigr) \\ \times \exp\Bigl\{\xi_b\Bigl(\frac{\partial\phi_m}{\partial\tau_b}z_m - \frac{\partial\phi_m^*}{\partial\tau_b}\frac{\partial}{\partial z_m}\Bigr)\Bigr\}\exp\Bigl(\frac{1}{2}z_m M_{mn} z_n\Bigr)\exp(-z_m^* z_m),$$

integrating over ξ, and using the property $M_{mn}\partial\phi_n^*/\partial\tau_b = 0$ we conclude that the functional integral is equal to zero.

Let us check now that integral (60) differs from zero. The commutation relations

$$[\hat{a}_\alpha, \hat{a}_\beta] = [\hat{a}_\alpha^+, \hat{a}_\beta^+] = \left[\hat{a}_\alpha, \frac{\partial\phi_m}{\partial\tau_b}\hat{\psi}_m^+ - \frac{\partial\phi_m^*}{\partial\tau_b}\hat{\psi}_m^-\right],$$

$$\left[\hat{a}_\alpha^+, \frac{\partial\phi_m}{\partial\tau_b}\hat{\psi}_m^+ - \frac{\partial\phi_m^*}{\partial\tau_b}\hat{\psi}_m^-\right] = 0, \qquad [\hat{a}_\alpha, \hat{a}_\beta^+] = \delta_{\alpha,\beta}$$

make this integral equal to

$$(\nu_{k+1})!(\nu_{k+2})!\cdots \int d^k\xi \Big(\widehat{\Phi}_0, \exp\Big(\frac{1}{2}\hat{\psi}_m^- M_{mn}^* \hat{\psi}_n^-\Big) \exp\Big(\xi_b \frac{\partial\phi_m}{\partial\tau_b}\hat{\psi}_m^+ - \frac{\partial\phi_m^*}{\partial\tau_b}\hat{\psi}_m^-\Big)\Big) \times \exp\Big(\frac{1}{2}\hat{\psi}_m^+ M_{mn}^* \hat{\psi}_n^+\Big)\widehat{\Phi}_0 .$$

It is easy to check that the last expression differs from zero. The corollary is proved.

Now we can prove Theorem 2. Let us check that

$$\Big(i\frac{\partial}{\partial t} - \frac{1}{\varepsilon}H(\sqrt{\varepsilon}\hat{\psi}^+, \sqrt{\varepsilon}\hat{\psi}^-)\Big)\widehat{\tilde{\Psi}}^\varepsilon(t) \underset{\varepsilon\to 0}{\longrightarrow} 0 . \tag{62}$$

The vector $\widehat{\tilde{\Psi}}^\varepsilon(t)$ may be presented as

$$\widehat{\tilde{\Psi}}^\varepsilon(t) = \int \frac{d\tau\chi(\tau,t)}{\varepsilon^{k/4}} \exp\Big(\frac{i}{\varepsilon}S(\tau,t) - \frac{1}{\varepsilon}\phi_j^*(\tau,t)\phi_j(\tau,t)\Big) \times \exp\Big(\frac{1}{\sqrt{\varepsilon}}\phi_j(\tau,t)\hat{\psi}_j^+\Big) \exp\Big(-\frac{1}{\sqrt{\varepsilon}}\phi_j^*(\tau,t)\hat{\psi}_j^-\Big) \times (\hat{a}_{k+1}^+)^{\nu_{k+1}}(\hat{a}_{k+2}^+)^{\nu_{k+2}}\cdots\widehat{\Phi}_{0,M(\tau,t)}, \qquad j = 1,2,\dots,$$

$$\chi(\tau,t) = \frac{f(\tau,t)}{(2\pi)^{k/2}} \frac{\sqrt{\det\frac{\partial\phi_j}{\partial\tau_a}(\tau,t)\,\frac{\partial\phi_j^*}{\partial\tau_b}(\tau,t)}}{\sqrt[4]{\det G^+(\tau,t)G(\tau,t)}}, \quad a,b = 1,\dots,k,\ j = 1,2,\dots . \tag{63}$$

Since

$$i\frac{\partial}{\partial t}\Big(\exp\Big(-\frac{1}{\varepsilon}\phi_j^*\phi_j\Big)\exp\Big(\frac{1}{\sqrt{\varepsilon}}\phi_j\hat{\psi}_j^+\Big)\exp\Big(-\frac{1}{\sqrt{\varepsilon}}\phi_j^*\hat{\psi}_j^-\Big)\Big) = \exp\Big(-\frac{1}{\varepsilon}\phi_j^*\phi_j\Big)\exp\Big(\frac{1}{\sqrt{\varepsilon}}\phi_j\hat{\psi}_j^+\Big)\exp\Big(-\frac{1}{\sqrt{\varepsilon}}\phi_j^*\hat{\psi}_j^-\Big) \times \Big(i\frac{\partial}{\partial t} - \frac{i}{\varepsilon}\dot{\phi}_j^*\phi_j + \frac{i}{\sqrt{\varepsilon}}(\dot{\phi}_j\hat{\psi}_j^+ - \dot{\phi}_j^*\hat{\psi}_j^-)\Big),$$

where $j = 1, 2, \ldots$, arguments t, τ of the functions ϕ_j^*, ϕ_j are omitted, we obtain

(64)
$$\begin{aligned}
&\Big(i\,\frac{\partial}{\partial t} - \frac{1}{\varepsilon}\,H(\sqrt{\varepsilon}\,\hat{\psi}^+, \sqrt{\varepsilon}\,\hat{\psi}^-)\Big)\widehat{\widetilde{\Psi}}^\varepsilon(t)\\
&\quad = \int \frac{d\tau}{\varepsilon^{k/4}} \exp\Big(\frac{i}{\varepsilon}\,S(\tau,t) - \frac{1}{2\varepsilon}\,\phi_j^*(\tau,t)\,\phi_j(\tau,t) + \frac{1}{\sqrt{\varepsilon}}\,(\phi_j(\tau,t)\,\hat{\psi}_j^+ - \phi_j^*(\tau,t)\,\hat{\psi}_j^-)\Big)\\
&\quad \times \Big(i\,\frac{\partial}{\partial t} + \frac{i}{\sqrt{\varepsilon}}\,(\dot{\phi}_j(\tau,t)\,\hat{\psi}_j^+ - \dot{\phi}_j^*(\tau,t)\,\hat{\psi}_j^-) - \frac{1}{\varepsilon}\,\frac{\partial S}{\partial t}\\
&\qquad - \frac{i}{\varepsilon} - \dot{\phi}_j^*(\tau,t)\,\phi_j(\tau,t) - \frac{1}{\varepsilon}\,H(\phi^* + \sqrt{\varepsilon}\,\hat{\psi}^+, \phi + \sqrt{\varepsilon}\,\hat{\psi}^-)\Big)\\
&\quad \times \chi(\tau,t)(\hat{a}_{k+1}^+)^{\nu_{k+1}}(\hat{a}_{k+2}^+)^{\nu_{k+2}} \cdots \widehat{\Phi}_{0,M(\tau,t)}, \qquad j = 1, 2, \ldots .
\end{aligned}$$

The conditions of the theorem imply that the norm of the vector

$$H(\phi^*(\tau,t) + \sqrt{\varepsilon}\,\hat{\psi}^+, \phi(\tau,t) + \sqrt{\varepsilon}\,\hat{\psi}^-)(\hat{a}_{k+1}^+)^{\nu_{k+1}}(\hat{a}_{k+2}^+)^{\nu_{k+2}} \cdots \widehat{\Phi}_{0,M(\tau,t)}$$

is uniformly bounded. Since the operator $H(\phi^* + \sqrt{\varepsilon}\,\hat{\psi}^+, \phi + \sqrt{\varepsilon}\,\hat{\psi}^-)$ is a polynomial in $\sqrt{\varepsilon}$ of the form

$$\begin{aligned}
&H(\phi^* + \sqrt{\varepsilon}\,\hat{\psi}^+, \phi + \sqrt{\varepsilon}\,\hat{\psi}^-)\\
&\quad = \sum_{n=0}^{s}\sum_{m=0}^{s} \varepsilon^{(m+n)/2} \frac{\partial^{m+n} H}{\partial\phi_{i_1}^* \cdots \partial\phi_{i_m}^*\,\partial\phi_{j_1} \cdots \partial\phi_{j_n}}\,\hat{\psi}_{i_1}^+ \cdots \hat{\psi}_{i_m}^+\,\hat{\psi}_{j_1}^- \cdots \hat{\psi}_{j_n}^-,\\
&\qquad\qquad i_1, \ldots, i_m, j_1, \ldots, j_n = 1, 2, \ldots,
\end{aligned}$$

the norms of the vectors

$$\frac{\partial^{m+n} H}{\partial\phi_{i_1}^* \cdots \partial\phi_{i_m}^*\,\partial\phi_{j_1} \cdots \partial\phi_{j_n}}\,\hat{\psi}_{i_1}^+ \cdots \hat{\psi}_{i_m}^+\,\hat{\psi}_{j_1}^- \cdots \hat{\psi}_{j_n}^-\,(\hat{a}_{k+1}^+)^{\nu_{k+1}}(\hat{a}_{k+2}^+)^{\nu_{k+2}} \cdots \widehat{\Phi}_{0,M(\tau,t)}$$

are also uniformly bounded. By Lemma 5, the right-hand side of formula (64) may be presented as

$$\begin{aligned}
&\int \frac{d\tau}{\varepsilon^{k/4}} \exp\Big(\frac{i}{\varepsilon}\,S(\tau,t) - \frac{1}{2\varepsilon}\,\phi_j^*(\tau,t)\,\phi_j(\tau,t) + \frac{1}{\sqrt{\varepsilon}}\,(\phi_j(\tau,t)\,\hat{\psi}_j^+ - \phi_j^*(\tau,t)\,\hat{\psi}_j^-)\Big)\\
&\quad \times \Big(i\,\frac{\partial}{\partial t} - \frac{1}{2}\,\frac{\partial^2 H}{\partial\phi_m^*\partial\phi_n^*}\,\hat{\psi}_m^+\hat{\psi}_n^+ - \frac{1}{2}\,\frac{\partial^2 H}{\partial\phi_m\partial\phi_n}\,\hat{\psi}_m^-\hat{\psi}_n^- - \frac{\partial^2 H}{\partial\phi_m^*\partial\phi_n}\,\hat{\psi}_m^+\hat{\psi}_n^-\chi(\tau,t)\Big)\\
&\quad \times (\hat{a}_{k+1}^+)^{\nu_{k+1}}(\hat{a}_{k+2}^+)^{\nu_{k+2}} \cdots \widehat{\Phi}_{0,M(\tau,t)} + O(\varepsilon^{1/2}), \qquad j, m, n = 1, 2, \ldots,
\end{aligned}$$

where we used the relations

$$\frac{\partial S}{\partial t} = -i\phi_l\dot{\phi}_l^* - H(\phi^*, \phi)$$

from the definition of S, and

$$i\dot{\phi}_l = \frac{\partial H}{\partial\phi_l^*}, \qquad -i\dot{\phi}_l^* = \frac{\partial H}{\partial\phi_l}, \qquad l = 1, 2, \ldots,$$

and also used equations (42).

It is easy to check that

$$(65)\quad \left[i\frac{\partial}{\partial t}-\widehat{H}_2,\hat{a}^+_\alpha\right]=\left[i\frac{\partial}{\partial t}-\widehat{H}_2,\hat{a}_\alpha\right]=\left[i\frac{\partial}{\partial t}-\widehat{H}_2,\frac{\partial\phi_j}{\partial\tau_a}\hat{\psi}^+_j-\frac{\partial\phi^*_j}{\partial\tau_a}\hat{\psi}^-_j\right]=0,$$

$$\alpha=k+1,k+2,\dots,\qquad a=1,\dots,k,\quad j=1,2,\dots,$$

$$\widehat{H}_2=\frac{1}{2}\frac{\partial^2 H}{\partial\phi^*_m\partial\phi^*_n}\hat{\psi}^+_m\hat{\psi}^+_n+\frac{1}{2}\frac{\partial^2 H}{\partial\phi_m\partial\phi_n}\hat{\psi}^-_m\hat{\psi}^-_n+\frac{\partial^2 H}{\partial\phi^*_m\partial\phi_n}\hat{\psi}^+_m\hat{\psi}^-_n,\qquad m,n=1,2,\dots.$$

Let us use Lemma 5, the commutation relations (65), and the functional integral presentation for the inner product ([**2**]). It is then easy to check that if for any z we have

(66)

$$\left(i\frac{\partial}{\partial t}-\frac{1}{2}\frac{\partial^2 H}{\partial\phi^*_m\partial\phi^*_n}z_m z_n-\frac{1}{2}\frac{\partial^2 H}{\partial\phi_m\partial\phi_n}\frac{\partial}{\partial z_m}\frac{\partial}{\partial z_n}-\frac{\partial^2 H}{\partial\phi^*_m\partial\phi_n}z_m\frac{\partial}{\partial z_n}\right)$$
$$\times\int d^k\xi\exp\left(-\frac{1}{2}\xi_b\xi_c\frac{\partial\phi_j}{\partial\tau_b}\frac{\partial\phi^*_j}{\partial\tau_c}\right)\exp\left(\xi_b\frac{\partial\phi_j}{\partial\tau_b}z_j\right)\exp\left(-\xi_b\frac{\partial\phi^*_j}{\partial\tau_b}\frac{\partial}{\partial z_j}\right)$$
$$\times\chi\exp\left(\frac{1}{2}z_m M_{mn}z_n\right)=0,\qquad m,n,j=1,2,\dots,\ b,c=1,\dots,k$$

(where the arguments τ, t of the functions M, χ, ϕ^*, ϕ and the arguments $\phi^*(\tau,t)$, $\phi(\tau,t)$ of the derivatives of H are omitted), then the norm of the vector

$$\left(i\frac{\partial}{\partial t}-\frac{1}{\varepsilon}H(\sqrt{\varepsilon}\hat{\psi}^+,\sqrt{\varepsilon}\hat{\psi}^-)\right)\widehat{\widetilde{\Psi}}^\varepsilon(t)$$

tends to zero as $\varepsilon\to 0$.

Formula (66) can be verified by calculating the Gaussian integral at ξ, using the presentations for χ, M, F, G. Therefore, property (62) is proved. Now let us estimate $\hat{\delta}^\varepsilon(t)$. Denote

$$\widehat{H}=\frac{1}{\varepsilon}H(\sqrt{\varepsilon}\hat{\psi}^+,\sqrt{\varepsilon}\hat{\psi}^-),\quad \hat{\kappa}^\varepsilon(t)=\widehat{\widetilde{\Psi}}^\varepsilon(t)-\widehat{\Phi}(t),\quad \hat{s}^\varepsilon(t)=\left(i\frac{\partial}{\partial t}-\widehat{H}\right)\widehat{\widetilde{\Psi}}^\varepsilon(t).$$

The function

$$\hat{\kappa}^\varepsilon(t)=\int_0^t dt'\exp(-i\widehat{H}(t-t'))\,\hat{s}^\varepsilon(t')$$

is the solution of the Cauchy problem

$$i\frac{\partial}{\partial t}\hat{\kappa}^\varepsilon(t)-\widehat{H}\hat{\kappa}^\varepsilon(t)=\hat{s}^\varepsilon(t),\qquad \hat{\kappa}^\varepsilon(0)=0.$$

Hence

$$\|\hat{\kappa}^\varepsilon(t)\|\leqslant\int_0^t\|\hat{s}^\varepsilon(t')\|\,dt'\xrightarrow[\varepsilon\to 0]{}0,$$

which implies

$$\|\hat{\delta}^\varepsilon(t)\|\xrightarrow[\varepsilon\to 0]{}0.$$

Theorem 2 is proved.

References

1. V. V. Belov and S. Yu. Dobrokhotov, Teoret. Mat. Fiz.; English transl., Theoret. and Math. Phys. **92** (1992), no. 2, 215–254.
2. F. A. Berezin, *Method of secondary quantization*, "Nauka", Moscow, 1986. (Russian)
3. N. N. Bogoliubov and D. V. Shirkov, *Introduction to the theory of quantized fields*, Interscience Publishers, N. Y., 1959.
4. P. A. M. Dirac, Proc. Roy. Soc. A **144** (1927), 234–262.
5. V. A. Fock, *Konfigurationsraum und zweite Quantelung*, Z. Phys. **75** (1932), 622–647.
6. ———, Soviet Phys. **6** (1934), 425.
7. J. Glimm and A. Jaffe, *Quantum Physics. A functional integral point of view*, N. Y., 1981.
8. V. Guillemin and S. Sternberg, *Geometric asymptotics*, Providence, 1977.
9. P. Jordan and E. Wigner, Zh. Phys. **47** (1928), 631–658.
10. M. V. Karasev and V. P. Maslov, *Nonlinear Poisson brackets. Geometry and quantization*, "Nauka", Moscow, 1991; English transl., Amer. Math. Soc., Providence, RI, 1993.
11. A. D. Krakhnov, Uspekhi Mat. Nauk **31** (1976), 217–218. (Russian)
12. L. D. Landau and E. M. Livshitz, *Quantum mechanics. Nonrelativistic theory*, Fourth edition, "Nauka", Moscow, 1989; English transl. of the first edition, Pergamon, London-Paris, 1958.
13. V. P. Maslov, *Operational methods*, "Nauka", Moscow, 1973; English transl., "Mir", Moscow, 1976.
14. ———, *Complex WKB method in nonlinear equations*, "Nauka", Moscow, 1977; Partial English transl., Birkhäuser, Basel, 1994.
15. V. P. Maslov and O. Yu. Shvedov, *Quantization in a neighborhood of classical solutions in the N-particle problem, and superfluidity*, Teoret. Mat. Fiz. **98** (1994), 266–288; English transl., Theoret. and Math. Phys. **98** (1994).
16. ———, *An asymptotic formula for the N-particle density function as $N \to \infty$ and a violation of the chaos hypothesis*, Russian J. Math. Phys. **2** (1994), 217–234.
17. ———, *The problem of chaos conservation in many-particle systems*, Dokl. Akad. Nauk **338** (1994), 15–18; English transl. in Phys. Dokl. **39** (1994).
18. V. P. Maslov and S. E. Tariverdiev, *The asymptotic behavior of the Kolmogorov–Feller equation for a system of a large number of particles*, Probability theory, mathematical statistics, theoretical cybernetics, vol. 19, VINITI, Moscow, 1982, pp. 85–125; English transl. in J. Soviet Math. **23** (1983), no. 5.
19. M. Schönberg, Nuovo Cimento **9** (1952), 1139.
20. ———, Nuovo Cimento **10** (1953), 419.
21. S. Schweber, *An introduction to relativistic quantum field theory*, Elmsford, New York, 1961.
22. B. Simon, *The $P(\phi)_2$ Euclidean (quantum) field theory*, Princeton Univ. Press, Princeton, NJ, 1974.

Translated by S. LANDO

Amer. Math. Soc. Transl.
(2) Vol. **177**, 1996

The Three-Body Problem in Radioactive Decay: The Case of One Atom and At Most Two Photons

R. Minlos and H. Spohn

ABSTRACT. We show that the spin-boson Hamiltonian with the number of photons restricted to either 0, 1, or 2 is an analog of the standard 3-body problem. We then use techniques of stationary scattering theory to prove the existence and completeness of wave operators for a sufficiently small coupling α. In particular, we prove that the Hamiltonian has a unique ground state separated by a gap from the purely absolutely continuous part of the spectrum.

§1. Introduction

Atoms decay to the ground state through the emission of electromagnetic radiation. To have a simple theoretical model, one takes into account the ground state and the excited state of the (bare) atom only and describes the coupling to the radiation field in the dipole approximation. Polarization then plays no particular role and one arrives at the so-called spin-boson Hamiltonian

$$H = \varepsilon\sigma_z + \int_{\mathbb{R}^\nu} \omega(k)\, a^*(k)\, a(k)\, d^\nu k + \alpha\sigma_x \int_{\mathbb{R}^\nu} \lambda(k)(a^*(k) + a(k))\, d^\nu k. \tag{1.1}$$

Here σ_z, σ_x are the Pauli spin matrices, $\varepsilon\sigma_z$ is the atomic part with energy levels $\pm\varepsilon$ ($\varepsilon > 0$), $\{a^*(k), a(k)\}$ is a Bose field with canonical commutation relations $[a(k), a^*(k)] = \delta(k - k')E$ (E is the identity operator), $\omega(k)$ is the dispersion of the free field; $\omega(k) = |k|$ for the photon case, but we shall allow more general dispersions as will be stated below; $\alpha\lambda(k)$ is the coupling between the atoms and the field modes. We think of the real function λ as being fixed. Then α is the coupling constant. Usually $\lambda(k) = (\omega(k))^{1/2}$ with some cut-off for large k. However, λ may be modified by atomic form factors. Thus it is again convenient to impose only some general conditions on $\lambda(k)$. The operator H acts in the Hilbert space $\mathbb{C}^2 \otimes \mathcal{F}_s$, where $\mathcal{F}_s$ is the bosonic Fock space. If

$$\int_{\mathbb{R}^\nu} \frac{\lambda^2(k)}{\omega(k)}\, d^\nu k < \infty,$$

1991 *Mathematics Subject Classification*. Primary 81V10; Secondary 81U10.

then H is bounded from below and essentially self-adjoint on the natural domain for H_0 (i.e., H with $\alpha = 0$). If

$$\int_{\mathbb{R}^\nu} \frac{\lambda^2(k)}{\omega(k)}\, d^\nu k < \infty,$$

then H has a unique ground state ([**1**]). Radioactive decay means that if initially the atom is excited, then after a long time the state of the system is the ground state for H plus one or several out-traveling photons. We could also imagine other initial states, e.g., a photon is incoming, scatters the atom, excites the atom, which then emits a few photons. Thus radioactive decay is only one part of the time asymptotics of $e^{-iHt}\Psi$ as $t \to \infty$, for general initial Ψ. Outgoing photons no longer interact with the atom. Thus we are in the framework of scattering theory. Compared to the standard N-particle Schrödinger problem, a simplifying feature is that photons interact only indirectly through the atom. On the other hand, the general dispersion $\omega(k)$ and the unbounded number N of photons cause some extra difficulties.

If in fact we restrict the number of photons to $N \leqslant 1$, then we have the analog of the 2-body problem. In particular, the interaction is of trace class. Our main achievement is the complete handling of the analog of the 3-body problem, namely the case $N \leqslant 2$, at least for small coupling. We shall use stationary methods in the spirit of Faddeev. Time-dependent methods were discussed in [**2**].

In the next section we formulate our main results and explain how the rest of the paper is organized.

§2. The Hamiltonian for $N \leqslant 2$ and statement of the main results

The Hilbert space of our system is

$$\mathcal{H} = \mathbb{C}^2 \otimes \{\mathbb{C} \oplus L_2(\mathbb{R}^\nu) \oplus L_2^{\mathrm{sym}}(\mathbb{R}^\nu \times \mathbb{R}^\nu)\},$$

where $L_2^{\mathrm{sym}}(\mathbb{R}^\nu \times \mathbb{R}^\nu)$ is the space of symmetric (square integrable) functions with inner product

$$(f, g) = \frac{1}{2} \int_{\mathbb{R}^\nu \times \mathbb{R}^\nu} f(k_1, k_2)\overline{g(k_1, k_2)}\, d^\nu k_1\, d^\nu k_2 .$$

The Hamiltonian of the system (for the case $N \leqslant 2$) is

$$\begin{aligned}
(HG)_0^{(\sigma)} &= \varepsilon\sigma G_0^{(\sigma)} + \alpha \int \lambda(q)\, G_1^{(-\sigma)}(q)\, d^\nu q,\\
(HG)_1^{(\sigma)}(k) &= (\varepsilon\sigma + \omega(k))\, G_1^{(\sigma)}(k) + \alpha\lambda(k)\, G_0^{(-\sigma)}\\
&\quad + \alpha \int_{\mathbb{R}^\nu} G_2^{(-\sigma)}(q, k)\, \lambda(q)\, d^\nu q, \qquad \sigma = \pm, \qquad (2.1)\\
(HG)_2^{(\sigma)}(k_1, k_2) &= (\omega(k_1) + \omega(k_2) + \sigma\varepsilon)\, G_2^{(\sigma)}(k_1, k_2)\\
&\quad + \alpha\lambda(k_1)\, G_1^{(-\sigma)}(k_2) + \alpha\lambda(k_2)\, G_1^{(-\sigma)}(k_1),
\end{aligned}$$

where

$$G = \{G_0^{(\sigma)}, G_1^{(\sigma)}(k), G_2^{(\sigma)}(k_1, k_2)\}.$$

Throughout the paper we make the following assumptions:

A.i) $\varepsilon > 0$.

A.ii) The dispersion ω is radial:
$\omega(k) = \overline{\omega}(|k|)$, $\overline{\omega}(0) = 0$, $\overline{\omega}(k) > 0$ for $k > 0$, $\overline{\omega}(k) \cong k$ close to $k = 0$, $\overline{\omega}(k)$ is strictly increasing: $\overline{\omega}'(k) \geqslant a_0 > 0$, and $\overline{\omega}(k)$ is smooth away from $k = 0$.

A.iii) $\lambda(k)$, $k \in \mathbb{R}^\nu$ is a smooth real function, rapidly decreasing as $|k| \to \infty$, and $\lambda(k) \cong |k|^n$ with $n > 0$, $2n + \nu \geqslant 4$. Furthermore, $\lambda(k)$ does not vanish in the annulus $\{k : 2\varepsilon - \delta \leqslant \omega(k) \leqslant 2\varepsilon + \delta\}$ with $\varepsilon > \delta > 0$.

A.iv) $0 < |\alpha| < \alpha_0$ with α_0 depending on ε, ω, and λ.

We note that the interaction cannot be of trace class. If one photon is close to the atom, then the interaction does not decay as the other photon is removed arbitrarily far away from the atom. This structure is similar to the 3-body problem.

Next we introduce the channel Hamiltonians. Here we have three channels of scattering:

I) one photon is outgoing and the other is bound to the atom;

II and III) both photons travel outwards, the atom is then uncoupled and has two eigenstates with eigenvalues $-\varepsilon$ (channel II) and $+\varepsilon$ (channel III).

There are transitions between channels I and II and there are no transitions between I and III or between II and III. This behavior is an artifact of our approximation. If N is unbounded, such a channel cannot exist. For channel I, the Hilbert space is $\widehat{\mathcal{H}}_1 = L_2(\mathbb{R}^\nu)$ and for channels II and III it is $\widehat{\mathcal{H}}_2 = \mathbb{C}^2 \otimes L_2^{\mathrm{sym}}(\mathbb{R}^\nu \times \mathbb{R}^\nu)$. We denote $\widehat{\mathcal{H}} = \widehat{\mathcal{H}}_1 \oplus \widehat{\mathcal{H}}_2$. The channel Hamiltonian $\widehat{H}$ on $\widehat{\mathcal{H}}_1$ is

$$(\widehat{H}f)_1(k) \equiv (\widehat{H}_1 f_1)(k) = (\omega(k) + E_0)\, f_1(k)\,. \tag{2.2}$$

Here E_0 is the ground state energy for the case $N \leqslant 1$ with Hamiltonian

$$\begin{aligned} (hg)_0^{(\sigma)} &= \varepsilon\sigma g_0^{(\sigma)} + \alpha \int \lambda(k)\, g_1^{(-\sigma)}(k)\, d^\nu k, \\ (hg)_1^{(\sigma)}(k) &= (\varepsilon\sigma + \omega(k))\, g_1^{(\sigma)}(k) + \alpha\lambda(k)\, g_0^{(-\sigma)} \end{aligned} \tag{2.3}$$

on $\mathbb{C}^2 \otimes \{\mathbb{C} \oplus L_2(\mathbb{R}^\nu)\}$. The Hamiltonian h has features of the Hamiltonian for the 2-body problem; its properties play an important role and will be discussed in §3. The channel Hamiltonian on $\widehat{\mathcal{H}}_2$ is

$$(\widehat{H}f)_2^{(\sigma)}(k_1,k_2) \equiv (\widehat{H}_2 f_2)^{(\sigma)}(k_1,k_2) = (\omega(k_1) + \omega(k_2) + \varepsilon\sigma)\, f_2^{(\sigma)}(k_1,k_2)\,. \tag{2.4}$$

We first state the spectral properties of H.

THEOREM 1. (i) *The spectrum of H equals $\{E_g\} \cup [E_0, \infty)$ with $E_g < E_0$. The eigenvalue E_g is nondegenerate and $\Psi_g \in \mathcal{H}$ is the corresponding eigenvector.*

(ii) *If $\mathcal{H}_{\mathrm{ac}}$ is the subspace of $\mathcal{H}$ orthogonal to Ψ_g, then $H|_{\mathcal{H}_{\mathrm{ac}}}$ is unitarily equivalent to $\widehat{H}$ on $\widehat{\mathcal{H}}$. In particular, H has no singular continuous spectrum.*

REMARK. By a very different technique this theorem is proved for arbitrary N in [**2**].

The large time asymptotics will be given in the two Hilbert space versions of scattering theory ([**3**]). Thus we must first define an embedding of $\widehat{\mathcal{H}}$ into $\mathcal{H}$. The obvious embedding is given by Ψ the ground state for h in (2.3)

$$J\colon \widehat{\mathcal{H}} \to \mathcal{H} \quad \text{with } J|_{\widehat{\mathcal{H}}_i} = J_i, \quad i = 1,2,$$

and for $F = (f_1, f_2) \in \widehat{\mathcal{H}}$

$$(2.5)\qquad \begin{gathered} (J_1 f_1)_0^{(\sigma)} = 0, \quad (J_1 f_1)_1^{(\sigma)}(k) = \Psi_0^{(\sigma)} f_1(k), \\ (J_1 f_1)_2^{(\sigma)}(k_1, k_2) = \Psi_1^{(\sigma)}(k_1) f_1(k_2) + \Psi_1^{(\sigma)}(k_2) f_1(k_1), \end{gathered}$$

$$(2.6)\qquad (J_2 f_2)_0^{(\sigma)} = 0, \quad (J_2 f_2)_1^{\sigma}(k) = 0, \quad (J_2 f_2)_2^{\sigma}(k_1, k_2) = f_2^{(\sigma)}(k_1, k_2).$$

It turns out to be more convenient to use a modified embedding, first introduced by Yafaev in [4] in the context of the 3-body problem. The Yafaev embedding is defined by

$$(2.7)\qquad \bar{J}_1 = J_1, \quad \bar{J}_2 = J_2 - J_1 J_1^* J_2$$

with $J_1^* : \mathcal{H} \to \widehat{\mathcal{H}}_1$ the adjoint operator to J_1. Explicitly we obtain

$$(2.8)\qquad \begin{aligned} (\bar{J}_2 f_2)_0^{(\sigma)} &= 0, \\ (J_2 f_2)_1^{(\sigma)}(k) &= -\Psi_0^{(\sigma)} \int_{\mathbb{R}^\nu} \overline{\Psi}_1^{(\sigma)}(q)\, f_2^{(\sigma)}(q, k)\, d^\nu q, \\ (J_2 f_2)_2^{(\sigma)}(k_1, k_2) &= f_2^{(\sigma)}(k_1, k_2) - \Psi_1^{(\sigma)}(k_1) \int_{\mathbb{R}^\nu} \overline{\Psi}_1^{(\sigma)}(q)\, f_2^{(\sigma)}(q, k_2)\, d^\nu q \\ &\quad - \Psi_1^{(\sigma)}(k_2) \int_{\mathbb{R}^\nu} \overline{\Psi}_1^{(\sigma)}(q)\, f_2^{(\sigma)}(q, k_1)\, d^\nu q. \end{aligned}$$

We introduce the wave operators

$$(2.9)\qquad W_- = \operatorname*{s-lim}_{t\to\infty} e^{iHt} J e^{-i\widehat{H}t}, \qquad \overline{W}_- = \operatorname*{s-lim}_{t\to\infty} e^{iHt} \bar{J} e^{-i\widehat{H}t}$$

and the inverse wave operator

$$(2.10)\qquad \overline{W}_+ = \operatorname*{s-lim}_{t\to\infty} e^{i\widehat{H}t} \bar{J}^* e^{-iHt} P_{\mathrm{ac}},$$

where P_{ac} is the projection on the space $\mathcal{H}_{\mathrm{ac}} \subset \mathcal{H}$. We state the existence and completeness of the wave operators W_-, $\overline{W}_-$, $\overline{W}_+$.

THEOREM 2. *The wave operators W_-, $\overline{W}_-$ exist and are equal:*

$$(2.11)\qquad W_- = \overline{W}_-.$$

Moreover, $\overline{W}_-$ is an isometry from $\widehat{\mathcal{H}}$ to $\mathcal{H}$.

COROLLARY. *Let $\mathcal{H}_{\mathrm{in}} = \operatorname{Ran} \overline{W}_-$. The space $\mathcal{H}_{\mathrm{in}}$ is an invariant subspace for H and $H|_{\mathcal{H}_{\mathrm{in}}}$ is unitarily equivalent to $\widehat{H}$.*

THEOREM 3. *The limit (2.10) exists, we have*

$$(2.12)\qquad \overline{W}_+ = (\overline{W}_-)^*,$$

and $\overline{W}_+$ is an isometry from $\mathcal{H}_{\mathrm{ac}}$ to $\widehat{\mathcal{H}}$.

COROLLARY. *We have $\mathcal{H}_{\mathrm{in}} = \mathcal{H}_{\mathrm{ac}}$, $\operatorname{Ran} \overline{W}_+ = \widehat{\mathcal{H}}$, and*

$$(2.13)\qquad \overline{W}_+ = (\overline{W}_-)^{-1}.$$

REMARK. The intertwining relation

$$HP_{\mathrm{ac}}\overline{W}_- = \overline{W}_-\widehat{H} \tag{2.14}$$

implies the second assertion of Theorem 1.

The remainder of the paper is organized as follows. In §3 we discuss the case $N \leqslant 1$ (2-body problem). In §4 we establish Theorem 2 using Cook's method. In §5 we investigate the resolvent $R_H(z)$ of operator H.

Finally in §6 we prove Theorem 3 by using the previous analysis of the resolvent $R_H(z)$.

§3. The one-boson system

The Hamiltonian h for such a system is defined by formula (2.3) on the space of pairs

$$g = \{g_0^{(\sigma)}, g_1^{(\sigma)}(k)\}. \tag{3.1}$$

LEMMA 1. *The operator h has a unique eigenvector Ψ (which is the ground state) with eigenvalue $E_0 < -\varepsilon$. The (normalized) eigenvector Ψ has the form*

$$\begin{gathered} \Psi_0^{(+)} = 0, \qquad \Psi_0^{(-)} = 1/s, \\ \Psi_1^{(+)}(k) = \frac{-\alpha\lambda(k)}{(\omega(k)+\varepsilon-E_0)\,s}, \qquad \Psi_1^{(-)}(k) = 0, \end{gathered} \tag{3.2}$$

where

$$s = \left(1 + \alpha^2 \int_{\mathbb{R}^\nu} \frac{\lambda^2(k)\,d^\nu k}{(\omega(k)+\varepsilon-E_0)^2}\right)^{1/2}.$$

PROOF. The equations for the eigenvalue E_0 and the eigenvector Ψ are of the form

$$\begin{gathered} (\varepsilon\sigma - E)\,\Psi_0^{(\sigma)} + \alpha \int_{\mathbb{R}^\nu} \lambda(q)\,\Psi_1^{(-\sigma)}(q)\,d^\nu q = 0, \qquad \sigma = \pm, \\ (\varepsilon\sigma + \omega(k) - E)\,\Psi_1^{(\sigma)}(k) + \alpha\lambda(q)\,\Psi_0^{(-\sigma)} = 0. \end{gathered}$$

After eliminating $\Psi_1^{(\sigma)}(k)$ from the second equations, we get two equations

$$\left(\sigma\varepsilon - E - \alpha^2 \int_{\mathbb{R}^\nu} \frac{\lambda^2(q)\,d^\nu q}{-\varepsilon\sigma + \omega(q) - E}\right)\Psi_0^{(\sigma)} = 0, \qquad \sigma = \pm.$$

Now we introduce two functions of z

$$m_\sigma(z) = \varepsilon\sigma - z - \alpha^2 \int_{\mathbb{R}^\nu} \frac{\lambda^2(q)\,d^\nu q}{-\varepsilon\sigma + \omega(q) - z}, \qquad \sigma = \pm. \tag{3.3}$$

Denote by Π_σ the complex plane with the cut $[-\sigma\varepsilon, \infty)$, together with both banks of the cut (see Figure 1).

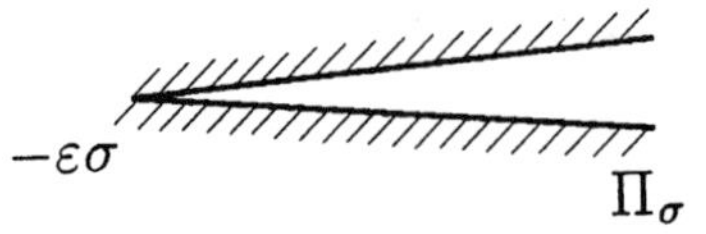

FIGURE 1

LEMMA 2. 1) *The function* $m_\sigma(z)$ *is defined and continuous on* Π_σ; *it is an analytic function outside the cut* $[-\varepsilon\sigma, \infty)$.

2) *The functions on the real axis*

$$m_\sigma^u(x) = m_\sigma(x + i0), \quad m_\sigma^d(x) = m_\sigma(x - i0), \qquad x \in \mathbb{R}^1 \tag{3.4}$$

(*coinciding outside the cut* Π_σ) *have* $2n + \nu - 3$ *continuous derivatives.*

3) *We have* $m_+(z) \neq 0$, $z \in \Pi_+$, *but the function* $m_-(z)$ *has a unique zero* E_0:

$$m_-(E_0) = 0, \qquad E_0 \text{ is real and } E_0 < -\varepsilon.$$

REMARK. Evidently E_0 is the unique eigenvalue of h and the form (3.2) of the eigenvector Ψ is found directly. Thus Lemma 1 follows from Lemma 2.

PROOF OF LEMMA 2. The integral

$$I_\sigma(z) \equiv \int_{\mathbb{R}^\nu} \frac{\lambda^2(q)\, d^\nu q}{-\sigma\varepsilon + \omega(q) - z}$$

is an analytic function of $z \in \mathbb{C} \setminus [-\varepsilon\sigma, \infty)$. Further, for $\operatorname{Im} z < 0$, we have

$$\begin{aligned} I_\sigma(z) &= i \int \lambda^2(q)\, d^\nu q \int_0^\infty e^{is(\omega(q) - \varepsilon\sigma - z)} ds \\ &= i \int_0^\infty e^{-is(z + \varepsilon\sigma)} ds \int_{\mathbb{R}^\nu} \lambda^2(q)\, e^{is\omega(q)} d^\nu q \\ &= i \int_0^\infty e^{-is(z + \varepsilon\sigma)} b(s)\, ds, \end{aligned} \tag{3.5}$$

where

$$b(s) = \int_{\mathbb{R}^\nu} \lambda^2(q)\, e^{i\omega(q)s} d^\nu q = \int_0^\infty h(r)\, e^{isr} dr; \tag{3.6}$$

here we denote

$$h(r) = \int_{\omega(q) = r} \lambda^2(q)\, d^\nu q = \frac{d}{dr} \int_{\omega(q) \leqslant r} \lambda^2(q)\, d^\nu q. \tag{3.7}$$

From our assumption about $\lambda(k)$ and $\omega(k)$, it follows that

$$|b(s)| < \frac{\text{const}}{|s|^k + 1}, \qquad k = 2n + \nu - 1. \tag{3.8}$$

Relations (3.8) and (3.5) imply that I_σ is continuous and bounded up to the lower bank of the cut Π_σ and the function $m_\sigma^d(x)$ has $2n + \nu - 3$ continuous derivatives. The case $\operatorname{Im} z > 0$ is considered in a similar way. Further, if $\operatorname{Im} z \neq 0$, we have $m_\sigma(z) \neq 0$, because a nonreal z cannot be an eigenvalue of a self-adjoint operator h.

Consider now the case of real $z = x$ and $\sigma = +$. Let $x < -\varepsilon$. Putting $x = -\varepsilon - y$, $y > 0$, we obtain

$$m_+(x) = \varepsilon - x - \alpha^2 \int_{\mathbb{R}^\nu} \frac{\lambda^2(q)\,d^\nu q}{-\varepsilon + \omega(q) - x} = 2\varepsilon + y - \alpha^2 \int \frac{\lambda^2(q)\,d^\nu q}{y + \omega(q)}.$$

The function

$$2\varepsilon + y - \alpha^2 \int_{\mathbb{R}^\nu} \frac{\lambda^2(q)\,d^\nu q}{y + \omega(q)}$$

for $y > 0$ is monotone increasing and positive for $y = 0$ (for small α). Hence $m_+(x) \neq 0$ by $x < -\varepsilon$. Consider the case of values $m_+(x \pm i0)$ on the banks of the cut $(-\varepsilon, \infty)$. Let $x = -\varepsilon + u$, $u > 0$, then

$$\begin{aligned} m_+(x - i0) &= 2\varepsilon - u - \alpha^2 \int_{\mathbb{R}^\nu} \frac{\lambda^2(q)\,d^\nu q}{\omega(q) - u - i0} \\ &= 2\varepsilon - u - \alpha^2 \Big(- i\pi h(u) + P \int_0^\infty \frac{h(r)}{r - u}\,dr \Big) \end{aligned} \tag{3.9}$$

(by Sokhotski's formula, see [**5**]). Now in the interval $|x - \varepsilon| < \delta$, i.e., $|u - 2\varepsilon| < \delta$, we have

$$m_+(x - i0) > \alpha^2 \pi \min_{2\varepsilon - \delta < u < 2\varepsilon + \delta} h(u) > \alpha^2 \text{const} > 0,$$

due to condition A.iii). Outside of this interval, we have

$$m_+(x - i0) > \delta + O(\alpha^2) > \delta/2$$

for small α. The case of the function $m_+(x + i0)$ is considered similarly.

REMARK. Condition A.iii) means that $h(r) > 0$ in the neighborhood of the value $r = 2\varepsilon$ is important, because if it is not satisfied, a real zero of $m_+(x \pm i0)$ may appear near $x = \varepsilon$. This means that the eigenvalue $E = \varepsilon_0$ of h_0 (by $\alpha = 0$) lying on the continuous spectrum of h_0 remains there for small $\alpha \neq 0$. This case is more complicated. Condition A.iii) provides a transition of the eigenvalue $E = \varepsilon$ to the resonance of h.

Now consider the case $\sigma = -$. Let $x < \varepsilon$. Again put $x = \varepsilon - y$, $y > 0$, obtaining

$$m_-(x) = -\varepsilon - x - \alpha^2 \int_{\mathbb{R}^\nu} \frac{\lambda^2(q)\,d^\nu q}{\varepsilon + \omega(q) - x} = -2\varepsilon + y - \alpha^2 \int_{\mathbb{R}^\nu} \frac{\lambda^2(q)\,d^\nu q}{y + \omega(q)}.$$

This function is monotone increasing for $y \geqslant 0$, negative at $y = 0$, and goes to $+\infty$ as $y \to \infty$. Hence there exists a unique zero y_0 of the equation

$$-2\varepsilon + y - \alpha^2 \int \frac{\lambda^2(q)\,d^\nu q}{y + \omega(q)} = 0.$$

Since the left-hand side of this equation is negative for $y = 2\varepsilon$, we have $y_0 > 2\varepsilon$ and $y_0 = 2\varepsilon + O(\alpha^2)$. Thus $E_0 = \varepsilon - y_0 < -\varepsilon$ and $E_0 = -\varepsilon + O(\alpha^2)$ is the unique zero of $m_+(x)$ for $x < -\varepsilon$. Now consider the values $m_-(x \pm i0)$ of the function m_- on the banks of the cut $[\varepsilon, \infty)$. Let $x = \varepsilon + u$, $u > 0$. Then

$$m_-(x \pm i0) = -2\varepsilon - u - \alpha^2 \int_{\mathbb{R}^\nu} \frac{\lambda^2(q)\,d^\nu q}{\omega(q) - u \mp i0}.$$

Therefore, $|m_-(x \pm i0)| > 2\varepsilon + O(\alpha^2)$ for small α. Lemma 2 is proved.

§4. Proof of Theorem 2 (Cook's method)

Let $W_t = e^{itH} J e^{-it\widehat{H}}$. Then

$$dW_t f/dt = ie^{itH}[HJ - J\widehat{H}]e^{-it\widehat{H}} f, \qquad f \in \widehat{\mathcal{H}}.$$

The following formulas hold:

a) for $f = f_1 \in \widehat{\mathcal{H}}_1$,

$$(4.1)\quad \begin{aligned} &(HJ - J\widehat{H})f = (HJ_1 - J_1\widehat{H}_1)f_1, \\ &((HJ_1 - J_1\widehat{H}_1)f_1)_0^{(\sigma)} = \begin{cases} \alpha\Psi_0^{(-)} \int_{\mathbb{R}^\nu} \lambda(k) f_1(k)\, d^\nu k, & \sigma = +, \\ 0, & \sigma = -, \end{cases} \\ &((HJ_1 - J_1\widehat{H}_1)f_1)_1^{(\sigma)}(k) = \begin{cases} 0, & \sigma = +, \\ \alpha\Psi_1^{(+)}(k) \int_{\mathbb{R}^\nu} \lambda(q) f_1(q)\, d^\nu q, & \sigma = -, \end{cases} \\ &((HJ_1 - J_1\widehat{H})f_1)_2^{(\sigma)}(k_1, k_2) = 0, \qquad \sigma = \pm; \end{aligned}$$

b) for $f = f_2 \in \widehat{\mathcal{H}}_2$

$$(4.2)\quad \begin{aligned} &(HJ - J\widehat{H})f = (HJ_2 - J_2\widehat{H}_2)f_2, \\ &((HJ_2 - J_2\widehat{H})f_2)_0^{(\sigma)} = 0, \\ &((HJ_2 - J_2\widehat{H})f_2)_1^{(\sigma)}(k) = \alpha \int \lambda(q) f_2^{(-\sigma)}(q, k)\, d^\nu q, \\ &((HJ_2 - J_2\widehat{H})f_2)_2^{(\sigma)}(k_1, k_2) = 0. \end{aligned}$$

Denote by $D_1 \subset L_2(\mathbb{R}^\nu)$ the set of smooth functions $f(k)$ rapidly decreasing at infinity such that

$$f(k) \cong |k|^n$$

near zero. Evidently D_1 is a dense set in $L_2(\mathbb{R}^\nu)$. Then for $f_1 \in D_1$

$$\begin{aligned} \left\| \frac{dW_t f_1}{dt} \right\|^2 &= \|(HJ_1 - J_1\widehat{H}_1)e^{-i\widehat{H}_1 t} f_1\|^2 \\ &= \alpha^2 |\Psi_0^{(-)}|^2 \left| \int_{\mathbb{R}^\nu} \lambda(q) e^{-i(\omega(q)+E_0)t} f_1(q)\, d^\nu q \right|^2 \\ &\quad + \alpha^2 \|\Psi_1^{(+)}\|^2 \left| \int \lambda(q) e^{-i(\omega(q)+E_0)t} f_1(q)\, d^\nu q \right|^2 \\ &< \frac{C(f_1)}{(|t|^k + 1)^2}, \qquad k = 2n + \nu - 1. \end{aligned}$$

Hence the existence of the first limit in (2.9) for $f_1 \in \widehat{\mathcal{H}}_1$ follows in the usual way. Now consider the set $D_2 \subset L_2^{\mathrm{sym}}(\mathbb{R}^\nu \times \mathbb{R}^\nu) \otimes \mathbb{C}^2$ of finite linear combinations

$$f_2^{(\sigma)}(k_1, k_2) = \sum_i C_i^{(\sigma)} f_i(k_1) f_i(k_2), \qquad \sigma = \pm,$$

where $f_i \in D_1$. Obviously D_2 is a dense set in $\widehat{\mathcal{H}}_2$. For $f_2 \in D_2$ we have

$$\left\|\frac{dW_t f}{dt}\right\| = \|(HJ_2 - J_2\widehat{H}_2)e^{-i\widehat{H}_2 t} f_2\| \leqslant \sum_{\sigma,i} |c_i^{(\sigma)}| \, \|f_i\| \left| \int \lambda(q)\, e^{-i\omega(q)t} f_i(q)\, d^\nu q \right| < \frac{c(f_2)}{|t|^k + 1}.$$

Hence the existence of the first limit in (2.9) for $f_2 \in \widehat{\mathcal{H}}_2$ follows. Now we prove the existence of the second limit in (2.9). We have

$$W_t - \overline{W}_t = e^{iHt}(J - \bar{J})e^{-i\widehat{H}t} \quad \text{and} \quad \|(W_t - \overline{W}_t)f\| = \|(J - \bar{J})e^{-i\widehat{H}t} f\|.$$

If $f = f_1 \in \widehat{\mathcal{H}}_1$, then $(J - \bar{J})e^{-i\widehat{H}t} f_1 = 0$ and

$$\lim_{t\to\infty} \overline{W}_t f_1 = \lim_{t\to\infty} W_t f_1 = W_- f_1. \tag{4.3}$$

For $f = f_2 \in \widehat{\mathcal{H}}_2$ we get $(J - \bar{J}) f_2 = J_1 J_1^* J_2 f_2$. Relation (2.8) implies

$$\begin{aligned}
&\|(J - \bar{J})e^{-i\widehat{H}_2 t} f_2\|^2 \\
&\leqslant |\Psi_0^{(-)}|^2 \left\| \int_{\mathbb{R}^\nu} \overline{\Psi}_1^{(+)}(q)\, e^{-i(\varepsilon+\omega(k)+\omega(q))t} f_2^{(+)}(q,k)\, d^\nu q \right\|^2_{L_2(\mathbb{R}^\nu)} \\
&\quad + 2\|\Psi_1^{(+)}\|^2 \left\| \int_{\mathbb{R}^\nu} \overline{\Psi}_1^{(+)}(q)\, e^{-i(\varepsilon+\omega(k)+\omega(q))} f_2^{+}(q,k)\, d^\nu q \right\|^2_{L_2(\mathbb{R}^\nu)}.
\end{aligned}$$

For $f \in D_2$, from the formula (3.2) for $\Psi_1^{(+)}(q)$, we find that

$$\left\| \int_{\mathbb{R}^\nu} \Psi_1^{(+)}(q)\, e^{i(\varepsilon+\omega(q)+\omega(k))t} f_2^{(+)}(q,k)\, d^\nu k \right\|_{L_2(\mathbb{R}^\nu)} \to 0$$

as $t \to \infty$. Hence (4.3) implies

$$\overline{W}_- = \operatorname*{s\text{-}lim}_{t\to\infty} \overline{W}_t = \operatorname*{s\text{-}lim}_{t\to\infty} W_t = W_-.$$

Now we prove that W_- is an isometry. For $f, g \in \widehat{\mathcal{H}}$ we have

$$\begin{aligned}
(W_- g, W_- f)_{\mathcal{H}} &= \lim_{t\to\infty} (W_t g, W_t f)_{\mathcal{H}} \\
&= \lim_{t\to\infty} (g, W_t^* W_t f)_{\mathcal{H}} = \lim_{t\to\infty} (g, e^{it\widehat{H}} J^* J e^{-it\widehat{H}} f)_{\mathcal{H}}.
\end{aligned}$$

Then

$$J^* J \approx \begin{pmatrix} J_1^* J_1 & J_1^* J_2 \\ J_2^* J_1 & J_2^* J_2 \end{pmatrix}.$$

This operator matrix corresponds to the decomposition $\widehat{\mathcal{H}} = \widehat{\mathcal{H}}_1 \oplus \widehat{\mathcal{H}}_2$. Let us calculate every element of this matrix. We have

$$\begin{aligned}(J_1^* J_1 f_1)_1(k) &= f_1(k) + \Psi_1^{(+)}(k) \int_{\mathbb{R}^\nu} \overline{\Psi}_1^{(+)}(q)\, f_1(q)\, d^\nu q,\\ (J_2^* J_1 f_1)_2^{(-)}(k_1, k_2) &= 0,\\ (J_2^* J_1 f_1)_2^{(+)}(k_1, k_2) &= \Psi_1^{(+)}(k_1)\, f_1(k_2) + \Psi_1^{(+)}(k_2)\, f_1(k_1),\\ (J_2^* J_2 f_2)_2^{(\sigma)}(k_1, k_2) &= f_2^{(\sigma)}(k_1, k_2),\\ (J_1^* J_2 f_2)_1(k) &= \int_{\mathbb{R}^\nu} \overline{\Psi}_1^{(+)}(q)\, f_2^{(+)}(q, k)\, d^\nu q.\end{aligned}$$

Further, for $f = (f_1, f_2^{(\sigma)}) \in \widehat{\mathcal{H}}$, we have

$$\begin{aligned}(e^{i\widehat{H}t} J^* J e^{-i\widehat{H}t} f)_1(k) &= f_1(k) + e^{i(\omega(k)+E_0)t} \Psi_1^{(+)}(k)\\ &\quad \times \int_{\mathbb{R}^\nu} \overline{\Psi}_1^{(+)}(q)\, e^{-i(\omega(q)+E_0)t} f_1(q)\, d^\nu q\\ &\quad + \int_{\mathbb{R}^\nu} e^{-it(\omega(q)+\varepsilon-E_0)} f_2^{(+)}(q, k)\, \overline{\Psi}_1^{(+)}(q)\, d^\nu q,\\ (e^{i\widehat{H}t} J^* J e^{-i\widehat{H}t} f)_2^{(-)}(k_1, k_2) &= f_2^{(-)}(k_1, k_2),\\ (e^{i\widehat{H}t} J^* J e^{-i\widehat{H}t} f)_2^{(+)}(k_1, k_2) &= f_2^{+}(k_1, k_2) + e^{i(\varepsilon-E_0)t}(e^{i\omega(k_1)t} \Psi_1^{(+)}(k_1)\, f_1(k_2)\\ &\quad + e^{i\omega(k_2)t} \Psi_1^{(+)}(k_2)\, f_1(k_1)).\end{aligned}$$

Finally,

$$\begin{aligned}&(g, e^{i\widehat{H}t} J^* J e^{-i\widehat{H}t} f)_{\widehat{\mathcal{H}}}\\ &\quad = (g, f)_{\widehat{\mathcal{H}}} + \int_{\mathbb{R}^\nu} e^{i\omega(k)t} g_1(k)\, \overline{\Psi}_1^{(+)}(k)\, d^\nu k \int_{\mathbb{R}^\nu} e^{i\omega(q)t}\, \overline{f_1(q)}\, \Psi_1^{(+)}(q)\, d^\nu q\\ &\qquad + \int_{\mathbb{R}^\nu \times \mathbb{R}^\nu} e^{it(\omega(q)+\varepsilon-E_0)} \big(g_1(k)\, \Psi_1^{+}(q)\, \overline{f_2^{(+)}(q, k)}\\ &\qquad\qquad + g_2^{+}(q, k)\, \Psi_1^{(+)}(q)\, \overline{f_1(k)}\big)\, d^\nu q\, d^\nu k.\end{aligned}$$

Hence, with the help of the above arguments for $g, f \in D_1 \oplus D_2 \subset \widehat{\mathcal{H}}$, we get

$$\lim_{t\to\infty} (g, e^{i\widehat{H}t} J^* J e^{-i\widehat{H}t} f)_{\widehat{\mathcal{H}}} = (g, f)_{\widehat{\mathcal{H}}}.$$

This means we have an isometry of W_-. Theorem 2 is proved.

§5. Investigation of the resolvent of H

To prove Theorem 3, we need to study the resolvent $R_H(z) = (H - zE)^{-1}$ of the operator H.

5.1. Let $F = F(z) = R_H(z)G$, $G \in \mathcal{H}$. This means that we must solve the equations

$$
\begin{aligned}
&(\varepsilon\sigma - z) F_0^{(\sigma)}(z) + \alpha \int_{\mathbb{R}^\nu} \lambda(q) F_1^{(-\sigma)}(q;z)\, d^\nu q = G_0^{(\sigma)}, \\
&(\varepsilon\sigma + \omega(k) - z) F_1^{(\sigma)}(k;z) + \alpha\lambda(k) F_0^{(-\sigma)}(z) \\
&\qquad + \alpha \int \lambda(q) F_2^{(-\sigma)}(q,k;z)\, d^\nu q = G_1^{(\sigma)}(k), \\
&(\varepsilon\sigma + \omega(k_1) + \omega(k_2) - z) F_2^{(\sigma)}(k_1,k_2;z) \\
&\qquad + \alpha(\lambda(k_1) F_1^{(-\sigma)}(k_2;z) \\
&\qquad + \lambda(k_2) F_1^{(-\sigma)}(k_1;z)) = G_2^{(\sigma)}(k_1,k_2), \qquad \sigma = \pm.
\end{aligned}
\tag{5.1}
$$

Below we shall investigate the system (5.1) for $G = \{G_0^{(\sigma)}, G_1^{(\sigma)}, G_2^{(\sigma)}\} \in \mathcal{H}$ such that $G_1^{(\sigma)} \in D_1$, $\sigma = \pm$, $G_2 = \{G_2^\sigma, \sigma = \pm\} \in D_2$. We have

$$
F_2^{(\sigma)}(k_1,k_2;z) = \frac{G_2^{(\sigma)}(k_1,k_2) - \alpha(\lambda(k_1) F_1^{(-\sigma)}(k_2;z) + \lambda(k_2) F_1^{(-\sigma)}(k_1;z))}{\omega(k_1) + \omega(k_2) + \sigma\varepsilon - z}. \tag{5.2}
$$

After the substitution of $F_2^{(-\sigma)}(k_1,k_2;z)$ in the second equation of system (5.1), we get

$$
\begin{aligned}
&\left(\varepsilon\sigma + \omega(k) - z - \alpha^2 \int_{\mathbb{R}^\nu} \frac{\lambda^2(q)\, d^\nu q}{-\varepsilon\sigma + \omega(q) + \omega(k) - z}\right) F_1^{(\sigma)}(k;z) \\
&\qquad - \alpha^2 \lambda(k) \int_{\mathbb{R}^\nu} \frac{\lambda(q) F_1^{(\sigma)}(q;z)\, d^\nu q}{\omega(q) + \omega(k) - \varepsilon\sigma - z} = \Gamma^{(\sigma)}(k;z),
\end{aligned}
\tag{5.3}
$$

where

$$
\Gamma^{(\sigma)}(k;z) = -\alpha\lambda(k) F_0^{(-\sigma)}(z) + G_1^{(\sigma)}(k) - \alpha \int_{\mathbb{R}^\nu} \frac{G_2^{(-\sigma)}(q,k)\lambda(q)\, d^\nu q}{-\varepsilon\sigma + \omega(q) + \omega(k) - z}. \tag{5.4}
$$

Denote

$$
\begin{aligned}
H_\sigma(k;z) &= \varepsilon\sigma + \omega(k) - z - \alpha^2 \int_{\mathbb{R}^\nu} \frac{\lambda^2(q)\, d^\nu q}{-\varepsilon\sigma + \omega(q) + \omega(k) - z} \\
&= m_\sigma(z - \omega(k)),
\end{aligned}
\tag{5.5}
$$

where the functions $m_\sigma(\zeta)$ were defined by (3.3). We have studied them earlier and found that $m_+(\zeta) \neq 0$, $\zeta \in \Pi_+$, $m_-(\zeta)$ has a unique zero $E_0 < -\varepsilon$ (Lemma 3.2), and E_0 is a simple zero of m_-. Consider the functions

$$
v_+(\zeta) = \frac{1}{m_+(\zeta)}, \quad \zeta \in \Pi_+, \quad v_-(\zeta) = \frac{1}{m_-(\zeta)} - \frac{c}{\zeta - E_0}, \qquad \zeta \in \Pi_-, \tag{5.6}
$$

where c is the residue of $1/m_-(\zeta)$ at its pole $\zeta = E_0$. From Lemma 3.2 it follows that bcth functions $v_\sigma(\zeta)$ are continuous on Π_σ respectively and are analytic outside the cut $(-\varepsilon\sigma, \infty)$. Moreover, the functions

$$
v_\sigma^{u,d}(x) = v_\sigma(x \pm i0) \quad (d \sim -, u \sim +), \quad x \in \mathbb{R}^1,
$$

have $2n + \nu - 3$ continuous derivatives. The next lemma gives a refinement of this.

LEMMA 1. *The following representations*

$$v_\sigma(\zeta) = \begin{cases} \int_0^\infty e^{-is\zeta} \varkappa_\sigma^d(s)\, ds, & \operatorname{Im}\zeta < 0 \text{ or } \zeta = x - i0,\ x \in \mathbb{R}^1, \\ \int_0^\infty e^{is\zeta} \varkappa_\sigma^u(s)\, ds, & \operatorname{Im}\zeta > 0 \text{ or } \zeta = x + i0,\ x \in \mathbb{R}^1, \end{cases} \tag{5.6a}$$

hold, where the functions $\varkappa_\sigma^{u,d}(s)$ *satisfy the estimates*

$$|\varkappa_+^{u,d}(s)| < \frac{\text{const}}{(\alpha^2 s)^k + 1}, \qquad |\varkappa_-^{u,d}(s)| < \frac{\text{const}}{s^k + 1}, \tag{5.7}$$

in which $k = 2n + \nu - 3$ *and the constants do not depend on* α.

PROOF. Let $\varkappa_\sigma^d(s)$ be the Fourier transform of $v_\sigma^d(x) = v_\sigma(x - i0)$,

$$\varkappa_\sigma^d(s) = \frac{1}{2\pi} P \int_{-\infty}^{\infty} v_\sigma^d(x)\, e^{ixs} dx, \tag{5.8}$$

where the integral is understood in the sense of the principal part (namely, as $\lim_{L\to\infty} \int_{-L}^{L} \cdots dx$). Because we have $v_\sigma^d(x) = \text{const}/x + O(1/x^2)$ for large x, the integral (5.8) converges for all s and is bounded. Since the function $v_\sigma^d(x)$ has an analytic continuation to the lower half-plane, we have $\varkappa_\sigma^d(s) = 0$ for $s < 0$. Now let us consider in detail the case $v_+^d(x) = 1/m_+(x - i0)$. Using representation (3.9) (which is true for all y), we get (putting $x = \varepsilon + y$)

$$\begin{aligned} \int_{-\infty}^{\infty} \frac{e^{isx} dx}{m_+(x - i0)} &= -e^{is\varepsilon} \int_{-\infty}^{\infty} \frac{e^{isy} dy}{y + \alpha^2(-i\pi h(2\varepsilon + y) + P\int_0^\infty \frac{h(r)\, dr}{r - y - 2\varepsilon})} \\ &= -e^{is\varepsilon} \int_{-\infty}^{\infty} \frac{e^{is\alpha^2 v} dv}{v + (-\pi i h(2\varepsilon + \alpha^2 v) + P\int_0^\infty \frac{h(r)\, dr}{r - 2\varepsilon - \alpha^2 v})} \\ &= e^{is\varepsilon} f_\alpha(\alpha^2 s), \end{aligned} \tag{5.9}$$

where $f_\alpha(\tau)$ is a bounded function satisfying the estimate

$$|f_\alpha(\tau)| < \frac{\text{const}}{\tau^k + 1} \tag{5.10}$$

with the constant not depending on α. Estimate (5.10) is obtained by integration by parts of the integral (5.9). Relation (5.10) implies estimate (5.7) for $\varkappa_+^d$. Now let us pass to the function $v_-^d(x)$. Note that

$$m'(x - i0) = -1 + O(\alpha^2) \quad \text{and} \quad m_-^{(p)}(x - i0) = O(\alpha^2), \quad p > 1.$$

Further, for $|x + \varepsilon| < 2\varepsilon$ the function $m_-(x)$ is analytic and therefore

$$\frac{1}{m_-(x)} - \frac{c}{x - E_0},$$

where $c = 1/m'_-(E_0) = -1 + O(\alpha^2)$ is an analytic function and its derivatives have order $O(\alpha^2)$. If $|x + \varepsilon| > 2\varepsilon$, both functions $1/m_-(x - i0)$ and $1/(x - E_0)$ have $2n + \nu - 3$ bounded and integrable derivatives. Taking the integral (5.8) by parts for $\sigma = -$, we get the second inequality (5.7). The function $\varkappa_\sigma^u(s)$ is considered in a similar way. From (5.6a) and (5.7), we find by the inverse Fourier transform that

$$v_\sigma^d(x) = \int_0^\infty \varkappa_\sigma^d(s)\, e^{-isx} ds.$$

Both sides of this formula can be continued to the lower half-plane, and we get (5.6a) for $\operatorname{Im}\zeta \leqslant 0$. In a similar way we can prove (5.6a) for $\operatorname{Im}\zeta \geqslant 0$. Lemma 1 is proved.

Now consider the equation (5.3) for $\sigma = -$:

$$H_-(k;z)\, F_1^{(-)}(k;z) - \alpha^2\lambda(k) \int_{\mathbb{R}^\nu} \frac{\lambda(q)\, F_1^{(-)}(q;z)\, d^\nu q}{\varepsilon + \omega(q) + \omega(k) - z} = \Gamma^{(-)}(k;z). \tag{5.10a}$$

After introducing the notation $\Phi^{(-)}(k;z) = H_-(k;z)\, F_1^{(-)}(k;z)$, we get

$$\Phi^{(-)}(k;z) - \alpha^2\lambda(k) \int_{\mathbb{R}^\nu} \frac{\lambda(q)\, \Phi^{(-)}(q;z)\, d^\nu q}{(\varepsilon + \omega(q) + \omega(k) - z)\, H_-(q;z)} = \Gamma^{(-)}(k;z). \tag{5.11}$$

Put

$$\frac{1}{H_-(q;z)} = \frac{1}{m_-(z - \omega(q))} = \frac{c}{z - \omega(q) - E_0} + v_-(z - \omega(q)).$$

We introduce two operators

$$\begin{aligned} (K_0\Phi)(k;z) &= -\lambda(k)\, c \int_{\mathbb{R}^\nu} \frac{\lambda(q)\, \Phi(q;z)\, d^\nu q}{(\varepsilon + \omega(q) + \omega(k) - z)(\omega(q) + E_0 - z)}, \\ (K_1\Phi)(k;z) &= \lambda(k) \int_{\mathbb{R}^\nu} \frac{\lambda(q)\, v_-(z - \omega(q))\, \Phi(q;z)\, d^\nu q}{\varepsilon + \omega(q) + \omega(k) - z}, \end{aligned} \tag{5.12}$$

acting in a certain space of functions $\Phi(q;z)$, $q \in \mathbb{R}^\nu$, $z \in \mathbb{C} \setminus [E_0;\infty)$, analytic in z, which we shall define below. Equation (5.11) can be written

$$\Phi^{(-)} - \alpha(K_0 + K_1)\, \Phi^{(-)} = \Gamma^{(-)}. \tag{5.12a}$$

5.2. Investigation of the operators K_0 and K_1 and a representation of $F(z)$. Consider the space $\mathcal{A}_-$ of functions of the form

$$\Phi(k;z) = \lambda(k)\, g(\omega(k);z), \tag{5.13}$$

where $g(r;z)$ is a continuous function on $\mathbb{R}^1 \times \widehat{\Pi}_-$, analytic in z outside the cut $[E_0,\infty)$ and continuous in $z \in \widehat{\Pi}_-$, where $\widehat{\Pi}_-$ is the complex plane with cut $[E_0;\infty)$ together with both its banks. Moreover, the function $g(r;z)$ has an analytic continuation in r to the upper half-plane for $\operatorname{Im} z \leqslant 0$ and to the lower half-plane in r for $\operatorname{Im} z \geqslant 0$. More exactly, let $g(r;z)$ possess the representations

$$g(r;z) = \begin{cases} \int_0^\infty \int_0^\infty e^{ir\xi - izt}\Psi^d(\xi,t)\, d\xi\, dt, & z \in \widehat{\Pi}_-,\ \operatorname{Im} z \leqslant 0,\ r \in \mathbb{R}^1, \\ \int_0^\infty \int_0^\infty e^{-ir\xi + izt}\Psi^u(\xi,t)\, dx\, dt, & z \in \widehat{\Pi}_-,\ \operatorname{Im} z \geqslant 0,\ r \in \mathbb{R}^1, \end{cases} \tag{5.14}$$

where both functions $\Psi^{u,d}(\xi,t)$ satisfy the estimate

$$\int_0^\infty |\Psi^{u,d}(\xi,t)|\, dt \leqslant \frac{c}{\xi^p + 1}; \tag{5.15}$$

here $p = k - 1 = 2n + \nu - 2$. Denote the space of functions $g(r; z)$ described above by $\eta^{(-)}$ and introduce a norm in $\eta^{(-)}$ by the formula

$$\|g\|_{\eta^{(-)}} = \inf c,$$

where c is any constant satisfying (5.15).

LEMMA 2. *The space $\mathcal{A}_-$ of functions $\Phi(k; z)$ of the form* (5.13), *with $g \in \eta^{(-)}$, is invariant with respect to the operators K_0 and K_1, and we have*

$$\max\{\|\widetilde{K}_0\|_{\eta^{(-)}}, \|\widetilde{K}_1\|_{\eta^{(-)}}\} < \infty,$$

where $\widetilde{K}_0$ and $\widetilde{K}_1$ are the operators in $\eta^{(-)}$ generated by K_0 and K_1 respectively.

PROOF. First consider K_0. Obviously for $\Phi(k; z)$ of the form (5.13) we have

$$(K_0\Phi)(k; z) = \lambda(k)\,\tilde{g}(\omega(k); z),$$

where

$$\begin{aligned} \tilde{g}(\rho; z) \equiv (\widetilde{K}_0 g)(\rho; z) &= \int_{\mathbb{R}^\nu} \frac{\lambda^2(q)\,g(\omega(q); z)\,d^\nu q}{(\omega(q) + \rho + \varepsilon - z)(\omega(q) + E_0 - z)} \\ &= \int_{-\infty}^{\infty} \frac{h(r)\,g(r, z)\,dr}{(r + \rho + \varepsilon - z)(r + E_0 - z)}; \end{aligned}$$

here $h(r) = 0$ if $r \leqslant 0$ and is defined by (3.7) if $r \geqslant 0$. Let $\operatorname{Im} z < 0$. Then

$$\begin{aligned} &\int_{-\infty}^{\infty} \frac{h(r)\,g(r; z)\,dr}{(r + \rho + \varepsilon - z)(r + E_0 - z)} \\ &\quad = -\int_{-\infty}^{\infty} b(u)\,du \int_{-\infty}^{\infty}\int_0^{\infty}\int_0^{\infty} e^{ir(u+\xi) - izt}\Psi^d(\xi, t)\,dr\,d\xi\,dt \\ &\qquad \times \int_0^{\infty}\int_0^{\infty} e^{i\eta(r+\rho+\varepsilon-z) + is(r+E_0-z)}\,d\eta\,ds \\ &\quad = -2\pi i \int_0^{\infty}\int_0^{\infty}\int_0^{\infty}\int_0^{\infty} e^{i\eta\rho - iz\tau + i\eta\varepsilon + isE_0} \\ &\qquad\qquad \times b(-\xi - \eta - s)\,\Psi^d(\xi, \tau - s - \eta)\,d\eta\,d\tau\,ds\,d\xi_0. \end{aligned}$$

Hence we see that the function $\tilde{g}(\rho; z)$ for $\operatorname{Im} z \leqslant 0$ has the representation (5.14) with

$$\widetilde{\Psi}^d(\eta, \tau) = -2\pi i e^{i\eta(\varepsilon - E_0)} \int_0^{\infty} d\xi \int_\eta^{\tau} dv\, b(-\xi - v)\,\Psi^d(\xi, \tau - v)\,e^{iE_0 v}. \tag{5.16}$$

Here $b(\,\cdot\,)$ is defined by formula (3.6). From (5.16) and estimates (5.15), (3.8), we find that

$$\begin{aligned} \int_0^{\infty} |\widetilde{\Psi}^d(\eta, \tau)|\,d\tau &\leqslant \|g\|_{\eta^{(-)}} \cdot \text{const} \int_0^{\infty} d\xi \int_\eta^{\infty} \frac{dv}{(\xi + v)^k + 1} \cdot \frac{1}{\xi^p + 1} \\ &< \text{const}\,\|g\|_{\eta^{(-)}} \int_0^{\infty} \frac{d\xi}{((\xi + \eta)^p + 1)(\xi^p + 1)} < \frac{\text{const}}{\eta^p + 1}\|g\|_{\eta^{(-)}}, \end{aligned}$$

where the constant is absolute. This inequality means that $K_0\Phi \in \mathcal{A}_-$ and $\|\widetilde{K}_0\|_{\eta^{(-)}} < \infty$. The operator K_1 is studied in a similar way. Lemma 2 is proved.

Note, however, that the right-hand side $\Gamma^{(-)}$ of equation (5.11) does not have the form (5.13). But we have the following

LEMMA 3. *If* $G_1^{(-)} \in D_1$, $G_2 \in D_2$, *then* $K_0\Gamma^{(-)} \in \mathcal{A}_-$, $K_1\Gamma^{(-)} \in \mathcal{A}_-$.

PROOF. Consider every term going to $\Gamma^{(-)}$.

$$\begin{aligned} 1.\quad (K_0\lambda)(k;z) &= c\lambda(k)\int_{\mathbb{R}^\nu} \frac{\lambda^2(q)\,d^\nu q}{(\omega(k)+\omega(q)+\varepsilon-z)(\omega(q)+E_0-z)} \\ &= \lambda(k)\,g(\omega(k);z), \end{aligned}$$

where

$$g(\rho;z) = c\int_0^\infty \frac{h(r)\,dr}{(\rho+r+\varepsilon-z)(r+E_0-z)}.$$

Let $\operatorname{Im} z < 0$. Then

$$\begin{aligned} &\int_0^\infty \frac{h(r)\,dr}{(\rho+r+\varepsilon-z)(r+E_0-z)} \\ &\quad = -\int_{-\infty}^\infty du \int_{-\infty}^\infty dr \int_0^\infty d\eta \int_0^\infty ds\, b(u)\, e^{iur+i\eta(r+\rho+\varepsilon-z)+is(r+E_0-z)} \\ &\quad = -2\pi i \int_0^\infty d\eta \int_\eta^\infty b(-\tau)\, e^{-i\tau z+i\eta\rho}\, d\tau. \end{aligned}$$

Thus

$$\Psi^d(\eta,\tau) = \begin{cases} 0, & \tau < \eta, \\ b(\tau)\, e^{i\eta(\varepsilon-E_0)+i\tau E_0}, & \tau \geqslant \eta. \end{cases}$$

Hence

$$\int_0^\infty |\Psi^d(\eta,\tau)|\, d\tau < \text{const}\int_\eta^\infty \frac{d\tau}{\tau^k+1} < \text{const}\,\frac{1}{\eta^p+1}.$$

The case $\operatorname{Im} z > 0$ is investigated similarly.

2. The second term $G_1^{(-)}(k) \in D_1$ is studied in a similar way, because this function has the same properties as $\lambda(k)$.

3. Now consider the third term going to $\Gamma^{(-)}(k;z)$. After applying K_0 to it, we get a function of the form (5.13), where

$$\begin{aligned} g(r;z) &= -\alpha c \int_{\mathbb{R}^\nu\times\mathbb{R}^\nu} \frac{G_2^{(+)}(q,k)\,\lambda(q)\,\lambda(k)\,d^\nu q\, d^\nu k}{(\varepsilon+\omega(k)+\omega(q)-z)(\omega(k)+E_0-z)(r+\omega(k)+\varepsilon-z)} \\ &= -\alpha c \int_{-\infty}^\infty\int_{-\infty}^\infty \frac{l(\rho_1,\rho_2)\,d\rho_1\,d\rho_2}{(\varepsilon+\rho_1+\rho_2-z)(r+\rho_2+\varepsilon-z)(\rho_2+E_0-z)}, \\ l(\rho_1,\rho_2) &= \int_{\substack{\omega(q)=\rho_1\\ \omega(k)=\rho_2}} G_2^{(+)}(q,k)\,\lambda(q)\,\lambda(k)\,d^\nu q\, d^\nu k. \end{aligned}$$

Again we take $\operatorname{Im} z < 0$ and get

$$\int_{-\infty}^{\infty}\int_{-\infty}^{\infty} \frac{l(\rho_1,\rho_2)\,d\rho_1\,d\rho_2}{(\varepsilon+\rho_1+\rho_2-z)(\varepsilon+r+\rho_2-z)(\rho_2+E_0-z)}$$
$$= \int_{-\infty}^{\infty}\int_{-\infty}^{\infty} d\rho_1\,d\rho_2 \int_{-\infty}^{\infty}\int_{-\infty}^{\infty} b(u_1,u_2)\,e^{iu_1\rho_1+iu_2\rho_2}\,du_1\,du_2$$
$$\times \int_0^{\infty}\int_0^{\infty}\int_0^{\infty} ds\,d\eta\,d\xi\; e^{is(\varepsilon+\rho_1+\rho_2-z)+i\eta(\varepsilon+r+\rho_2-z)+i\xi(\rho_2+E_0-z)}$$
$$= -4\pi^2 \int_0^{\infty} d\eta \int_{\eta}^{\infty} d\tau \int_0^{\tau-\eta} ds\, b(-s,-\tau)\,e^{-iz\tau+i\eta\varepsilon+i(\tau-\eta-s)E_0+i\eta r},$$

where $b(u_1,u_2)$ is the Fourier transform of $l(\rho_1,\rho_2)$. If $G_2 \in D_2$, then, as above, we have

$$|b(u_1,u_2)| < \frac{\text{const}}{(|u_1|^k+1)(|u_2|^k+1)}.$$

Thus

$$\Psi^d(\eta,\tau) = \begin{cases} 0, & \tau < \eta, \\ \int_0^{\tau-\eta} b(-s,\tau)e^{-isE_0}\,ds \cdot e^{i\eta(\varepsilon-E_0)+i\tau E_0}, & \tau \geqslant \eta. \end{cases}$$

Hence

$$\int_0^{\infty} |\Psi^d(\eta,\tau)|\,d\tau \leqslant \text{const} \int_{\eta}^{\infty} \frac{d\tau}{\tau^k+1} \leqslant \frac{\text{const}}{\eta^p+1}.$$

The case $\operatorname{Im} z > 0$ is studied similarly. Finally, the case of the operator K_1 is analogous. Lemma 3 is proved.

Let us turn to equation (5.12a) and look for its solution in the form

$$\Phi^{(-)} = \Gamma^{(-)} + \widetilde{\Phi}^{(-)}.$$

For $\widetilde{\Phi}^{(-)}$ we obtain the equation

$$\widetilde{\Phi}^{(-)} - \alpha^2(K_0+K_1)\,\widetilde{\Phi}^{(-)} = \alpha^2(K_0+K_1)\Gamma^{(-)}.$$

From Lemmas 2 and 3 it follows that for small α there exists a unique solution of this equation that belongs to the class $\mathcal{A}_-$. Thus the solution of equation (3.12a) has the form

$$\Phi^{(-)}(k;z) = -\alpha\lambda(k)\,F_0^{(+)}(z) + \alpha^3\Delta_-(k;z)\,F_0^{(+)}(z) + \widetilde{\Gamma}^{(-)}(k;z) + \widetilde{\Phi}^{(-)}(k;z),$$

where $\Delta_-(k;z)$ and $\widehat{\Phi}^{(-)}(k;z)$ are functions from the class $\mathcal{A}_-$; here $\widetilde{\Gamma}^{(-)}(k;z)$ has the form

$$\widetilde{\Gamma}^{(-)}(k;z) = G_1^{(-)}(k) - \alpha \int_{\mathbb{R}^\nu} \frac{G_2^+(q,k)\,\lambda(q)\,d^\nu q}{\varepsilon+\omega(k)+\omega(q)-z}. \tag{5.17}$$

Hence the solution of equation (5.10a) is

$$\begin{aligned} F_1^{(-)}(k;z) &= \frac{\Phi^{(-)}(k;z)}{H_-(k;z)} \\ &= F_0^{(+)}(z)\,\frac{-\alpha\lambda(k)+\alpha^3\Delta_-(k;z)}{H_-(k;z)} + \frac{\widetilde{\Gamma}^{(-)}(k;z)}{H_-(k;z)} + \frac{\widetilde{\Phi}^{(-)}(k;z)}{H_-(k;z)}. \end{aligned} \tag{5.18}$$

Repeating the above arguments, we see that a solution of equation (5.3) for $\sigma = +$ is of a similar form:

$$(5.19)\qquad F_1^{(+)}(k;z) = F_0^{(-)}(z)\,\frac{-\alpha\lambda(k) + \alpha^3\Delta_+(k,z)}{H_+(k;z)} + \frac{\widetilde{\Gamma}^{(+)}(k;z)}{H_+(k;z)} + \frac{\widetilde{\Phi}^{(+)}(k;z)}{H_+(k;z)},$$

where $\Delta_+(k;z)$ and $\widetilde{\Gamma}^+(k;z)$, $\widetilde{\Phi}^{(+)}(k;z)$ are defined as in the previous case and Δ_+, $\widetilde{\Phi}^{(+)} \in \mathcal{A}_+$ (the class $\mathcal{A}_+$ is also defined in a similar way). In addition

$$(5.20)\qquad \begin{aligned} \Delta_+(k;z) &= \lambda(k)\int_{\mathbb{R}^\nu} \frac{\lambda^2(q)\,d^\nu q}{(-\varepsilon + \omega(k) + \omega(q) - z)\,H_+(q;z)} + O(\alpha^2) \\ &= \lambda(k)\int_{\mathbb{R}^\nu} \frac{\lambda^2(q)\,d^\nu q}{(-\varepsilon + \omega(q) + \omega(k) - z)(\varepsilon + \omega(q) - z)} + O(\alpha^2). \end{aligned}$$

Substituting $F_1^{(+)}(k;z)$ and $F_1^{(-)}(k;z)$ into the first equation of system (5.1), we get

$$M_\sigma(z)\,F_0^{(\sigma)}(z) = D_\sigma(z), \qquad \sigma = \pm,$$

where

$$(5.21)\qquad \begin{aligned} M_\sigma(z) &= \varepsilon\sigma - z - \alpha^2\int_{\mathbb{R}^\nu} \frac{\lambda^2(q)\,d^\nu q}{H_{-\sigma}(q,z)} + \alpha^4\int_{\mathbb{R}^\nu} \frac{\Delta_{-\sigma}(q,z)\,\lambda(q)\,d^\nu q}{H_{-\sigma}(q;z)}, \\ D_\sigma(z) &= G_0^{(\sigma)} - \alpha\left(\int_{\mathbb{R}^\nu} \frac{(\widetilde{\Gamma}^{(-\sigma)}(q;z) + \widetilde{\Phi}^{(-\sigma)}(q,z))\,\lambda(q)\,d^\nu q}{H_{-\sigma}(q;z)}\right). \end{aligned}$$

5.3. Investigation of the functions $M_\sigma(z)$ and $D_\sigma(z)$. Complete structure of the resolvent. Here we study both the pairs of functions $\{M_\sigma(z), D_\sigma(z)\}$ and their ratio $F_0^{(\sigma)}(z)$. Let be $\widehat{\Pi}_+ = \Pi_+$ and $\widehat{\Pi}_-$ be the complex plane with the cut $[E_0;\infty)$ together with both banks of this cut. Consider the following spaces $\mathcal{B}_\sigma$ of functions defined on $\widehat{\Pi}_\sigma$ respectively:

1) $f \in \mathcal{B}_\sigma$ is a continuous (bounded) function on $\widehat{\Pi}_\sigma$ and is analytic outside the cut;
2) the restrictions of f to the upper (f^u) and lower (f^d) half-planes have the following representations:

$$(5.21\text{a})\qquad \begin{aligned} f^d(z) &= \int_0^\infty \Psi^d(t)\,e^{-izt}\,dt, &\qquad \operatorname{Im} z \leqslant 0, \\ f^u(z) &= \int_0^\infty \Psi^u(t)\,e^{izt}\,dt, &\qquad \operatorname{Im} z \geqslant 0, \end{aligned}$$

where

$$\int_0^\infty |\Psi^{(u,d)}(t)|\,dt < \infty.$$

We introduce a norm on $\mathcal{B}_\sigma$ by putting

$$\|f\| = \max\{\|\Psi^u\|_{L_1(\mathbb{R}^1_+)}, \|\Psi^u\|_{L_1(\mathbb{R}^1_+)}\}.$$

Obviously $\max_{z\in\widehat{\Pi}_\sigma} |f(z)| \leqslant \|f\|$. Moreover, $\mathcal{B}_\sigma$ is a Banach algebra with respect to pointwise multiplications of functions. This follows from the formula

$$f_1^d(z)\, f_2^d(z) = \int_0^\infty \Psi^d(\tau)\, e^{-iz\tau}\, d\tau,$$

where

$$\Psi^d(\tau) = \int_0^\tau \Psi_1^d(t)\, \Psi_2^d(\tau - t)\, dt \quad \text{and} \quad \|\Psi^d\|_{L_1(\mathbb{R}_+^1)} \leqslant \|\Psi_1^d\|_{L_1(\mathbb{R}_+^1)} \|\Psi_2^d\|_{L_1(\mathbb{R}_+^1)},$$

and similar relations for f_1^u and f_2^u.

Let $\widehat{\mathcal{B}}$ be the algebra obtained from $\mathcal{B}_\sigma$ by adding the unit $\mathbb{I}$, i.e., the set of functions $\hat{f}(z) = c + f(z)$, $c \in \mathbb{C}^1$, $f \in \mathcal{B}_\sigma$. Let $\|\hat{f}\|_{\widehat{\mathcal{B}}} = |c| + \|f\|_{\mathcal{B}_\sigma}$. Obviously $\mathcal{B}_\sigma \subset \widehat{\mathcal{B}}_\sigma$ is an ideal of $\widehat{\mathcal{B}}_\sigma$.

LEMMA 4. 1) *We have* $D_\sigma(z) \in \widehat{\mathcal{B}}_{-\sigma}$, $\sigma = \pm$.

2) *The functions* $M_\sigma(z)$ *are continuous on* $\widehat{\Pi}_{-\sigma}$ *respectively and analytic outside the cut.*

3) *We have* $M_+(z) \neq 0$, $z \in \widehat{\Pi}_-$ *and* $1/M_+(z) \in \mathcal{B}_-$. *Hence* $F_0^+(z) = D_+(z)/M_+(z) \in \mathcal{B}_-$.

4) *The function* $M_-(z)$ *has a unique zero* $E_1 \in \widehat{\Pi}_+$, E_1 *is real and* $E_1 < E_0$. *This value is the unique eigenvalue of the operator* H; *let* Ψ_g *be the corresponding eigenvector. If the vector* $G = \{G_0^{(\sigma)}, G_1^{(\sigma)}, G_2^{(\sigma)}\}$, $G_1^{(\sigma)} \in D_1$, $G_2 \in D_2$ *is orthogonal to* Ψ_g, *then the function* $D_-(z)$ *vanishes at* $z = E_1$: $D_-(E_1) = 0$ *and*

$$F_0^{(-)}(z) = D_-(z)/M_-(z) \in \mathcal{B}_+. \tag{5.22}$$

PROOF. 1) Let us consider the case $D_+(z)$ and study each integral in (5.21). We have

$$\int_{\mathbb{R}^\nu} \frac{\widetilde{\Gamma}^{(-)}(q;z)\,\lambda(q)\, d^\nu q}{H_-(q;z)} = -c \int_{\mathbb{R}^\nu} \frac{\widetilde{\Gamma}^{-}(q;z)\,\lambda(q)\, d^\nu q}{\omega(q) + E_0 - z} + \int_{\mathbb{R}^\nu} \widetilde{\Gamma}^{(-)}(q;z)\, v_-(z - \omega(q))\, d^\nu q.$$

Then the first term will be

(5.23)
$$\int_{\mathbb{R}^\nu} \frac{\widetilde{\Gamma}^{(-)}(q;z)\,\lambda(q)\, d^\nu q}{\omega(q) + E_0 - z} = \int \frac{G_1^{(-)}(q)\, d^\nu q}{\omega(q) + E_0 - z} - \alpha \int_{\mathbb{R}^\nu\times\mathbb{R}^\nu} \frac{G_2^+(q,k)\,\lambda(q)\,\lambda(k)\, d^\nu q\, d^\nu k}{(\varepsilon + \omega(q) + \omega(k) - z)(\omega(k) + E_0 - z)}.$$

Further

$$\int_{\mathbb{R}^\nu} \frac{G_1^{(-)}(q)\,\lambda(q)\, d^\nu q}{\omega(q) + E_0 - z} = \int_{-\infty}^{\infty} \frac{l(r)\, dr}{r + E_0 - z},$$

where

$$l(r) = \begin{cases} \int_{\omega(q)=r} G_1^{(-)}(q)\,\lambda(q)\, d^\nu q, & r \geqslant 0, \\ 0, & r < 0. \end{cases}$$

For $\operatorname{Im} z < 0$ we get

$$\int_{-\infty}^{\infty} \frac{l(r)\,dr}{r+E_0-z} = \int_{-\infty}^{\infty} dr \int_0^{\infty} ds\, l(r)\, e^{isr+isE_0-isz} = \int_0^{\infty} \tilde{l}(s)\, e^{is(E_0-z)} ds,$$

where $\tilde{l}(s) = \int_0^{\infty} l(r)\, e^{isr} dr$. For $G_1^{(-)} \in D_1$, we have $|\tilde{l}(s)| < \text{const}/(|s|^k+1)$ and for $k > 1$, we have $\tilde{l}(s) \in L_1(\mathbb{R}^1_+)$. The case $\operatorname{Im} z \geqslant 0$ is similar. For $G_2^{(+)}(k,q) = f(k)f(q)$, $f \in D_1$, we can write the second integral in (5.23) as

$$\int_{\mathbb{R}^\nu} f(q)(K_0 f)(q;z)\, d^\nu q.$$

As we saw before, for $f \in D_1$ we have $(K_0 f)(q;z) = \lambda(q)\, g(\omega(q);z)$, where $g \in \eta^{(-)}$. It is easy to check that

$$\int_{\mathbb{R}^\nu} \lambda(q)\, f(q)\, g(\omega(q);z)\, d^\nu q \in \mathcal{B}_-\,.$$

Thus

$$\int_{\mathbb{R}^\nu} \frac{\widetilde{\Gamma}^-(q;z)\, d^\nu q}{\omega(q)+E_0-z} \in \mathcal{B}_-\,.$$

Similarly

$$\int_{\mathbb{R}^\nu} \widetilde{\Gamma}^{(-)}(q;z)\, v_-(z-\omega(q))\, d^\nu q \in \mathcal{B}_-\,.$$

Finally, in a similar way we can prove that

$$\int_{\mathbb{R}^\nu} \frac{\widetilde{\Phi}^{(-)}(k;z)}{H_-(k;z)}\, \lambda(k)\, d^\nu k \in \mathcal{B}_-\,.$$

Thus $D_+(z) \in \widehat{\mathcal{B}}_-$. The proof for $D_-(z)$ is the same.

2) Analogously we can prove that

$$\int_{\mathbb{R}^\nu} \frac{\lambda^2(q)\, d^\nu q}{H_{-\sigma}(q;z)} \in \mathcal{B}_{-\sigma} \quad \text{and} \quad \int_{\mathbb{R}^\nu} \frac{\Delta_{-\sigma}(q;z)\,\lambda(q)\, d^\nu q}{H_{-\sigma}(q;z)} \in \mathcal{B}_{-\sigma}\,.$$

This implies assertion 2) concerning the functions M_σ.

3) Now consider the case of $M_+(z)$. Again $M_+(z)$ has no nonreal zeros. Let z be real and $z < E_0$: $z = E_0 - x$, $x > 0$. Then

(5.24)

$$\begin{aligned}\widehat{M}_+(z) &\equiv \varepsilon - z - \alpha^2 \int_{\mathbb{R}^\nu} \frac{\lambda^2(q)\, d^\nu q}{H_-(q;z)} \\ &= \varepsilon - E_0 + x - \alpha^2 \int_{\mathbb{R}^\nu} \frac{\lambda^2(q)\, d^\nu q}{-\varepsilon - E_0 + \omega(q) + x - \alpha^2 \int_{\mathbb{R}^\nu} \frac{\lambda^2(q')\, d^\nu q'}{\varepsilon - E_0 + \omega(q) + \omega(q') + x}}\,.\end{aligned}$$

The right-hand side is a monotone increasing function of $x \geqslant 0$ and for $x = 0$ is equal to $\varepsilon - E_0 + O(\alpha^2) > (3/2)\varepsilon$ for small α. Hence $M_+(z) = \widehat{M}_+(z) + O(\alpha^4)$ for $z < E_0$. Let be $z > E_0$ and consider $\widehat{M}_+(z - i0)$. Then:

a) let $|\varepsilon - z| > 2k\alpha^2$, where

$$k = \max_{z \in \widehat{\Pi}_-} \int_{\mathbb{R}^\nu} \frac{\lambda^2(q)\, d^\nu q}{H_-(q;z)};$$

in this case $|\widehat{M}_+(z - i0)| > k\alpha^2$;

b) let $|\varepsilon - z| < 2k\alpha^2$; in this case by putting $\varepsilon - z = y$ we obtain

$$\widehat{M}_+(z - i0) = y - \alpha^2 \int_{\mathbb{R}^\nu} \frac{\lambda^2(q)\, d^\nu q}{-2\varepsilon + \omega(q) + y + i0 - \alpha^2 \int_{\mathbb{R}^\nu} \frac{\lambda^2(q')\, d^\nu q'}{\omega(q)+\omega(q')+y+i0}}$$
$$= y - \alpha^2 \int_0^\infty \frac{h(r)\, dr}{-2\varepsilon + r + y + i0 - \alpha^2 \int_0^\infty \frac{h(r')\, dr'}{r+r'+y+i0}}.$$

Denote

$$\tau(r) = \int_0^\infty \frac{h(r')\, dr'}{r + r' + y} \quad \text{for } r > |y|.$$

We can write

$$\widehat{M}_+(z - i0) = y - \alpha^2 \Big[\int_{|r-2\varepsilon|<\delta/2} \dots dr + \int_{|r-2\varepsilon|>\delta/2} \dots dr \Big]. \tag{5.25}$$

First consider the integral $\int_{|r-2\varepsilon|<\delta/2} \dots dr$ in (5.25). We have

$$\int_{2\varepsilon-\delta/2}^{2\varepsilon+\delta/2} \frac{h(r)\, dr}{r - 2\varepsilon + y + i0 - \alpha^2\tau(r)} = \int_{\gamma_1}^{\gamma_2} \frac{\widetilde{h}(\rho)\, d\rho}{\rho - 2\varepsilon + y + i0},$$

where $\rho = r - \alpha^2\tau(r)$, $d\rho = (1 - \alpha^2\tau'(r))\, dr$, $\widetilde{h}(\rho) = h(r)/(1 - \alpha^2\tau'(r))$, $\gamma_1 = \rho|_{r=2\varepsilon-\delta/2}$, $\gamma_2 = \rho|_{r=2\varepsilon+\delta/2}$. Hence by Sokhotski's formula,

$$\operatorname{Im} \int_{\gamma_1}^{\gamma_2} \frac{\widetilde{h}(\rho)\, d\rho}{\rho - 2\varepsilon + y + i0} = \pi\widetilde{h}(2\varepsilon - y) > \text{const} > 0$$

by condition A.iii). Let us pass to the second integral in (5.25) and consider two cases:

i) $y > 0$; then the denominator of the integrand in this integral is real and does not vanish; therefore the second integral is real;

ii) $y < 0$ (and, as before, $|y| < 2k\alpha^2$); then

$$\operatorname{Im} \int_0^\infty \frac{h(r')\, dr'}{r + r' + y + i0} = \begin{cases} 0, & r > |y|, \\ \pi h(-r - y), & 0 < r < |y|. \end{cases}$$

By condition A.iii), we have

$$h(-r - y) \sim |y + r|^{2n+\nu-1} = O(\alpha^{4n+2\nu-2}).$$

Hence the imaginary part of the second integral in (5.25) by $y < 0$, $|y| = O(\alpha^2)$ has order $O(\alpha^{4n+2\nu})$ and consequently in both cases i) and ii)

$$|\widehat{M}_+(z - i0)| > \text{const}\, \alpha^2.$$

Therefore,

$$|M_+(z - i0)| = |\widehat{M}_+(z - i0) + O(\alpha^4)| > \text{const}\, \alpha^2.$$

A similar estimate is true for $M_+(z + i0)$. Thus $M_+(z) \neq 0$, $z \in \widehat{\Pi}_-$. Now let us show that $1/M_+(z) \in \mathcal{B}_-$. The continuity of $M_+(z)$ on $\widehat{\Pi}_-$ and its analyticity outside the cut is obvious. We must establish representations (5.21a). Consider the Fourier transformation

$$I_+^d(s) = \int_{-\infty}^{\infty} \frac{e^{izs}\,dz}{\widehat{M}_+(z-i0)}$$
$$= \int_{-\infty}^{\infty} \frac{e^{izs}dz}{\varepsilon - z + i0 - \alpha^2 \displaystyle\int_0^\infty \frac{h(r)\,dr}{-\varepsilon + r - z + i0 - \alpha^2 \int_0^\infty \frac{h(r')\,dr'}{\varepsilon + r + r' - z + i0}}}.$$

Setting $z = \varepsilon - \alpha^2 u$, we get

$$I_+^d(s) = e^{i\varepsilon s} \int_{-\infty}^{\infty} \frac{e^{-iu\alpha^2 s}du}{u + i0 - \displaystyle\int_0^\infty \frac{h(r)\,dr}{-2\varepsilon + r + \alpha^2 u + i0 - \alpha^2 \int_0^\infty \frac{h(r')\,dr'}{r + r' + \alpha^2 u + i0}}}$$

with the help of the above reasoning (see the derivative in the first estimate (5.7)), we again get

$$|I_+^d(s)| < \text{const}\,/((s\alpha^2)^k + 1),$$

where the constant does not depend on α. A similar estimate is true for $I_+^u(s)$. Therefore, $1/\widehat{M}_+(z) \in \mathcal{B}_-$ and

$$\|1/\widehat{M}_+\|_{\mathcal{B}_-} < \text{const}\,/\alpha^2. \tag{5.26}$$

Further

$$M_+(z) = \widehat{M}_+(z)\left(1 + \frac{\alpha^4}{\widehat{M}_+(z)} \int_{\mathbb{R}^\nu} \frac{\Delta_-(q;z)\,\lambda(q)\,d^\nu q}{H_-(q;z)}\right). \tag{5.27}$$

Since

$$\int_{\mathbb{R}^\nu} \frac{\Delta_-(q;z)\,\lambda(q)\,d^\nu q}{H_-(q;z)} \in \mathcal{B}_- \quad \text{and} \quad \left\|\int_{\mathbb{R}^\nu} \frac{\Delta_-(q;z)\,\lambda(q)\,d^\nu q}{H_-(q;z)}\right\|_{\mathcal{B}_-} < K$$

uniformly with respect to α, it follows from (5.26) that

$$\left\|\frac{\alpha^4}{\widehat{M}_+(z)} \int_{\mathbb{R}^\nu} \frac{\Delta_-(q;z)\,\lambda(q)\,d^\nu q}{H_-(q;z)}\right\| < \alpha^2\,\text{const}$$

and hence

$$\left(1 + \frac{\alpha^4}{\widehat{M}_+(z)} \int_{\mathbb{R}^\nu} \frac{\Delta_+(q;z)\,\lambda(q)\,d^\nu q}{H_-(q;z)}\right)^{-1} \in \mathcal{B}_-.$$

This inclusion and (5.27) imply that $1/M_+ \in \mathcal{B}_-$.

4) Now consider the function $M_-(z)$. Again $M_-(z) \neq 0$ if $\operatorname{Im} z \neq 0$. Let

$$\widehat{M}_-(z) = -\varepsilon - z - \alpha^2 \int_{\mathbb{R}^\nu} \frac{\lambda^2(q)\,d^\nu q}{H_+(q;z)}.$$

First consider the case $z < -\varepsilon$, $z = -\varepsilon - x$, $x > 0$. Then

$$\begin{aligned}\widehat{M}_-(z) &= x - \alpha^2 \int_{\mathbb{R}^\nu} \frac{\lambda^2(q)\, d^\nu q}{2\varepsilon + x + \omega(q) - \alpha^2 \int_{\mathbb{R}^\nu} \frac{\lambda^2(q')\, d^\nu q'}{\omega(q)+\omega(q')+x}} \\ &= x - \alpha^2 \int \frac{\lambda^2(q)\, d^\nu q}{2\varepsilon + x + \omega(q)} \\ &\quad - \alpha^4 \int_{\mathbb{R}^\nu}\int_{\mathbb{R}^\nu} \frac{d^\nu q\, d^\nu q'}{(2\varepsilon + x + \omega(q))^2(\omega(q) + \omega(q') + x)} + O(\alpha^6).\end{aligned}$$

This relation and (5.20) imply that

$$\begin{aligned}M_-(z) =& \widehat{M}_-(z) + \alpha^4 \int_{\mathbb{R}^\nu} \frac{\Delta_+(q;z)\lambda(q)\, d^\nu q}{H_+(q;z)} \\ =& x - \alpha^2 \int_{\mathbb{R}^\nu} \frac{\lambda^2(q)\, d^\nu q}{2\varepsilon + x + \omega(q)} \\ & - \alpha^4 \int_{\mathbb{R}^\nu}\int_{\mathbb{R}^\nu} \frac{\lambda^2(q)\lambda^2(q')}{\omega(q) + \omega(q') + x} \\ & \times \left[\frac{1}{(2\varepsilon + x + \omega(q))^2} - \frac{1}{(\omega(q) + 2\varepsilon + x)(\omega(q') + 2\varepsilon + x)}\right] \\ & \times d^\nu q\, d^\nu q' + O(\alpha^6).\end{aligned}$$

Because the integral coefficient of α^4 can be written in the form

$$\int_{\mathbb{R}^\nu}\int_{\mathbb{R}^\nu} \frac{\lambda^2(q)\lambda^2(q')}{\omega(q) + \omega(q') + x}\left[\frac{1}{2}\frac{1}{(2\varepsilon + x + \omega(q))^2} + \frac{1}{2}\frac{1}{(2\varepsilon + x + \omega(q'))^2} - \frac{1}{(2\varepsilon + x + \omega(q))(2\varepsilon + x + \omega(q'))}\right] d^\nu q\, d^\nu q',$$

the inequality

$$\frac{1}{2}\left(\frac{1}{a^2} + \frac{1}{b^2}\right) > \frac{1}{ab}$$

yields

$$M_-(z) < x - \alpha^2 \int \frac{\lambda^2(q)\, d^\nu q}{2\varepsilon + x + \omega(q)} = m_-(z). \tag{5.28}$$

From (5.21) it follows that

$$M_-'(z) = -1 + O(\alpha^2) < 0. \tag{5.29}$$

From (5.28) and (5.29) we see that $M_-(z)$ has a unique simple zero $z = E_1$ with $E_1 < E_0$, $E_0 - E_1 \sim O(\alpha^4)$. Obviously E_1 is an eigenvalue of H and the corresponding normalized eigenvector Ψ_g has the components

$$(\Psi_g)_0^{(-)} = 1/T, \qquad (\Psi_g)_0^{(+)}, \qquad (\Psi_g)_1^{(-)}(k) = 0,$$
$$(\Psi_g)_1^{(+)}(k) = \frac{-\alpha\lambda(k) + \alpha^3\Delta_+(k, E_1)}{H_-(k, E_1)T},$$
$$(\Psi_g)_2^{(-)}(k_1, k_2) = -\frac{\alpha(\lambda(k_1)(\Psi_g)_1^{(+)}(k_2) + \lambda(k_2)(\Psi_g)_2^{(+)}(k_1))}{-\varepsilon + \omega(k_1) + \omega(k_2) - E_1},$$
$$(\Psi_g)_2^{(+)}(k_1, k_2) = 0,$$

where T is the normalization factor. If a vector $G \in \mathcal{H}$ is orthogonal to Ψ_g, then the function $D_+(z) = D_+(z; G)$ vanishes at $z = E_1$. Indeed, for any $G \in \mathcal{H}$, the component

$$(R_H(z)G)_0^{(+)} = F_0^{(+)}(z; G) = D_+(z; G)/M_+(z)$$

has a pole at $z = E_1$. On the other hand, it is known that the residue of the resolvent $R_H(z)$ at isolated points of the spectrum E_1 is P_{Ψ_g} (the projector into the subspace $\{c\Psi_g\}$). Hence it follows that

$$D_+(z; G) = 0,$$

if $P_{\Psi_g}G = 0$. Thus $F_0^{(+)}(z)$ is an analytic function outside the cut $(-\varepsilon, \infty)$. Now let us show that $M_-(z + i0) \neq 0$ by $z > -\varepsilon$. Indeed, if $|-\varepsilon - z| < \varepsilon$, we have $\varepsilon - z + \omega(q) > 0$, and we get

$$M_-(z) = -\varepsilon - z - \alpha^2 \int_{\mathbb{R}^\nu} \frac{\lambda^2(g)\, d^\nu q}{\varepsilon - z + \omega(q)} + O(\alpha^4) = m_-(z) + O(\alpha^4).$$

Since $m_-(z) > \alpha^2 \operatorname{const}$, $(z > -\varepsilon)$, we have $|M_-(z \pm i0)| > \alpha^2 \operatorname{const}$ for $-\varepsilon < z < 0$. If $z > 0$, then $|M_-(z \pm i0)| > \varepsilon + O(\alpha^2) > \varepsilon/2$. Thus, the functions $M_-(z \pm i0)$ have no zero for $z > -\varepsilon$. Now consider the function $L(z) = M_-(z)/(E_1 - z)$. Evidently this function is continuous on $\widehat{\Pi}_+$, analytic outside the cut $(-\varepsilon, \infty)$ and $L(z) \neq 0$, $z \in \widehat{\Pi}_+$. Let us show now that $L(z) = 1 + N(z) \in \widehat{\mathcal{B}}_+$, where

$$N(z) = \frac{-\varepsilon - E_1 - \alpha^2 \int [\lambda^2(q) + \alpha^2\Delta_+(q; z)](H_+(q; z))^{-1} d^\nu q}{E_1 - z}.$$

Since $N(z)$ is continuous on $\widehat{\Pi}_+$ and analytic outside the cut $[-\varepsilon; \infty)$, it is sufficient to establish representations (5.21a). Now let us consider $L(x \pm i0) = 1 + N(x \pm i0)$ for real x and show that the Fourier transforms of $N(x \pm i0)$ belong to $L_1(\mathbb{R}^1)$. Let $\varphi_1(x) \geqslant 0$, $\varphi_2(x) \geqslant 0$ be two smooth functions (see Figure 2) such that

$$\varphi_1(x) + \varphi_2(x) = 1,$$
$$\operatorname{supp} \varphi_1 \subset [E_1 - \delta, E_1 + \delta], \qquad E_1 + \delta < -\varepsilon,$$
$$\operatorname{supp} \varphi_2 \subset (-\infty, E_1 - \delta') \cup (E_1 + \delta', \infty), \qquad \delta' < \delta.$$

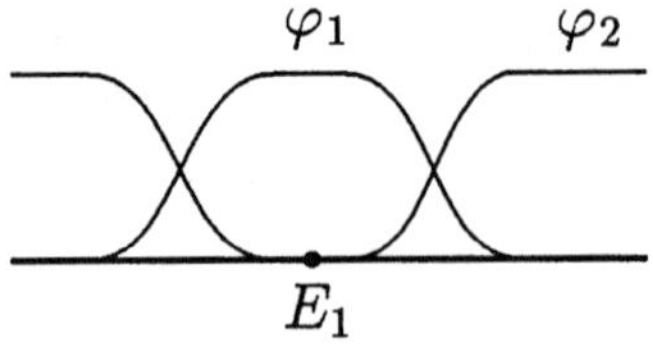

FIGURE 2

Then

$$\varphi_1(x)\,L(x+i0) = \varphi_1(x) + \varphi_1(x)\,N(x+i0)$$

are sufficiently smooth finite functions and hence their Fourier transforms belong to $L_1(\mathbb{R}^1)$, i.e., the Fourier transforms of $\varphi_1(x)\,N(x \pm i0)$ belong to $L_1(\mathbb{R}^\nu)$. Further

$$N(x \pm i0) = N_1(x \pm i0) + N_2(x \pm i0),$$

where

$$N_1(x \pm i0) = \frac{-\varepsilon - E_1 - \alpha^2 \int_{\mathbb{R}^\nu} \lambda^2(q)(H_+(q; x \pm i0))^{-1} d^\nu q}{E_1 - x},$$

$$N_2(x \pm i0) = \frac{\alpha^4 \int_{\mathbb{R}^\nu} \Delta_+(q; x \pm i0)\,\lambda(q)(H_+(q; x \pm i0))^{-1} d^\nu q}{E_1 - x}.$$

The functions $\varphi_2(x)\,N_1(x \pm i0)$ are sufficiently smooth functions (on $\operatorname{supp}\varphi_2$) and their Fourier transforms (which must be understood in the sense of principal values) belong to $L_1(\mathbb{R}^1)$. Since the Fourier transforms of the functions $\int_{\mathbb{R}^\nu} \Delta_+(q; x \pm i0)(H_+(q; x \pm i0))^{-1} d^\nu q$ belong to $L_1(\mathbb{R}^1)$, they retain that property after multiplication by the smooth function $\varphi_2(x)/(E_1 - x)$. Thus the Fourier transforms of $N(x \pm i0)$ belong to $L_1(\mathbb{R}^1)$. This establishes representations (5.21a). Thus $L(z) \in \widehat{\mathcal{B}}_+$. Hence it follows that

$$1/L(z) \in \widehat{\mathcal{B}}_+ .$$

Indeed, $1/L(z)$ is continuous on $\widehat{\Pi}_+$ and analytic outside the cut $[-\varepsilon, \infty)$. By Wiener's famous theorem (see [**6**])

$$1/L(x \pm i0) = 1 + K(x \pm i0),$$

where $K(x \pm i0)$ have integrable Fourier transforms.

Repeating the above arguments, we can prove that $D_-(z)/(E_1 - z) \in \mathcal{B}_+$ and consequently $F_0^{(-)}(z) \in \mathcal{B}_+$. Lemma 4 is proved.

Now we state the results of this section in the form the following theorem.

THEOREM 4. *Let $G = \{G_0^{(\sigma)}, G_1^{(\sigma)}, G_2^{(\sigma)}\}$ satisfy $G_1^\sigma \in D_1$, $G_2 = \{G_2^{(\sigma)}\} \in D_2$ and $G \in \mathcal{H}_{\rm ac} = \{c\Psi_g\}$, where Ψ_g is the unique eigenvector of H. Then*

$$F = F(z) = R_H(z)G = \{F_0^{(\sigma)}(z), F_1^{(\sigma)}(k; z), F_2^{(\sigma)}(k_1, k_2; z)\} \in \mathcal{H},$$
$$z \notin [E_0, \infty)$$

possess the following properties.

1) *All components are analytic on* $z \notin [E_0, \infty)$. *The components* $F_0^{(\sigma)}(z)$ *and* $F_1^{(\sigma)}(k, z)$ *(for any fixed* k*) have continuations by continuity to the banks of the cut* $[E_0; +\infty)$.

2) *We have*

$$F_0^{(\sigma)}(z) \in \mathcal{B}_{-\sigma}, \quad \sigma = \pm; \tag{5.30}$$

3) *We have*

$$F_1^{(\sigma)}(k; z) = \frac{F_0^{(-\sigma)}(z)[-\alpha\lambda(k) + \alpha^3\Delta_\sigma(k; z)] + \widetilde{\Gamma}^{(\sigma)}(k; z) + \widetilde{\Phi}^{(\sigma)}(k; z)}{H_\sigma(k; z)}, \tag{5.31}$$

where

$$\begin{gathered} \widetilde{\Gamma}^{(\sigma)}(k; z) = G_1^{(\sigma)}(k) - \alpha \int_{\mathbb{R}^\nu} \frac{G_2^{(-\sigma)}(q, k)\,\lambda(q)\, d^\nu q}{-\varepsilon\sigma + \omega(k) + \omega(q) - z}, \\ \Delta_\sigma(k; z) \in \mathcal{A}_\sigma, \quad \widetilde{\Phi}^{(\sigma)}(k; z) \in \mathcal{A}_\sigma. \end{gathered} \tag{5.32}$$

4) *We have*

$$\begin{aligned} &F_2^{(\sigma)}(k_1, k_2, z) \\ &\quad = \frac{G_2^{(\sigma)}(k_1, k_2) - \alpha(\lambda(k_1) F_1^{(-\sigma)}(k_2; z) + \lambda(k_2) F_1^{(-\sigma)}(k_1; z))}{\varepsilon\sigma + \omega(k_1) + \omega(k_2) - z}. \end{aligned} \tag{5.33}$$

Here $\widehat{\Pi}_\sigma$ *is the complex plane with the cuts* $[E_0, \infty)$ *for* $\sigma = -$ *and* $[-\varepsilon, \infty)$ *for* $\sigma = +$ *together with their banks. The spaces* $\mathcal{A}_\sigma$ *and* $\mathcal{B}_\sigma$ *are defined in* 5.2 *and* 5.3*, respectively.*

§6. Proof of Theorem 3

Denote $\overline{W}_t^{\,\mathrm{inv}} = e^{it\widehat{H}} \bar{J}^* e^{-itH} P_{\mathcal{H}_{\mathrm{ac}}}$. Then

$$d\overline{W}_t^{\,\mathrm{inv}} G/dt = i\,(e^{it\widehat{H}}(\widehat{H}\bar{J}^* - \bar{J}^* H)e^{-itH})P_{\mathcal{H}_{\mathrm{ac}}} G.$$

Again, we must calculate

$$\widehat{H}\bar{J}^* - \bar{J}^* H = (\widehat{H}_1 \bar{J}_1^* - \bar{J}_1^* H) + (\widehat{H}_2 \bar{J}_2^* - \bar{J}_2^* H).$$

The following formulas hold:

a)

$$\begin{aligned} &((\widehat{H}_1 \bar{J}_1^* - \bar{J}_1^* H)G)_1(k) \\ &\quad = -\alpha\lambda(k)\Big[G_0^{(+)}\overline{\Psi}_0^{(-)} + \int_{\mathbb{R}^\nu} G_1^{(-)}(q)\overline{\Psi}_1^{(+)}(q)\, d^\nu q\Big], \end{aligned} \tag{6.1}$$

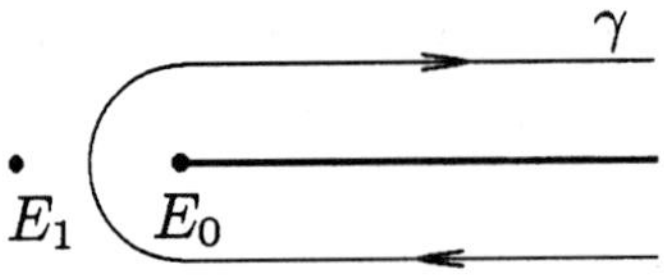

FIGURE 3

b)

$$
(6.2)\quad
\begin{aligned}
&((\widehat{H}_2 \bar{J}_2^* - \bar{J}_2^* H)G)_2^{(-)}(k_1, k_2) = -\alpha(\lambda(k_1)G_1^{(+)}(k_2) + \lambda(k_2)G_1^{(+)}(k_1)),\\
&((\widehat{H}_2 \bar{J}_2^* - \bar{J}_2^* H)G)_2^{(+)}(k_1, k_2)\\
&\quad = -\alpha\lambda(k_1)\Big[G_1^{(-)}(k_2)\|\Psi_1^{(+)}\|^2 - \Psi_0^{(-)}\int_{\mathbb{R}^\nu}\overline{\Psi}_1^{(+)}(q)\,G_2^{(+)}(q, k_2)\,d^\nu q\\
&\qquad\qquad - \Psi_1^{(+)}(k_2)\Big(\overline{\Psi}_0^{(-)}G_0^{(+)} + \int_{\mathbb{R}^\nu}\overline{\Psi}_1^{(+)}(q)\,G_1^{(-)}(q)\,d^\nu q\Big)\Big]\\
&\quad - \alpha\lambda(k_2)\Big[\qquad k_2 \leftrightarrows k_1 \qquad\Big],
\end{aligned}
$$

where the expression in the second square brackets is the same as in the first, with k_1 and k_2 interchanged. As above, we must estimate the norm $\|(\widehat{H}\bar{J}^* - \bar{J}^*H)e^{-itH}P_{\mathcal{H}_{ac}}G\|_{\widehat{\mathcal{H}}}$ for $G = \{G_0^{(\sigma)}, G_1^{(\sigma)}, G_2^{(\sigma)}\}$, $G_1^{(\sigma)} \in D_1$, $G_2 \in D_2$. We write

$$
e^{-itH}P_{\mathcal{H}_{ac}} = \frac{1}{2\pi i}\int_\gamma e^{-izt}R_H(z)\,dz,
$$

where $R_H(z)$ is a resolvent of H and γ is a contour on the complex plane that encircles the continuous spectrum of H (clockwise), so that the eigenvalue E_1 lies outside of γ (see Figure 3).

1. First, we must estimate the norm

$$
(6.3)\quad
\begin{aligned}
&\|((\widehat{H}_1\bar{J}_1^* - \bar{J}_1^*H)e^{-iHt}G)_1\|_{L_2(\mathbb{R}^\nu)}\\
&\quad = \left\| -\alpha\lambda(k)\int_\gamma\Big(F_0^{(+)}(z)\overline{\Psi}_0^{(-)} + \int_{\mathbb{R}^\nu}\overline{\Psi}_1^{(+)}(q)\,F_1^{(-)}(q;z)\,d^\nu q\Big)e^{-izt}dz\right\|\\
&\quad = |\alpha|\|\lambda\|_{L_2(\mathbb{R}^\nu)}\Big|\overline{\Psi}_0^{(-)}\int_\gamma F_0^{(+)}(z)e^{-izt}dz\\
&\qquad + \int_\gamma e^{-izt}dz\int_{\mathbb{R}^\nu}\overline{\Psi}_1^{(+)}(q)\,F_1^{(-)}(q;z)\,d^\nu q\Big|.
\end{aligned}
$$

Since both functions $F_0^{(+)}(z)$ and $\int_{\mathbb{R}^\nu}\overline{\Psi}_1^{(+)}(q)\,F_1^{(-)}(q,z)\,d^\nu q$ are analytic outside the cut $[E_0;\infty)$ and decreasing at ∞, the integral along the contour γ is equal to the sum of the two integrals

$$
\int_\gamma \dots\, dz = \int_{-\infty+i0}^{+\infty+i0} \dots\, dx + \int_{+\infty-i0}^{-\infty-i0} \dots\, dx\,.
$$

That is,

$$(6.4)\quad \int_\gamma F_0^{(+)}(z)e^{-izt}dz = \int_{-\infty}^{\infty} F_0^{(+)}(x+i0)e^{-ixt}dx - \int_{-\infty}^{\infty} F_0^{(+)}(x-i0)e^{-ixt}dx$$

and similarly for the integral

$$\int_\gamma e^{-izt}dz\left(\int_{\mathbb{R}^\nu} \overline{\Psi}_1^{(+)}(q)\,F_1^{(-)}(q;z)\,d^\nu q\right).$$

Since $F_0^{(+)} \in \mathcal{B}_-$, both integrals in (6.4) as functions of t belong to $L_1(\mathbb{R}^1)$. Let us show that

$$(6.5)\quad \int_{\mathbb{R}^\nu} \overline{\Psi}_1^{(+)}(q)\,F_1^{(-)}(q;z)\,d^\nu q \in \mathcal{B}_-\,.$$

To that end, we represent $F_1^{(-)}(q;z)$ in the form (5.31) and prove (6.3) for every term (5.31), using the above arguments (see the proof of the first assertion of Lemma 5.4). Hence we see that $G_1^{(-)} \in D_1$, $G_2 \in D_2$, $G \in \mathcal{H}_{\rm ac}$, imply

$$\|((\widehat{H}_1\overline{J}_1^* - \overline{J}_1^*H)\,e^{-itH}G)_1\|_{L_2(\mathbb{R}^\nu)} \in L_1(\mathbb{R}^1)$$

(as a function of t). Hence the following limit exists:

$$(6.6)\quad \lim_{t\to\infty}(\overline{W}_t^{\rm inv}G)_1 = (\overline{W}_+G)_1 \in \widehat{\mathcal{H}}_1\,.$$

2. Let us pass to the estimate of the norms

$$\|((\widehat{H}_2\overline{J}_2^* - \overline{J}_2^*H)\,e^{-iHt}G)_2^{(\sigma)}\|_{L_2^{\rm sym}(\mathbb{R}^\nu\times\mathbb{R}^\nu)}, \qquad \sigma = \pm,$$

for $G_1^{(\sigma)} \in D_1$, $G_2 \in D_2$, $G \in \mathcal{H}_{\rm ac}$.

a) $$\begin{aligned}&\|((\widehat{H}_2\overline{J}_2^* - \overline{J}_2^*H)\,e^{-iHt}G)_2^{(-)}\|_{L_2^{\rm sym}(\mathbb{R}^\nu\times\mathbb{R}^\nu)}\\ &= \left\| -\alpha\left(\lambda(k_1)\int_\gamma F_1^{(+)}(k_2;z)\,e^{-izt}dz \right.\right.\\ &\qquad\left.\left. + \lambda(k_2)\int_\gamma F_1^{(+)}(k_1;z)\,e^{-izt}dz\right)\right\|_{L_2^{\rm sym}(\mathbb{R}^\nu\times\mathbb{R}^\nu)}\\ &\leqslant |\alpha|\|\lambda\|_{L_2(\mathbb{R}^\nu)}\left\|\int F_1^{(+)}(k_2;z)\,e^{-izt}dz\right\|_{L_2(\mathbb{R}^\nu)}.\end{aligned}$$

Again, using the decomposition (5.31) of $F_1^{(+)}$, let us consider all its terms.

1. We have

$$\int_\gamma F_0^{(+)}(z)\,\frac{-\alpha\lambda(k)+\alpha^3\Delta_+(k,z)}{H_+(k_2;z)}\,e^{-izt}dz$$
$$= -\alpha\lambda(k)\Big[\int_{-\infty}^{\infty}\Big(F_0^{(+)}(x+i0)\,\frac{e^{-ixt}}{H_+(k_2;x+i0)}$$
$$- F_0^{(+)}(x-i0)\,\frac{e^{-ixt}}{H_+(k_2,x-i0)}\Big)dx\Big] \tag{6.7}$$
$$+\alpha^3\lambda(k)\Big[\int_{-\infty}^{\infty}\Big(F_0^{(+)}(x+i0)\,\frac{g_+(\omega(k_2);x+i0)}{H_+(k_2;x+i0)}$$
$$- F_0^{(+)}(x-i0)\,\frac{g_+(\omega(k_2);x-i0)}{H_+(k_2;x-i0)}\Big)e^{-ixt}dx\Big].$$

Here we have used the representation

$$\Delta_+(k;z)=\lambda(k)\,g_+(\omega(k);z),\qquad g_+(r,z)\in\eta^{(+)}.$$

Using the above arguments it is easy to prove that $(H_+(k;z))^{-1}\in\mathcal{B}_+$, $g_+(\omega(k);z)\in\mathcal{B}_+$ for any $k\in\mathbb{R}^\nu$ and

$$\max_k\{\|(H_+(k;z))^{-1}\|_{\mathcal{B}_+},\|g_+(\omega(k);z)\|_{\mathcal{B}_+}\}<\infty.$$

Hence

$$\Big\|\lambda(k)\int_{-\infty}^{\infty}F_0^{(+)}(x+i0)\,\frac{e^{-ixt}}{H_+(k;x+i0)}\,dx\Big\|_{L_2(\mathbb{R}^\nu)}\in L_1(\mathbb{R}^1)$$

(as a function of t). The same is true for the other integrals in (6.7).

2. We have

$$\Big\|\int_\gamma\frac{\widetilde{\Gamma}^{(+)}(k;z)}{H_+(k;z)}\,e^{-izt}dz\Big\|_{L_2(\mathbb{R}^\nu)}$$
$$\leqslant\Big\|G_1^{(+)}(k)\int_\gamma\frac{e^{-izt}}{H_+(k;z)}\,dz\Big\|_{L_2(\mathbb{R}^\nu)} \tag{6.8}$$
$$+\alpha\Big\|\int_\gamma\Big(\int_{\mathbb{R}^\nu}\frac{G_2^{(-)}(q,k)\,\lambda(q)\,d^\nu q}{-\varepsilon+\omega(k)+\omega(q)-z}\Big)\frac{e^{-izt}dz}{H_+(k;z)}\Big\|_{L_2(\mathbb{R}^\nu)}.$$

The first term in the right-hand side of (6.8) is treated as above, and the estimate of the second one is obtained if we note that for $G_2^{(-)}(q,k)=f(q)\,f(k)$, $f\in D_1$,

$$\int_{\mathbb{R}^\nu}\frac{G_2^{(-)}(q,k)\,\lambda(q)\,d^\nu q}{-\varepsilon+\omega(k)+\omega(q)-z}=f(k)\int_{\mathbb{R}^\nu}\frac{f(q)\,\lambda(q)\,d^\nu q}{-\varepsilon+\omega(k)+\omega(q)-z},$$

and

$$\int_{\mathbb{R}^\nu}\frac{f(q)\,\lambda(q)\,d^\nu q}{-\varepsilon+\omega(q)+\omega(k)-z}\in\mathcal{B}_+$$

for any $k\in\mathbb{R}^\nu$, while

$$\max_{k\in\mathbb{R}^\nu}\Big\|\int_{\mathbb{R}^\nu}\frac{f(q)\,\lambda(q)\,d^\nu q}{-\varepsilon+\omega(k)+\omega(q)-z}\Big\|_{\mathcal{B}_+}<\infty.$$

Hence

$$\left\| \int_\gamma \left(\int_{\mathbb{R}^\nu} \frac{G_2^{(-)}(q,k)\,\lambda(q)\,d^\nu q}{-\varepsilon + \omega(q) + \omega(k) - z} \right) \frac{e^{-izt} dz}{H_+(k;z)} \right\|_{L_2(\mathbb{R}^\nu)} \in L_1(\mathbb{R}^1).$$

The same is true for any $G_2 \in D_2$.

3. The third term of $F_1^{(+)}$ is studied as above and we get

$$\left\| \int_\gamma \frac{\widetilde{\Phi}_+(k;z)}{H_+(k;z)} e^{-izt} dz \right\|_{L_2(\mathbb{R}^\nu)} \in L_1(\mathbb{R}^1).$$

Now consider the norm

(6.9)

$$\begin{aligned} &\|((\widehat{H}_2 \bar{J}^* - \bar{J}^* H) e^{-iHt} G)_2^{(+)}\|_{L_2^{\mathrm{sym}}(\mathbb{R}^\nu \times \mathbb{R}^\nu)} \\ &\quad \leqslant \left\| \int_\gamma e^{-izt} dz \Big(- \alpha\lambda(k_1) \Big[\|\Psi_1^{(+)}\|^2 F_1^{(-)}(k_2;z) \right. \\ &\qquad\qquad \left. - \Psi_0^{(-)} \int \overline{\Psi}_1^{(+)}(q)\, F_2^{(+)}(k_2,q;z)\, d^\nu q \Big] \Big) \right\|_{L_2(\mathbb{R}^\nu \times \mathbb{R}^\nu)} \\ &\quad + |\alpha| \|\lambda\|_{L_2(\mathbb{R}^\nu)} \|\Psi_1^{(+)}\|^2_{L_2(\mathbb{R}^\nu)} \\ &\qquad \times \left| \int_\gamma \left(\overline{\Psi}_0^{(-)} F_0^{(+)}(z) + \int_{\mathbb{R}^\nu} \overline{\Psi}_1^{(+)}(q)\, F_1^{(-)}(q;z)\, d^\nu q \right) e^{-izt} dz \right|. \end{aligned}$$

Notice that

(6.10)

$$\begin{aligned} &\|\Psi_1^{(+)}\|^2 F_1^{(-)}(k_2;z) - \Psi_0^{(-)} \\ &\quad \times \int_{\mathbb{R}^\nu} \overline{\Psi}_1^{(+)}(q) \Big[\frac{G_2^{(+)}(q,k_2)}{\varepsilon + \omega(k_2) + \omega(q) - z} \\ &\qquad\qquad - \frac{\alpha\lambda(q)\, F_1^{(-)}(k_2;z) - \alpha\lambda(k_2)\, F_1^{(-)}(q;z)}{\varepsilon + \omega(k_2) + \omega(q) - z} \Big]\, d^\nu q \\ &= \left(\|\Psi_1^{(+)}\| + \alpha\Psi_0^{(-)} \int_{\mathbb{R}^\nu} \overline{\Psi}_1^{(+)}(q)\, \frac{\lambda(q)\, d^\nu q}{\varepsilon + \omega(k_2) + \omega(q) - z} \right) F_1^{(-)}(k_2;z) \\ &\quad - \Psi_0^{(-)} \int_{\mathbb{R}^\nu} \overline{\Psi}_1^{(+)}(q)\, \frac{G_2^{(+)}(q,k_2)\, d^\nu q}{\varepsilon + \omega(k_2) + \omega(q) - z} \\ &\quad + \alpha\lambda(k_2)\, \Psi_0^{(-)} \int_{\mathbb{R}^\nu} \frac{\overline{\Psi}_1^{(+)}(q)\, F_1^{(-)}(q;z)\, d^\nu q}{\varepsilon + \omega(k_2) + \omega(q) - z}. \end{aligned}$$

We can write the first term in (6.10) using the decomposition (5.31) of $F_1^{(-)}(k;z)$:

$$\begin{aligned} &\frac{1}{H_-(k_2;z)} \left(\|\Psi_1^{(+)}\|^2 + \alpha\Psi_0^{(-)} \int_{\mathbb{R}^\nu} \overline{\Psi}_1^{(+)}(q)\, \frac{\lambda(q)\, d^\nu q}{\varepsilon + \omega(k_2) + \omega(q) - z} \right) \\ &\qquad \times F_0^{(-)}(z) (\alpha\lambda(k_2) + \alpha^3 \Delta_-(k_2,z) + \widetilde{\Gamma}^{(-)}(k_2;z) + \widetilde{\Phi}_-(k_2;z)). \end{aligned}$$

Denote the following function by μ:

$$\frac{1}{m_-(\zeta)} \left(\|\Psi_1^{(+)}\|^2 + \alpha\Psi_0^{(-)} \int \overline{\Psi}_1^{(+)}(q)\, \frac{\lambda(q)\, d^\nu q}{\varepsilon + \omega(q) - \zeta} \right) \equiv \mu(\zeta).$$

From (3.2) it is follows that

$$\alpha\Psi_0^{(-)}\int_{\mathbb{R}^\nu}\overline{\Psi}_1^{(+)}(q)\,\frac{\lambda(q)\,d^\nu q}{\varepsilon+\omega(q)-\zeta}\bigg|_{\zeta=E_0}=-\|\Psi_1^{(+)}\|^2$$

and hence the function $\mu(\zeta)$ is analytic outside the cut $(\varepsilon;+\infty)$. Repeating the previous arguments, we can prove that $\mu(\zeta)\in\mathcal{B}_-$ and

$$\|\mu(z-\omega(k))\|_{\mathcal{B}_-}<\text{const} \tag{6.10a}$$

uniformly in $k\in\mathbb{R}^\nu$. Therefore, reasoning as above, we can prove that

$$\bigg\|\int_\gamma \mu(z-\omega(k_2))\big[F_0^{(-)}(z)(-\alpha\lambda(k)+\alpha^3\Delta_-(k_2;z)) + \widetilde{\Gamma}^{(-)}(k_2;z)+\widetilde{\Phi}^{(-)}(k_2;z)\big]\,e^{-izt}dz\bigg\|_{L_2(\mathbb{R}^\nu)}\in L_1(\mathbb{R}^1). \tag{6.10b}$$

Further,

$$\bigg\|\int_\gamma\bigg(\int_{\mathbb{R}^\nu}\overline{\Psi}_1^{(+)}(q)\,\frac{G_2^+(q,k)\,d^\nu q}{\varepsilon+\omega(k_2)+\omega(q)-z}\bigg)e^{-izt}dz\bigg\|_{L_2(\mathbb{R}^\nu)}\in L_1(\mathbb{R}^1)$$

for $G_2\in D_2$, and the same is true for

$$\bigg\|\lambda(k_2)\int_\gamma\bigg(\int_{\mathbb{R}^\nu}\frac{\overline{\Psi}_1^{(+)}(q)\,F_1^{(-)}(q;z)\,d^\nu q}{\varepsilon+\omega(q)+\omega(k_2)-z}\bigg)e^{-izt}dz\bigg\|_{L_2(\mathbb{R}^\nu)}.$$

Finally, it follows from previous arguments that

$$\int_\gamma\bigg(\overline{\Psi}_0^{(-)}F_0^{(+)}(z)+\int_{R^\nu}\overline{\Psi}_1^{(+)}(q)\,F_1^{(-)}(q;z)\,d^\nu q\bigg)e^{-izt}dz\in L_1(\mathbb{R}^\nu).$$

Thus,

$$\|((\widehat{H}_2\bar{J}^*-\bar{J}_2^*H)e^{-iHt}G)_2^{(\sigma)}\|_{L_2^{\text{sym}}(\mathbb{R}^\nu\times\mathbb{R}^\nu)}\in L_1(\mathbb{R}^1),\qquad \sigma=\pm,$$

for $G=(G_0^{(\sigma)},G_1^{(\sigma)},G_2^{(\sigma)})\in\mathcal{H}_{\text{ac}}$ and $G_1^{(\sigma)}\in D_1$, $G_2\in D_2$. This means that the following limit exists

$$\overline{W}_+=\operatorname*{s-lim}_{t\to\infty}\overline{W}_t^{\text{inv}}.$$

Now we must prove that $\overline{W}_+$ is isometric, i.e.,

$$(\overline{W}_+G,\overline{W}_+G')_{\widehat{\mathcal{H}}}=(G,G')_{\mathcal{H}}. \tag{6.11}$$

We have

$$\begin{aligned}(\overline{W}_+G,\overline{W}_+G')&=\lim_{t\to\infty}(\overline{W}_t^{\text{inv}}G,\overline{W}_t^{\text{inv}}G')_{\widehat{\mathcal{H}}}=\lim_{t\to\infty}(e^{-iHt}G,\bar{J}\bar{J}^*e^{-itH}G')_{\mathcal{H}}\\ &=(G,G')_{\mathcal{H}}+\lim_{t\to\infty}(e^{-iHt}G,(\bar{J}\bar{J}^*-E)e^{-iHt}G')_{\mathcal{H}}.\end{aligned}$$

Now assertion (6.11) follows from the following lemma.

LEMMA 1. *If $G\in\mathcal{H}_{\text{ac}}$ and $G_1^{(\sigma)}\in D_1$, $G_2\in D_2$, then*

$$\|(\bar{J}\bar{J}^*-E)e^{-iHt}G\|_{\mathcal{H}}\in L_1(\mathbb{R}^1)$$

as a function of t.

PROOF. First we must calculate the action $\bar{J}\bar{J}^*\colon \mathcal{H} \to \mathcal{H}$. The following formulas hold:

(6.12)

$$(\bar{J}\bar{J}^*)_0^{(\sigma)} = 0, \quad \sigma = \pm,$$

$$(\bar{J}\bar{J}^*G)_1^{(+)}(k) = 0,$$

$$\begin{aligned}(\bar{J}\bar{J}^*G)_1^{(-)}(k) &= |\Psi_0^{(-)}|^2(1 + \|\Psi_1^{(+)}\|^2)G_1^{(-)}(k) + \Psi_0^{(-)}\|\Psi_1^{(+)}\|^2 \int_{\mathbb{R}^\nu} G_2^{(+)}(q,k)\overline{\Psi}_1^{(+)}(q)\, d^\nu q \\ &\quad + \Psi_0^{(-)}\Psi_1^{(+)}(k) \int_{\mathbb{R}^\nu}\int_{\mathbb{R}^\nu} G_2^{(+)}(q,q')\,\overline{\Psi}_1^{(+)}(q)\,\overline{\Psi}_1^{(+)}(q')\, d^\nu q\, d^\nu q' \\ &\quad + |\Psi_0^{(-)}|^2\Psi_1^{(+)}(k)\int_{\mathbb{R}^\nu} \overline{\Psi}_1^{(+)}(q)\,G_1^{(-)}(q)\, d^\nu q,\end{aligned}$$

$$\begin{aligned}(\bar{J}\bar{J}^*G)_2^{(+)}(k_1,k_2) &= G_2^{(+)}(k_1,k_2) - \Psi_1^{(+)}(k_1)\Big[|\Psi_0^{(-)}|^2 \int_{\mathbb{R}^\nu} G_2^{(+)}(q,k_2)\,\overline{\Psi}_1^{(+)}(q)\, d^\nu q \\ &\qquad\qquad - \overline{\Psi}_0^{(-)}\|\Psi_1^{(+)}\|^2 G_1^{(-)}(k_2)\Big] \\ &\quad - \Psi_1^{(+)}(k_2)\Big[\qquad k_1 \leftrightarrows k_2 \qquad\Big] \\ &\quad + 2\Psi_1^{(+)}(k_1)\,\Psi_1^{(+)}(k_2)\Big[\overline{\Psi}_0^{(-)}\int_{\mathbb{R}^\nu}\overline{\Psi}_1^{(+)}(q)\,G_1^{(-)}(q)\, d^\nu q \\ &\qquad\qquad + \int_{\mathbb{R}^\nu}\int_{\mathbb{R}^\nu} G_2^{(+)}(q,q')\,\overline{\Psi}_1^{(+)}(q)\,\overline{\Psi}_1^{(+)}(q')\, d^\nu q\, d^\nu q'\Big],\end{aligned}$$

$$(\bar{J}\bar{J}^*G)_2^{(-)}(k_1,k_2) = G_2^{(-)}(k_1,k_2).$$

From (6.12) if $G \in \mathcal{H}_{\mathrm{ac}}$ and $G_1^\sigma \in D_1$, $\sigma = \pm$, $G_2 \in D_2$, we get

a) $$((\bar{J}\bar{J}^* - E)e^{-itH}G)_0^{(\sigma)} = -\int_\gamma e^{-izt}F_0^{(\sigma)}(z)\, dz, \qquad \sigma = \pm,$$

b) $$((\bar{J}\bar{J}^* - E)e^{-itH}G)_1^{(+)}(k) = -\int_\gamma e^{-izt}F_1^{(+)}(k;z)\, dz,$$

c) $$\begin{aligned}((\bar{J}\bar{J}^* - E)e^{-itH}G)_1^{(-)}(k) &= \int_\gamma e^{-izt}dz\Big\{[|\Psi_0^{(-)}|^2(1 + \|\Psi_1^{(+)}\|^2) - 1]\,F_1^{(-)}(k,z) \\ &\quad + \Psi_0^{(-)}\|\Psi_1^{(+)}\|^2\int_{\mathbb{R}^\nu} F_2^{(+)}(q,k;z)\,\overline{\Psi}_1^{(+)}(q)\, d^\nu q \\ &\quad + \Psi_1^{(+)}(k)\Big(\Psi_0^{(-)}\int_{\mathbb{R}^\nu}\int_{\mathbb{R}^\nu}\overline{\Psi}_1^{(+)}\overline{\Psi}_1^{(+)}(q')\,F_2^{(+)}(q,q',z)\, d^\nu q\, d^\nu q' \\ &\qquad + |\Psi_0^{(-)}|^2\int_{\mathbb{R}^\nu}\overline{\Psi}_1^{(+)}(q)\,F_1^{(-)}(q;z)\, dq\Big)\Big\},\end{aligned}$$

d) $((\bar{J}\bar{J}^* - E)e^{-iHt}G)_2^{(-)}(k_1, k_2) = 0,$

e) $((\bar{J}\bar{J}^* - E)e^{-iHt}G)_2^{(+)}(k_1, k_2)$

$$= -\int_\gamma e^{-izt}dz\Big\{\Psi_1^{(+)}(k_1)\Big[\Psi_0^{(-)}\int_{\mathbb{R}^\nu} F_2^{(+)}(q, k_2; z)\overline{\Psi}_1^{(+)}(q)\,d^\nu q - \|\Psi_1^{(+)}\|^2\overline{\Psi}_0^{(-)}F_1^{(-)}(k; z)\Big] + \Psi_1^{(+)}(k_2)\,[\,k_1 \leftrightarrows k_2\,] - 2\Psi_1^{(+)}(k_1)\Psi_1^{(+)}(k_2) \times \Big[\overline{\Psi}_0^{(-)}\int_{\mathbb{R}^\nu}\overline{\Psi}_1^{(+)}(q)F_1^{(-)}(q; z)\,d^\nu q + \int_{\mathbb{R}^\nu}\int_{\mathbb{R}^\nu} F_2^{(+)}(q, q'; z)\overline{\Psi}_1^{(+)}(q)\overline{\Psi}_1^{(+)}(q')\,d^\nu q\,d^\nu q'\Big]\Big\}.$$

Now let us study each term in a)–e).

a) $\int_\gamma e^{-izt}F_0^{(\sigma)}(z)\,dz \in L_1(\mathbb{R}^1)$, as we saw before.

b) $\|\int_\gamma e^{-izt}F_1^{(+)}(k, z)\,dz\|_{L_2(\mathbb{R}^\nu)} \in L_1(\mathbb{R}^1)$, as we also saw before.

c) We can write the integrand in c) in the following way:

$$\Big[|\Psi_0^{(-)}|^2 + |\Psi_0^{(-)}|^2\|\Psi_1^{(+)}\|^2 - 1 + \Psi_0^{(-)}\|\Psi_1^{(+)}\|^2\int_{\mathbb{R}^\nu}\frac{\overline{\Psi}_1^{(+)}(q)(-\alpha\lambda(q))\,d^\nu q}{\varepsilon + \omega(q) + \omega(k) - z}\Big]F_1^{(-)}(k; z)$$
$$+ \Psi_0^{(-)}\|\Psi_1^{(+)}\|^2\int_{\mathbb{R}^\nu}\frac{G_2^{(+)}(q; k)\overline{\Psi}_1^{(+)}(q)\,d^\nu q}{\varepsilon + \omega(q) + \omega(k) - z}$$
$$- \alpha\lambda(k)\Psi_0^{(-)}\|\Psi_1^{(+)}\|^2\int_{\mathbb{R}^\nu}\frac{\overline{\Psi}_1^{(+)}(q)F_1^{(-)}(q; z)}{\varepsilon + \omega(q) + \omega(k) - z}\,d^\nu q$$
$$+ \Psi_1^{(+)}(k)\Big[\Psi_0^{(-)}\Big(\int_{\mathbb{R}^\nu\times\mathbb{R}^\nu}\overline{\Psi}_1^{(+)}(q)\overline{\Psi}_1^{(+)}(q')\frac{G_2^{(+)}(q, q')\,d^\nu q\,d^\nu q'}{\varepsilon + \omega(q) + \omega(q') - z} - 2\int_{\mathbb{R}^\nu\times\mathbb{R}^\nu}\frac{\alpha\lambda(q)F_1^{(-)}(q'; z)\overline{\Psi}_1^{(+)}(q)\overline{\Psi}_1^{(+)}(q')\,d^\nu q\,d^\nu q'}{\varepsilon + \omega(q) + \omega(q') - z}\Big) + |\Psi_0^{(-)}|^2\int_{\mathbb{R}^\nu}\overline{\Psi}_1^{(+)}(q)F_1^{(-)}(q; z)\,d^\nu q\Big].$$

Denote by $\chi(\zeta)$ the function

$$\chi(\zeta) = |\Psi_0^{(-)}|^2 + |\Psi_0^{(-)}|^2\|\Psi_1^{(+)}\|^2 - 1 + \Psi_0^{(-)}\|\Psi_1^{(+)}\|^2\int\frac{\overline{\Psi}^{(+)}(q)(-\alpha\lambda(q))\,d^\nu q}{\varepsilon + \omega(q) - \zeta}.$$

Obviously $\chi(\zeta) \in \widehat{\mathcal{B}}_-$ and (3.2) implies $\chi(E_0) = 0$. Hence, similarly to (6.10a),

$$\chi(z - \omega(k))/H_-(k; z) \in \mathcal{B}_- \quad \text{for any } k \in \mathbb{R}^\nu$$

and

$$\max_{k\in\mathbb{R}^\nu}\Big\|\frac{\chi(z - \omega(k))}{H_-(k; z)}\Big\|_{\mathcal{B}_-} < \infty.$$

Hence, as we saw before (see (6.10b)),

$$\left\| \int_\gamma \chi(z-\omega(k))\, F_1^{(-)}(k;z)\, e^{-izt} dz \right\|_{L_2(\mathbb{R}^\nu)} \in L_1(\mathbb{R}^1). \tag{6.13}$$

For the other terms in c), similar inclusions can be proved as above.

e) Again we can write

$$\begin{aligned}
&|\Psi_0^{(-)}|^2 \int_{\mathbb{R}^\nu} F_2^{(+)}(q;k_2;z)\,\overline{\Psi}_1^{(+)}(q)\, d^\nu q - \|\Psi_1^{(+)}\|^2 \overline{\Psi}_0^{(-)} F_1^{(-)}(k_2;z) \\
&= F_1^{(-)}(k_2;z)\left(|\Psi_0^{(-)}|^2 \int_{\mathbb{R}^\nu} \frac{-\alpha\lambda(q)\,\overline{\Psi}_1^{(+)}(q)\, d^\nu q}{\omega(k_2)+\omega(q)+\varepsilon - z} - \|\Psi_1^{(+)}\|^2 \overline{\Psi}_0^{(-)} \right) \\
&\quad + |\Psi_0^{(-)}|^2 \int_{\mathbb{R}^\nu} \frac{G_2^{(+)}(q,k_2)\,\overline{\Psi}_1^{(+)}(q)\, d^\nu q}{\omega(q)+\omega(k_2)+\varepsilon - z} \\
&\quad - \alpha\lambda(k_2) \int_{\mathbb{R}^\nu} \frac{F_1^{(-)}(q;z)\overline{\Psi}_1^{(+)}(q)\, d^\nu (q)}{\omega(q)+\omega(k_2)+\varepsilon - z}.
\end{aligned} \tag{6.14}$$

Again we consider the function

$$\pi(\zeta) = \overline{\Psi}_0^{(-)} \left(\Psi_0^{(-)} \int_{\mathbb{R}^\nu} \frac{-\alpha\lambda(q)\,\overline{\Psi}_1^{(+)}(q)\, d^\nu q}{\omega(q)+\varepsilon-\zeta} - \|\Psi_1^{(+)}\|^2 \right) \in \mathcal{B}_- .$$

From (3.2) we see that $\pi(E_0) = 0$ and therefore $\pi(z-\omega(k))/H_-(k,z) \in \mathcal{B}_-$ for any k and, as above (see (6.14)),

$$\left\| \int_\gamma \pi(z-\omega(k))\, F_1^{(-)}(k;z) e^{-izt} dz \right\|_{L_2(\mathbb{R}^\nu)} \in L_1(\mathbb{R}^1).$$

The other terms in e) and two last terms in (6.14) are treated as above. Finally,

$$\|((\bar{J}\bar{J}^* - E)\, e^{-iHt} G)_2^{(\sigma)}\|_{L_2^{\mathrm{sym}}(\mathbb{R}^\nu\times\mathbb{R}^\nu)} \in L_1(\mathbb{R}^1).$$

Lemma 1 is proved.

From the obvious relations

$$\begin{aligned}
&\lim_{t\to\infty} (e^{-iHt}G, (.J\bar{J}^* - E)\, e^{-iHt}G') \\
&\qquad = \lim_{T\to\infty} \frac{1}{T}\int_0^T (e^{-iHt}G, (\bar{J}\bar{J}^* - E)\, e^{-iHt}G')\, dt, \\
&\left| \int_0^T (e^{-iHt}G, (\bar{J}\bar{J}^* - E)\, e^{-iHt}G')\, dt \right| \leqslant \|G_1\| \int_0^T \|(\bar{J}\bar{J}^* - E)\, e^{-iHt}G'\|\, dt \\
&\qquad = O(1), \qquad T\to\infty,
\end{aligned}$$

for $G \in \mathcal{H}_{\mathrm{ac}}$, $G' \in \mathcal{H}_{\mathrm{ac}}$, $(G_1')^{(\sigma)} \in D_1$, $G_2' \in D_2$, we get (6.11). Theorem 3 and hence Theorem 1 are proved.

CONCLUDING REMARKS. 1. The operator H in $\mathcal{H}$ has four invariant subspaces: $\mathcal{H}_0$, $\mathcal{H}_1$, and $\mathcal{H}_2^{(\sigma)}$, $\sigma = \pm$:

$$\begin{aligned} \mathcal{H}_0 &= \{c\Psi_g\} \quad \text{is a one-dimensional proper subspace,} \\ \mathcal{H}_1 &= W_- \widehat{\mathcal{H}}_1, \\ \mathcal{H}_2^{(\sigma)} &= W_- \widehat{\mathcal{H}}^{(\sigma)}, \qquad \sigma = \pm. \end{aligned}$$

Here $\widehat{\mathcal{H}}_2^{(\sigma)} \subset \widehat{\mathcal{H}}_2$ is the subspace of vectors $\{G_2^{(\sigma)}\} \in \widehat{\mathcal{H}}_2$ with $G_2^{(-\sigma)} = 0$. The spectrum of H in $\mathcal{H}_0$ is the eigenvalue E_1, in $\mathcal{H}_1$ it is the Lebesgue spectrum filling the semiaxis $[E_0, \infty)$ and in $\mathcal{H}_2^{(\sigma)}$ there are also Lebesgue spectra filling $[\sigma\varepsilon; \infty)$. Note that there is an obvious decomposition of $\mathcal{H}$ into two subspaces invariant with respect to H, $\mathcal{H} = \mathcal{H}_- \oplus \mathcal{H}_+$, where

$$\begin{aligned} \mathcal{H}_- &= \{G_0^{(+)} = 0, G_1^{(-)} = 0, G_2^{(+)} = 0\}, \\ \mathcal{H}_+ &= \{G_0^{(-)} = 0, G_1^{(+)} = 0, G_2^{(-)} = 0\}. \end{aligned}$$

From our constructions it follows that

$$\mathcal{H}_- = \mathcal{H}_0 \oplus \mathcal{H}_1 \oplus \mathcal{H}_2^{(-)}, \qquad \mathcal{H}_+ = \mathcal{H}_2^{(+)}.$$

2. Physically $\mathcal{H}_0$ describes the state in which all bosons (0, 1 or 2) are bounded with spin. In addition, if the spin is directed down ($\sigma = -$), then there is an even number of bosons and if up ($\sigma = +$), there is one boson. The space $\mathcal{H}_1$ describes the states in which there is one free boson with momentum k and the other bosons (0 or 1) are bounded with spin. The energy of such states is $\omega(k) + E_0$; as before, the direction of the spin is defined by the number of bosons as in the previous case. The spaces $\mathcal{H}_2^{(\sigma)}$, $\sigma = \pm$, describe the states of two "dressing" bosons with momenta k_1 and k_2 and the energy of such states is equal to $\omega(k_1) + \omega(k_2) + \varepsilon\sigma$.

3. Although we proved the existence of wave operators and $\overline{W}_-$, $\overline{W}_+$, we use them only as a tool for the spectral analysis of H. However, scattering theory for our system can be developed in more detail.

4. An interesting problem is to get similar results for any α, without any assumption about its smallness.

Acknowledgments. One of authors (R. M.) thanks the F. W. Fund for Russian Mathematicians, the Russian Fundamental Research Fund (grant No. 93-011-1470), and the AMS Fund for fSU Aid for financial support, and the Physics Department of Munich University for its warm hospitality.

References

1. H. Spohn, *Ground state of the spin-boson Hamiltonian*, Comm. Math. Phys. **123** (1989), 277–304.
2. M. Huebner and H. Spohn, *Spectral properties of spin-boson Hamiltonian*, Ann. Inst. H. Poincaré Phys. Théor. **62** (1995), no. 3, 289–323.
3. M. Reed and B. Simon, *Methods of modern mathematical physics. Scattering theory*, vol. 3, Academic Press, New York–San Francisco–London, 1979.
4. D. R. Yafaev, *On the theory of Multichannel Scattering in a pair of spaces*, Teoret. Mat. Fiz. **37** (1978), 48–57; English transl. in Theoret. and Math. Phys.

5. M. A. Lavrent′ev and B. W. Shabat, *Methods of the theory of functions of complex variables* (1958), Gostekhizdat, Moscow. (Russian)
6. N. Wiener and H. R. Pitt, *On absolutely convergent Fourier–Stieltjes transform*, Duke Math. J. **4** (1938), 420–436.

INSTITUTE FOR PROBLEMS OF INFORMATION TRANSMISSION OF THE RUSSIAN ACADEMY OF SCIENCES, 19 ERMOLOVA STR. MOSCOW, 103051, RUSSIA

THEORETISCHE PHYSIC, LUDWIG-MAXIMILLIANS-UNIVERSITÄT, THERESIENSTR. 37, D-80333 MÜNCHEN, GERMANY

Translated by THE AUTHORS

Amer. Math. Soc. Transl.
(2) Vol. **177**, 1996

The Moduli Space of Instantons in $N = 2$ Supersymmetrical σ-Models

M. I. Monastyrsky and S. M. Natanzon

Introduction

In this paper we study the moduli space of instantons in the $N = 2$ supersymmetrical σ-model defined on a general Riemann surface where the field variables take values in the sphere. Our main result is the description of the connectivity components of the moduli space and the determination of new topological invariants of superinstantons. This article is a natural extension of [**1**].

It is a pleasure to publish this paper in the volume dedicated to the memory of Felix Berezin, a remarkable mathematician and one of the inventors of supermathematics. He was also interested in the instanton problem ([**2**]).

§1. The classical σ-model on Riemann surfaces

Let us collect some facts about two-dimensional σ-models that will be useful for our presentation (see [**3–5**]).

Let M^2 be a compact Riemannian two-dimensional real manifold equipped with a metric g and N a Kähler manifold with metric h. We shall call M^2 the *physical space* and N the *target space*. If $\varphi\colon M^2 \to N$ is a smooth map, then the differential $d\varphi\colon T_xM^2 \to T_{\varphi(x)}N$ is defined for all $x \in M^2$. For any map (field) φ we can define the Lagrangian density $\mathcal{L}(\varphi)(x) = \|d\varphi\|^2/2$, where $\|d\varphi\|$ is the norm of differential $d\varphi$. In local coordinates

$$\mathcal{L}(\varphi) = \frac{1}{2} g^{\mu\nu} \frac{d\varphi^a}{dx^\mu} \frac{d\varphi^b}{dx^\nu} h^{ab}. \tag{1.1}$$

The corresponding action is

$$S(\varphi) = \frac{1}{2} \int_{M^2} \|d\varphi\|^2 d\mu(x), \tag{1.2}$$

where $d\mu(x)$ is the canonical volume measure. The solutions of the equation

$$\delta S = 0 \tag{1.3}$$

1991 *Mathematics Subject Classification*. Primary 81T60; Secondary 32C11; 32G81.

are called *harmonic maps*[1]. It is well known that in the $O(3)$ σ-model all solutions of (1.3) can be split into holomorphic and antiholomorphic parts. With some precautions, this result is valid if we replace the physical space S^2 by a general compact Riemann surface M^2. There are two ways to prove this result.

The first way is to generalize the notion of topological charge and to get an estimate of energy from below. Let us give the outline of this approach (see the details in [**3**]). We use the standard notions of complex differential geometry ([**4**]); e.g., we use freely, depending on the context, two presentations of a Riemann surface: as a compact two-dimensional orientable manifold or as a one-dimensional complex curve. Let J_M be the operator of a (almost) complex structure consistent with the metric g:

$$J_M\colon TM \to TM, \qquad J_M^2 = -1\,.$$

The operator J_M may be extended by complex linearity to

$$J_M^{\mathbb{C}}\colon T^{\mathbb{C}}M \to T^{\mathbb{C}}M, \qquad (J_M^{\mathbb{C}})^2 = -1\,.$$

Since $(J_M^{\mathbb{C}})^2 = -1$, there is a direct sum decomposition $T^{\mathbb{C}}M = T^{1,0}M \oplus T^{0,1}M$, where $T^{1,0}M$ and $T^{0,1}M$ are the eigenbundles corresponding to the eigenvalues i and $-i$, respectively. The differential d of a smooth map $\varphi\colon M \to N$ can be extended by complex linearity to $d^{\mathbb{C}}\varphi\colon T^{\mathbb{C}}M \to T^{\mathbb{C}}N$ with canonical decomposition $d^{\mathbb{C}}\varphi = \partial\varphi + \overline{\partial}\varphi$, where

$$\partial\varphi\colon T^{1,0}M \to T^{1,0}N, \qquad \overline{\partial}\varphi\colon T^{0,1}M \to T^{0,1}N\,.$$

A map φ is (anti) holomorphic iff $\partial\varphi(\overline{\partial}\varphi) = 0$. Using these derivatives it is possible to represent

$$\mathcal{L}(\varphi) = \mathcal{L}^{1,0}(\varphi) + \mathcal{L}^{0,1}\varphi = \|\partial\varphi\|^2 + \|\overline{\partial}\varphi\|^2, \qquad S(\varphi) = S^{1,0}(\varphi) + S^{0,1}(\varphi)\,.$$

Just as in the usual $O(3)$ case, it is possible to introduce the topological charge Q of the field φ

$$Q = \frac{1}{2}\int \varphi^*(\omega), \tag{1.4}$$

where ω is the Kähler form on N. The following estimate is valid:

$$S \geqslant 2Q\,. \tag{1.5}$$

The estimate (1.5) is exactly similar to the $O(3)$ σ-model and the Yang–Mills action. The equality $S = 2Q$ leads to the conditions $\overline{\partial}\varphi = 0$ or $\partial\varphi = 0$. We call such equations *self-dual* or *anti-self-dual* and the corresponding solutions *instantons* or *anti-instantons*. Another way to study equation (1.3) is based on the estimation of the expression $(\varphi^*h)^{2,0}$ and belongs to Eells–Wood ([**8, 4**]). They proved the following result.

[1]Harmonic maps were studied in mathematics long before they attracted the attention of specialists in field theory ([**6**]). Certain kinds of chiral models appeared earlier in the theory of nematic liquid crystals ([**7**]).

THEOREM. *If* $\varphi\colon (M^2, g) \to (N^2, h)$ *is a harmonic map between compact orientable surfaces and if*

$$\chi(M^2) + |\deg(\varphi)\chi(N^2)| > 0, \tag{1.6}$$

then φ *is holomorphic or antiholomorphic. Here* $\chi(M^2) = 2 - 2p$, $\chi(N^2) = 2 - 2q$ *are the Euler characteristics of* M^2 *and* N^2, *and* $\deg(\varphi)$ *the Brouwer degree of* φ.

In our case, where $\chi(N^2) = 2$, condition (1.6) shows that if the map φ is harmonic and $|\deg\varphi| \geqslant p$, then φ is holomorphic or antiholomorphic.

It is important to stress that in the space of all harmonic mappings $\mathcal{H}$ there exists an open set Ω formed by functions in general position. The closure of Ω, denoted by $\overline{\Omega}$, coincides with $\mathcal{H}$. The space Ω consists of Morse type functions, i.e., mappings with branching points of second degree with distinct critical values. This class of functions is intimately related to the moduli space of instantons in general position. The complete moduli space of instantons $\mathcal{H}$ can be obtained by the completion procedure, i.e., by adding to the space Ω the component consisting of the class of functions with a lesser number of critical values.

These arguments lead us to introduce the natural category of spaces on which our σ-model can be determined. This category consists of a triple (M^2, N^2, φ), where M^2 and N^2 are compact Riemann surfaces with punctures, and φ is a map without branching points. For the spaces M^2 and N^2, we assume the limitations following from (1.6). These mappings may be extended as Morse mappings to the compactification $\overline{M}^2$ of the surface M^2. This generalization does not really extend our class of σ-models in the ordinary (bosonic) case, since it is possible to glue all punctures, but provide a new interesting topological characteristic of instantons in the supercase.

§2. $N = 2$ Super-Riemann surfaces

It is possible to define ordinary Riemann surfaces as one-dimensional complex manifolds by using charts and transition functions on the one hand and by the uniformization procedure on the other. Both methods admit supersymmetrical generalizations. The first approach yields a more transparent generalization of the σ-model to the supercase and the second one is useful in the description of the structure moduli space.

A. Let $L = L(K)$ be the Grassmann algebra over the field K generated by the infinite set $1, l_1, \dots, l_n, \dots$. Each element $a \in L$ can be represented as the sum of an element $a_0 \in K$ and a finite number of elements $a_{i_1 i_2 \dots i_p} l_{i_1} \wedge l_{i_2} \wedge \dots \wedge l_{i_p}$, where $p > 0$, $a_{i_1 \dots i_p} \in K$, and $i_1 > i_2 > \dots > i_p$. The mapping $a \mapsto a^{\#} = a_0$ defines an epimorphism $\# L \to K$. The monomial $l_{i_1} \wedge l_{i_2} \wedge \dots \wedge l_{i_p}$ is called *even* or *odd* depending on the parity of p. The monomial 1 is regarded as even. Linear combinations of even (odd) monomials with coefficients in the field K form the set $L_0(K)$ of *even elements* or the set $L_1(K)$ of *odd elements* of the algebra $L(K)$. The field K will be assumed complex or real. Let $K^{(n|m)}$ be the linear superspace of all sequences $(z_1, \dots, z_n \mid \theta_1, \dots, \theta_m)$, where $z_i \in L_0(K)$ and $\theta_i \in L_1(K)$. Let $\#(z_1, \dots, z_n \mid \theta_1, \dots, \theta_m) = (z_1^{\#}, \dots, z_n^{\#}) \subset K^n$. The set $(z_1^{\#}, \dots, z_n^{\#})$ is said to be the *body* of the superspace. Now we can define the $N = 2$ Riemann supersurfaces by using the coordinate approach (see [**9**] for details). An $N = 2$

super-Riemann surface M_S^2 is described locally by a complex coordinate z and Grassmann coordinates θ^1, θ^2 satisfying $\{\theta^1, \theta^2\} = 0$. Let $Z = (z, \theta^1, \theta^2)$. An arbitrary function $f(Z)$ has the expansion

$$f(z, \theta^1, \theta^2) = f_0(z) + \theta^i f_i(z) + (1/2)\theta^i\theta^j f_{ij}(z) \qquad (i, j = 1, 2)\,. \tag{2.1}$$

We consider the case where the field K is complex. The superconformal derivatives are

$$D_i = \frac{\partial}{\partial\theta^i} + \theta^i \frac{\partial}{\partial z}, \qquad i = 1, 2\,. \tag{2.2}$$

The summation convention is used with the $O(2)$-invariant metric δ_{ij}

$$\{D_i, D_j\} = \delta_{ij}\partial_z\,. \tag{2.3}$$

The transformation $Z \to \widetilde{Z}(Z)$ is supperconformal iff

$$D_i\widetilde{z} = \widetilde{\theta}_j D_i\widetilde{\theta}^j \tag{2.4}$$

and the transformation law of D_i is

$$D_i = (D_i\widetilde{\theta}^j)\widetilde{D}_j\,. \tag{2.5}$$

It follows from (2.5) that the composition of superconformal mappings is also a superconformal mapping. Now we can give the following definition.

DEFINITION. An $N = 2$ *super-Riemann surface* M_S^2 is a supermanifold (z, θ^1, θ^2) covered by coordinate neighborhoods $\{U_\alpha\}$ with local coordinates $(Z_\alpha = (z_\alpha^1, \theta_\alpha^1, \theta_\alpha^2)$ patched together by transition functions satisfying (2.5)). This definition is valid for an arbitrary even M.

B. The uniformized super-Riemann surfaces ($N = 2$). An $N = 2$ super-Riemann surface can be defined in terms of a covering space (extended half upper plane) and a discrete covering group ([**9, 10**]). Let us introduce the necessary notions. The superspace $H = \{z \mid \theta_1, \theta_2) \in \mathbb{C}^{1/2} \mid \operatorname{Im} z^\# > 0\}$ is called the $N = 2$ *upper super-half-plane.* The full group of automorphisms of $H \to \operatorname{Aut}(H)$ is the group $OSP(2|2)$ ([**9, 10**]). The projection $\#H$: $H^\#$ determines $A^\# = \#(A) \subset \operatorname{Aut}(H^\#)$, where

$$A^\# z = \frac{a^\# z + b^\#}{c^\# z + d^\#}, \qquad a^\#, b^\#, c^\#, d^\# \in \mathbb{R},\ \det A^\# = 1\,.$$

We say that $\Gamma \subset \operatorname{Aut}(H)$ is a $N = 2$ super-Fuchsian group if $\Gamma^\#$ is a Fuchsian group and $\#\Gamma \to \Gamma^\#$ is an isomorphism.

DEFINITION. An $N = 2$ *super-Riemann surface* is the space $M = H/\Gamma$, where $\Gamma \subset \operatorname{Aut}_0(H)$ is an $N = 2$ super-Fuchsian group consisting of hyperbolic or parabolic automorphisms. The projection $\#H \to H^\#$ determines the projection $\#M \to M^\#$ and the Riemann surface $H^\#/\Gamma^\#$ (the body of M).

As usual, we call the automorphism A *hyperbolic* (*parabolic*) if we have $|a^\# + d^\#| > 2$ ($|a + d| = 2$, respectively). Each hyperbolic automorphism A is conjugate to the automorphism

$$(z \mid \theta^1, \theta^2) \mapsto (\lambda z \mid h^{11}\theta^1 + h^{12}\theta_2, h^{21}\theta^1 + h^{22}\theta^2),$$

where $\lambda^{\#} > 1$, and each parabolic A is conjugate to

$$(z \mid \theta^1, \theta^2) \mapsto (z + 1 \mid h^{11}\theta^1 + h^{12}\theta^2 + \xi_1, h^{21}\theta^1 + h^{22}\theta^2 + \xi_2).$$

Using the relations between elements of the group $OSP(2|2)$, it is possible to obtain the following: $(h^{11}h^{22})^{\#}, (h^{12}h^{21})^{\#} \geqslant 0$ and either $(h^{11})^{\#} = (h^{22})^{\#} = 0$ or $(h^{12})^{\#} = (h^{21})^{\#} = 0$.

Now we define the functions $\Omega_1, \Omega_2 \colon \mathrm{Aut}_0(H) \to \{0,1\}$ ([**11**]).

1) $\Omega_1(A) = 1$, $\Omega_2(A) = 1$ if $(h^{11})^{\#}, (h^{22})^{\#} > 0$, $(h^{12})^{\#} = (h^{21})^{\#} = 0$;
2) $\Omega_1(A) = 0$, $\Omega_2(A) = 0$ if $(h^{11})^{\#}, (h^{22})^{\#} < 0$, $(h^{12})^{\#} = (h^{21})^{\#} = 0$;
3) $\Omega_1(A) = 1$, $\Omega_2(A) = 0$ if $(h^{11})^{\#} = (h^{22})^{\#} = 0$, $(h^{12})^{\#} = (h^{21})^{\#} > 0$;
4) $\Omega_1(A) = 0$, $\Omega_2(A) = 1$ if $(h^{11})^{\#} = (h^{22})^{\#} = 0$, $(h^{12})^{\#} = (h^{21})^{\#} < 0$.

If the supergroup $\mathrm{Aut}\,(H)$ consists only of superhyperbolic and superparabolic transformations, then the functions Ω_i induce functions $\widetilde{\omega}_i \colon \pi_1(M^{\#}) \to Z_2(0,1)$, which in turn generate the Arf (or spinoral) functions

$$\omega_i \colon H^1(M^{\#}, Z_2) \to Z_2$$

with the characteristic property $\omega(a+b) = \omega(a) + \omega(b) + a \cap b$. Here a, b are simple contours, $a, b \in \pi_1(M^{\#})$, and $a \cap b$ denotes the intersection index. Let $\Gamma \subset \mathrm{Aut}_0(H)$ and $\psi_\gamma \colon \Gamma \to \pi_1(M^{\#}, m_0)$ and $\psi_h \colon \pi_1(M^{\#}, m_0) \to H_1(M^{\#}, Z_2)$ be the natural projections. For $a \in H_1(M^{\#}, Z_2)$, we put $\omega_i(a) = \Omega_i(A)$, where $A \in \Gamma$ is such that $\psi_\gamma(A) \in \pi_1(M^{\#}, m_0)$ may be represented by a simple contour and $\psi_h\psi_\gamma(A) = a$. The conjugation classes of parabolic automorphisms belonging to superfuchsian groups correspond to punctures of the Riemann surface $M^{\#}$. The pair $(\omega_1(A), \omega_2(A))$ is called the *(topological) type* of the puncture. The connectivity components of the moduli space of $N = 2$ supersurfaces are determined by the topological type of the pair (ω_1, ω_2).

We need the following definition ([**11**]).

DEFINITION. Two pairs (ω_1, ω_2), (ω_1', ω_2') defined on $M^{\#}$ and $M^{\#'}$, respectively, are called *topologically equivalent* if there exists a diffeomorphism $M^{\#} \to M^{\#'}$ transforming (ω_1, ω_2) into (ω_1', ω_2'). In particular, the connectivity components for the space of the supercovering of the supersphere with punctures are determined by the numbers $n_{\alpha\beta}$ equal to the number of punctures a such that $\omega_1(a) = \alpha$ and $\omega_2(a) = \beta$, where α, $\beta \in Z_2(0,1)$. We use these notions in §3, where we study the moduli space of superinstantons.

C. Now we turn to the $N = 2$ supersymmetrical σ-model. All physical concepts encountered here can be found in [**12**].

The general form of the superfield action is

$$S(\Phi) = \int_{M^2} dz\, d\bar{z}\, d\theta_1\, d\theta_2\, g_{ij} h^{\mu\nu} (D\Phi^\mu)(D\Phi^\nu).$$

Here $\Phi^\mu(z, \theta_1, \theta_2)$ takes values in the supersphere. The fields Φ^μ have the form (2.1) and the superconformal derivatives D_i are defined by (2.2). We define the *superinstanton* as the solution of $\delta S = 0$ with fixed topological charge Q defined by (1.4). The main point of this subsection is to demonstrate that the general superinstanton solution is a superconformal mapping of a super-Riemann surface M^2 with punctures to the supersphere with punctures S^2. First we mention the paper [**1**], where we prove this fact in the case when M^2 is replaced by S^2. This result, based on previous papers ([**13–15**]), is valid with standard changes due to the more general supersurface considered. It is important to stress that we solve our extremal problem in the class of superfunctions with a numerical part that can be extended to the whole class of smooth mappings of compact surfaces to the sphere.

REMARK. We can consider the supersymmetric σ-model as a kind of superstring model with fixed gauge. In this case we can interpret superinstantons as certain boson-fermion excitations of the string.

§3. Moduli space of $N = 2$ superinstantons

The classification problem of superinstantons in $N = 2$ σ-models reduces to the classification problem of $N = 2$ supercovers[2] $f\colon M \to S^2$, where M is a super Riemann surface and S^2 is the supersphere.

Let us consider a super-Fuchsian group Γ such that H/Γ is the supersphere and choose a subgroup $\overline{\Gamma} \subset \Gamma$ of index n. Then $M = H/\overline{\Gamma}$ is a super-Riemann surface and the imbedding $\overline{\Gamma} \hookrightarrow \Gamma$ induces a supercovering $M \to S^2$ of degree n. The supercovering f induces the covering $f^{\#}\colon M^{\#} \to S^{2\#}$. It is possible to prove that the function $f^{\#}$ has no branching points of even degree (see [**1**]). This means that the class of coverings of super-Riemann surfaces without punctures excludes mappings in a general position. It follows from the Eells–Wood theorem that mappings satisfying (1.6) can be regarded as holomorphic coverings (without branching points) of the sphere with punctures. The extensions of these coverings give us functions in a general position.

The superanalog of this covering is the supercovering $\widetilde{f}\colon M \to S$ with the following properties:

1) the body part $\widetilde{f}^{\#}$ has no singular points on M;
2) $\widetilde{f}^{\#}$ maps the Riemann surface with punctures $M^{\#}$ to the sphere with punctures $S^{2\#}$ and extends to a holomorphic covering in general position.

We call such coverings *elementary*. Let us consider a pair of super-Fuchsian groups $\overline{\Gamma} \subset \Gamma$ related to the elementary covering $\widetilde{f}\colon M \to S$. This pair is isomorphic to the pair $\overline{\Gamma}^{\#} \subset \Gamma^{\#}$. The group $\overline{\Gamma}^{\#}$ is determined uniquely by the special algebraic conditions on the generators of $\Gamma^{\#}$ ([**16**]).

Thus, the topological invariants of the supercovering $\widetilde{f}$ fit to the invariants of the group Γ and, therefore, are determined by the topological invariants of S, i.e., the number of punctures $n_{\alpha\beta}$ of each topological type.

Let us turn to the construction of the moduli space $M(\xi)$ $(H/\Gamma \in M(\xi))$ of elementary supercoverings of type $(p, n_{00}, n_{01}, n_{10}, n_{11})$. Let us consider the space $\widetilde{T}(\xi)$ consisting of the sets of parabolic shifts $(c^{\alpha\beta}_{i_{\alpha\beta}})$, where $\alpha, \beta = 0, 1$, $i = 1, \dots, n_{\alpha\beta}$, with normalization condition $\prod_{i,\alpha,\beta} c_i^{\alpha\beta} = 1$. Following [**11**], we identify, in two steps, equivalent points in the space $\widetilde{T}(\xi)$ to get $M(\xi)$. First we consider the space $T(\xi)$, the set of orbits of the global group of automorphisms acting on $\widetilde{T}(\xi)$, $T(\xi) = \widetilde{T}(\xi)/\operatorname{Aut}(H)$.

[2]We omit the expression $N = 2$ supersymmetry below, since only this type of symmetry is considered in the present section.

PROPOSITION 1. *$T(\xi)$ is diffeomorphic to $R^{(3m-8+b/4m-8)}/\mathbb{Z}_2 \times \mathbb{Z}_2 \times \mathbb{Z}_2$, where $b = 0$ if $m_{01} = m_{10} = 0$, and $b = 1$ if $m_{01} + m_{10} > 0$. The group $\overline{\Gamma}$ stems from the system of generators $d^j_{\alpha\beta}$ that can be represented by $c_i^{\alpha\beta}$ in a special way* ([**16**]). *The moduli space $M(\xi)$ is the coset space $T(\xi)/\operatorname{Mod}$, where $\operatorname{Mod} = \{\varkappa \in \operatorname{Aut}(\Gamma)\, \varkappa\overline{\Gamma} = \overline{\Gamma}\}$, and $\varkappa c_i^{\alpha\beta} = \gamma_k c_{jk}^{\alpha\beta} \gamma_k^{-1}$, $\gamma_k \in \Gamma$.*

These considerations lead us to the following result.

PROPOSITION 2. *The space $M(\xi)$ is the superanalog of the space of the superinstantons in general position. The space of all instantons with fixed topological charge Q and genus p can be obtained by the "fusion" of punctures. Only fusions of punctures of the same topological types (α, β) are possible.*

We give an outline of the proof.

First of all note that if the degree of the covering decreases after the fusion procedure, then it is possible to glue a new puncture. The necessary condition for this gluing is the following:

$$m_{01} + m_{11} = m_{10} + m_{11} = 0 \bmod 2\,.$$

If the degree of the covering after the fusion remains the same, then the degree of covering at the point is 3. The topological type of this point coincides with the type of its image on S and is equal to $(0, 0)$.

Thus the dimension of the space of degenerate instantons $\partial M(\xi)$ is the dimension of $M(\xi)$ minus 3. These components constitute the natural boundary of the space of instantons in general position.

§4. Conclusion

The main result of this paper elucidates the role of punctures of the supersurface in the structure of the moduli space of superinstantons. There exist very interesting unsolved problems related to the moduli space of $N = 2$ superinstantons. One of the tempting problems in this direction is to find the connection between this space and the moduli space of solitons (monopoles) in the $N = 2$ super-σ-model in $2+1$-dimensional space ([**17, 18**]).

It would also be valuable to extend our approach to the four-dimensional Yang–Mills model. One of the natural ways in this direction is to generalize the construction of Atiyah ([**19**]) to the supercase.

Acknowledgements. The main results of this paper were reported on the seminars at the Enrico Fermi Institute, University of Chicago, and Northeastern University in Boston (February 1994) when the first author was the guest of these universities. M. M. is indebted to P. Freund, J. Gauntlett, J. Harvey, M. Shubin, and all other participants of the seminars for valuable conversations and suggestions.

This work was supported in part by the Russian Foundation for Basic Research and the Soros Foundation.

References

1. M. I. Monastyrsky and S. M. Natanzon, *The moduli space of superconformal instantons in sigma models*, Modern Phys. Lett. A **6** (1991), 1787–1796.

2. F. A. Berezin, *Instantons and Grassmannians*, Funktsional. Anal. i Prilozhen. **13** (1979), no. 2, 75–76; English transl. in Functional Anal. Appl. **13** (1979), no. 2.
3. T. P. Killingbeck, *Nonlinear σ-models on compact Riemann surfaces*, Comm. Math. Phys. **100** (1985), 481–493.
4. J. Eells and L. Lemaire, *A report on harmonic maps*, Bull. London Math. Soc. **10** (1978), 1–68.
5. M. I. Monastyrsky, *Topology of gauge fields and condensed matter*, Plenum Press, NY, 1993.
6. F. B. Fuller, *Harmonic mappings*, Proc. Nat. Acad. Sci. U.S.A. **40** (1954), 987–991.
7. C. W. Oseen, *The theory of liquid crystals*, Trans. Faraday Soc. **29** (1933), 883–899.
8. J. Eells and J. C. Wood, *Restrictions on harmonic maps of surfaces*, Topology **15** (1976), 263–266.
9. J. D. Cohn, $N = 2$ *super-Riemann surfaces*, Nuclear Phys. B **284** (1987), 349–364.
10. Yu. Manin, *Superalgebraic curves and quantum strings*, Trudy Mat. Inst. Steklov. **183** (1990), 126–138; English transl. in Proc. Steklov. Inst. Math. **1994**, no. 4.
11. S. M. Natanzon, *Moduli spaces of Riemann $N = 1$ and $N = 2$ superalgebras*, J. Geom. Phys. **12** (1993), 35–54.
12. M. B. Green, J. H. Schwarz, and E. Witten, *Superstring theory*, Cambridge Univ. Press, Cambridge, 1987.
13. B. Zumino, *Supersymmetry and Kähler manifolds*, Phys. Lett. B **87** (1993), no. 3, 203–206.
14. L. Alvarez-Gaumé and D. Z. Freedman, *Geometrical structure and ultraviolet finiteness in the supersymmetric σ-model*, Comm. Math. Phys. **80** (1981), 443–451.
15. P. J. Ruback, *Sigma model solitons and their moduli space metrics*, Comm. Math. Phys. **116** (1988), 645–658.
16. S. M. Natanzon, *Uniformization of the spaces of meromorphic functions*, Dokl. Akad. Nauk SSSR **287** (1986), no. 5, 1058–1061; English transl., Soviet Math. Dokl. **33** (1986), 487–490.
17. J. P. Gauntlett, *Low-energy dynamics of supersymmetric solitons*, Nuclear Phys. B **400** (1993), 103–105.
18. ———, *Low-energy dynamics of $N = 2$ supersymmetric monopoles*, Nuclear Phys. B **411** (1994), 443–460.
19. M. F. Atiyah, *Instantons in two and four dimensions*, Comm. Math. Phys. **93** (1984), 437–451.

Institute of Theoretical and Experimental Physics Moscow, 117259 Russia

Moscow Independent University, Moscow, Russia
E-mail address: Monastyrsky@vaxitep.itep.msk.su

Translated by THE AUTHORS

Amer. Math. Soc. Transl.
(2) Vol. **177**, 1996

Superanalogs of Symplectic and Contact Geometry and Their Applications to Quantum Field Theory

Albert Schwarz

ABSTRACT. The paper contains a short review of the theory of symplectic and contact manifolds and of the generalization of this theory to the case of supermanifolds. It is shown that this generalization can be used to obtain some important results in quantum field theory. In particular, regarding N-superconformal geometry as a particular case of contact complex geometry, one can better understand $N = 2$ superconformal field theory and its relationship to topological conformal field theory. Odd symplectic geometry constitutes a mathematical basis of the Batalin–Vilkovisky procedure for the quantization of gauge theories.

The exposition is mostly based on published papers. However, the paper also contains a review of some unpublished results (in the section devoted to the axiomatics of $N = 2$ superconformal theory and to topological quantum field theory).

§0. Introduction

It is a great pleasure for me to publish my paper in this volume. F. A. Berezin made extremely important contributions to mathematical physics in many different directions. However, the most significant part of his heritage is related to the idea that the theory of fermions becomes very similar to the theory of bosons, once the usual functions ("functions of commuting variables") are replaced by "functions of anticommuting variables" (elements of a Grassmann algebra). He was the first who realized that along with the standard algebra, analysis, and geometry, one can construct algebra and analysis of functions depending not only on commuting, but also on anticommuting variables, and develop the geometry of manifolds with commuting and anticommuting coordinates. These ideas have found very important applications to physics. They were used to analyze a new kind of symmetry. This symmetry, mixing bosons and fermions, is called supersymmetry; therefore the corresponding mathematical concepts are also provided with the prefix "super".

1991 *Mathematics Subject Classification.* Primary 58A50; Secondary 32G99, 32L99, 81E30.
Research supported in part by NSF grant No. DMS-9201366.

The present paper is entirely based on such concepts. It begins with a brief introduction to the ideas of supergeometry. The main part of the paper contains a short review of the theory of symplectic and contact manifolds and of the generalization of this theory to the case of supermanifolds. It is shown that this generalization can be used to obtain some important results in quantum field theory. In particular, regarding N-superconformal geometry as a particular case of contact complex geometry, one can better understand $N = 2$ superconformal field theory and its relationship to topological conformal field theory. Odd symplectic geometry constitutes a mathematical basis of the Batalin–Vilkovisky procedure for the quantization of gauge theories.

Our exposition is mainly based on the papers [**1–7**]. However, the paper also contains a review of some unpublished results (in the section devoted to the axiomatics of $N = 2$ superconformal theory and to topological quantum field theory).

§1. Supergeometry

A smooth m-dimensional manifold can be defined as an object obtained from domains in $\mathbb{R}^m$ glued together by means of smooth transformations. This definition can be formulated in a purely algebraic way. Namely, one can identify a domain $U \subset \mathbb{R}^m$ with the algebra $C^\infty(U)$ of all smooth functions on U and a smooth map of U into V with a homomorphism of $C^\infty(V)$ into $C^\infty(U)$. This algebraic construction can be generalized as follows. By definition, we identify an $(m|n)$-dimensional *superdomain* U_n with the $\mathbb{Z}_2$-graded algebra $C^\infty(U) \otimes \Lambda_n$, where U is a domain in $\mathbb{R}^m$ and Λ_n is the Grassmann algebra with n generators $\xi^1, \dots, \xi^n$. In particular, if $U = \mathbb{R}^m$ we obtain a superdomain denoted by $\mathbb{R}^{m|n}$. One says that $\mathbb{R}^{m|n}$ is an $(m|n)$-dimensional linear *superspace*. A map of U_n into $V_{n'}$, where U is a domain in $\mathbb{R}^m$, V is a domain in $\mathbb{R}^{m'}$, is defined as an even homomorphism of $C^\infty(V) \otimes \Lambda_{n'}$ into $C^\infty(U) \otimes \Lambda_n$. (We say that an operator acting on $\mathbb{Z}_2$-graded spaces is *even* if it is parity-preserving and *odd* if it is parity-reversing.) Elements of the algebra $C^\infty(U) \otimes \Lambda_n$ can be written as formal linear combinations

$$F = \sum_k \sum_{i_1,\dots,i_k} f_{i_1,\dots,i_k}(x)\xi^{i_1} \cdots \xi^{i_k},$$

where $f_{i_1,\dots,i_k}(x)$ are smooth functions on U and $\xi^i\xi^j = -\xi^j\xi^i$. Without loss of generality we assume that the coefficients $f_{i_1,\dots,i_k}$ are antisymmetric with respect to all permutations of $i_1, \dots, i_k$. It is convenient to consider elements of $C^\infty(U) \otimes \Lambda_n$ as functions depending on commuting variables $(x^1, \dots, x^m) \in U$ and anticommuting variables $\xi^1, \dots, \xi^n$. (In other words, elements of $C^\infty(U) \otimes \Lambda_n$ are regarded as functions on the superdomain U_n with m commuting coordinates $x^1, \dots, x^m$ and n anticommuting coordinates $\xi^1, \dots, \xi^n$). A map of a superdomain U_n with coordinates $x^1, \dots, x^m, \xi^1, \dots, \xi^n$ into a superdomain $V_{n'}$ with coordinates $\tilde{x}^1, \dots \tilde{x}^{m'}, \tilde{\xi}^1, \dots, \tilde{\xi}^{n'}$ can be specified by the formulas

$$(1) \qquad \tilde{x}^i = a^i(x^1, \dots, x^m, \xi^1, \dots, \xi^n), \qquad \tilde{\xi}^j = \alpha^j(x^1, \dots, x^m, \xi^1, \dots, \xi^n),$$

where a^i, $1 \leqslant i \leqslant m'$, and α^j, $1 \leqslant j \leqslant n'$ are even and odd elements of $C^\infty(U) \otimes \Lambda_n$. (It is easy to check that the substitution of a^i and α^j into the functions of variables $\tilde{x}^i$, $\tilde{\xi}^j$ determines a homomorphism from $C^\infty(V) \otimes \Lambda_{n'}$ to $C^\infty(U) \otimes \Lambda_n$ if the functions $a^i(x^1, \dots, x^m, 0, \dots, 0)$, $1 \leqslant i \leqslant m'$ determine a map of the domain $U \subset \mathbb{R}^m$ into the domain $V \subset \mathbb{R}^{m'}$.)

Now one can define an $(m|n)$-dimensional *supermanifold* as an object glued together from $(m|n)$-dimensional superdomains by invertible maps. The definition of superdomain and supermanifold given above is completely algebraic. In this definition a supermanifold "has no points". However, if M is a supermanifold and Λ is an arbitrary Grassmann algebra, one can construct the set M_Λ of Λ-points of M. In the particular case when M is the superdomain U_n, we define M_Λ as the set of all rows $(x^1, \dots, x^m, \xi^1, \dots, \xi^n)$, where $\xi^1, \dots, \xi^n$ are arbitrary odd elements of Λ and $x^1, \dots, x^m$ are even elements of Λ satisfying $(m(x^1), \dots, m(x^m)) \in U$ (here m is the standard homomorphism of Λ onto R.) It is easy to check that a map of superdomains generates a map of the corresponding sets of Λ-points. Using this remark, one can define the set M_Λ for an arbitrary supermanifold M. To every parity-preserving homomorphism $\rho\colon \Lambda \to \Lambda'$ of Grassmann algebras, one can assign a map $\tilde{\rho}\colon M_\Lambda \to M_{\Lambda'}$ in a natural way: if $\rho\colon \Lambda \to \Lambda'$ and $\rho'\colon \Lambda' \to \Lambda''$ are two parity-preserving homomorphisms of Grassmann algebras then $\widetilde{\rho\rho'} = \tilde{\rho} \cdot \tilde{\rho}'$. In other words, a supermanifold M determines a functor on the category of Grassmann algebras with values in the category of sets. (If M is an $(m|n)$-dimensional manifold and Λ is a Grassmann algebra with l generators, one can consider M_Λ as a smooth manifold of dimension $2^{l-1}(m+n)$. Therefore, one can also say that a supermanifold determines a functor on the category of Grassmann algebras with values in the category of smooth manifolds.) The language of Λ-points is often very convenient.

We define an algebraic operation on the supermanifold M as a map $M \times M \to M$. It is easy to see that such an operation in M naturally determines a map $M_\Lambda \times M_\Lambda \to M_\Lambda$, i.e., an operation on M_Λ. (Here we have in mind binary operations; however, one can consider operations with an arbitrary number of arguments in the same way.) Let us suppose that M is equipped with a binary operation and a unary operation in such a way that M_Λ is a Lie group with respect to the corresponding operations in M_Λ for every Λ. (The operations are regarded as multiplication and taking the inverse, respectively.) We say then that M is provided with a structure of a *Lie supergroup.* (In this way a Lie supergroup determines a functor on the category of Grassmann algebras with values in the category of Lie groups.) The notions of *super Lie algebra, supercommutative algebra,* etc., can be defined in similar way.

We shall define a superspace as a functor on the category of Grassmann algebras taking values in the category of sets (*morphisms* in the category of Grassmann algebras are parity-preserving homomorphisms). Similarly, a *supergroup* can be defined as a functor on the category of Grassmann algebras taking values in the category of groups. One can define in a natural way the notion of action of a supergroup on a superspace and the notion of the orbit superspace of such an action. It is clear that a supermanifold can be considered as a superspace and a Lie supergroup can be regarded as a supergroup. However, if a Lie supergroup acts on a supermanifold, the corresponding orbit superspace is not necessarily a supermanifold.

Almost all notions of algebra, geometry, and analysis can be formulated for supermanifolds. For example, the (Berezin) integral of the function (1) over U_n is defined by the formula

(2) $$\int_{U_n} F d^n x\, d^n \xi = n! \int_U f_{1,\ldots,n}(x)\, d^n x\,.$$

A (left) derivation in $C^\infty(U) \otimes \Lambda_n$ can be defined as a linear operator D satisfying the condition

$$D(uv) = Du \cdot v + (-1)^{\varepsilon(D)\varepsilon(u)} u \cdot Dv\,.$$

(Here $\varepsilon(u)$ denotes the parity of u and $\varepsilon(D)$ is the parity of D.) The left derivative $\partial_l/\partial\xi^i$ with respect to an anticommuting variable ξ^i is defined as the odd derivation obeying

$$\frac{\partial_l}{\partial \xi^i}\, \xi^j = \delta_i^j\,, \qquad \frac{\partial_l}{\partial \xi^i}\, f(x) = 0\,.$$

It is easy to check that derivations in $C^\infty(U) \otimes \Lambda_n$ can be identified with first order differential operators, i.e., operators of the form

$$A^i(x,\xi)\, \frac{\partial}{\partial x^i} + \alpha^j(x,\xi)\, \frac{\partial_l}{\partial \xi^j}\,.$$

The definitions of right derivation and right derivative $\partial_r/\partial\xi^i$ are similar. In order to define a differential form on a superdomain U_n with commuting coordinates $(x^1,\ldots,x^m) \in U$ and anticommuting coordinates $\xi^1,\ldots,\xi^n$, we consider a function of the commuting variables $x^1,\ldots,x^m,\tilde{\xi}^1,\ldots,\tilde{\xi}^n$ and the anticommuting variables $\xi^1,\ldots,\xi^n,\tilde{x}^1,\ldots,\tilde{x}^m$. Such a function determines a *k-form* if it is a polynomial of degree k with respect to the variables $\tilde{x}^1,\ldots,\tilde{x}^m,\tilde{\xi}^1,\ldots,\tilde{\xi}^n$. The variables $\tilde{x}^i$, $\tilde{\xi}^j$ can be identified with differentials dx^i, $d\xi^j$ (note that the parity of the differential is opposite to the parity of corresponding variable). The generalization of the de Rham differential is defined by the formula

$$d = \sum_i \tilde{x}^i\, \frac{\partial}{\partial x^i} + \sum_j \tilde{\xi}^j\, \frac{\partial_l}{\partial \xi^j}\,.$$

All notions defined above for superdomains are invariant with respect to invertible transformations and therefore our definitions work also for supermanifolds. The only exception is the Berezin integral. As in the case of the ordinary integral, the integrand acquires a factor equal to the determinant of the Jacobian matrix of the variable change. Of course, the determinant here should be understood as the superdeterminant (Berezinian).

For a supermanifold M, one can introduce the notion of tangent bundle TM and cotangent bundle T^*M. One can consider also the tangent and cotangent bundles with the parity of fibers reversed; we shall use the notations ΠTM and ΠT^*M in this case. Note that a section of the bundle TM or of the bundle ΠTM can be regarded as a vector field on M; a section of T^*M or of ΠT^*M is a 1-form on M. A k-form on M can be regarded as a function on ΠTM.

We shall freely use the supergeneralizations of the standard notions of algebra and analysis.

§2. Symplectic supermanifolds

We can apply the standard definition of symplectic manifold to the case of supermanifolds. Namely, a symplectic structure on a supermanifold M can be specified by means of an even nondegenerate closed 2-form

$$\omega = dz^a \omega_{ab}(z)\, dz^b. \tag{3}$$

Here z^a are (even and odd) coordinates on M. As usual, ω is *closed* if $d\omega = 0$ and *nondegenerate* if the (super) matrix ω_{ab} is invertible. If the form ω in the definition above is odd, we say that it specifies an *odd symplectic structure*; the manifolds equipped with an odd symplectic structure (odd symplectic manifolds or P-manifolds) will be considered at the end of the paper. *Symplectomorphisms* (canonical transformations) of a symplectic manifold M are defined as transformations of M preserving the form ω (i.e., f is a symplectomorphism if $f^*\omega = \omega$). Locally a 2-form ω specifying a symplectic structure in an appropriate coordinate system can be written as

$$\omega_0 = \sum_{1\leqslant i\leqslant n} dp_i\, dx^i + \sum_{1\leqslant j\leqslant m} (d\xi^j)^2. \tag{4}$$

(Here $x^1, \dots, x^n, p_1, \dots, p_n$ are even coordinates, $\xi^1, \dots, \xi^m$ are odd coordinates.) Therefore, one can define a symplectic supermanifold as a supermanifold glued together from superdomains in $\mathbb{R}^{2n|m}$ by means of transformations preserving ω_0 (canonical transformations).

One can define the Poisson bracket of two functions F, G on the symplectic manifold M by the formula

$$\{F, G\} = \frac{\partial_r F}{\partial z^a}\, \omega^{ab}(z)\, \frac{\partial_l G}{\partial z^b}, \tag{5}$$

where ω^{ab} stands for the matrix inverse to ω_{ab}.

The space $C(M)$ of functions on M can be regarded as a $\mathbb{Z}_2$-graded Lie algebra (super Lie algebra) with respect to this operation. To every function H we can assign a vector field K_H on M by the formula

$$K_H^a = \omega^{ab}(z)\, \frac{\partial_l H}{\partial z^b}\,. \tag{6}$$

The corresponding first order differential operator $\widehat{K}_H$ acts in the following way:

$$\widehat{K}_H F = \{F, H\}\,. \tag{7}$$

Vector fields obtained by means of this construction (Hamiltonian vector fields) preserve the form ω; in other words they can be regarded as infinitesimal symplectomorphisms. The map $H \to K_H$ is a homomorphism of the (super)Lie algebra $C(M)$ into the (super)Lie algebra of vector fields (of first order differential operators).

If, in the definition of symplectic (super)manifold, we omit the requirement that the form ω be nondegenerate, we obtain the definition of *presymplectic manifold.* If ω is degenerate, there exist vectors ξ^l satisfying $\omega_{kl}(z)\xi^l = 0$. Such vectors are called *null vectors of the form* ω. The set W_z of null vectors at a point $z \in M$ can be regarded as linear subspace of the tangent space T_z to M at z.

We want to construct a symplectic manifold M' corresponding to the presymplectic manifold M. To this end, we identify the point $z \in M$ with the point $z + dz$ if dz is a null vector of ω (i.e., $\omega_{kl}\, dz^l = 0$). To be more rigorous, we assume that W_z is a linear superspace whose dimension does not depend on $z \in M$; this dimension will be denoted by $(d_1|d_2)$. By the Frobenius theorem, for every point $z \in M$ we can construct, at least locally, a $(d_1|d_2)$-dimensional manifold R satisfying the following conditions:

1) $z \in R$,

2) the tangent space to R at an arbitrary point $u \in R$ coincides with W_u.

(The subspaces W_z determine a $(d_1|d_2)$-dimensional distribution W on the manifold M. This distribution is integrable; this means that for every two vector fields $\xi_1(z) \in W_z$, $\xi_2(z) \in W_z$ their commutator also belongs to W_z. By the Frobenius theorem, for every point $z \in M$ we can construct an integral manifold R of the distribution W containing the point z.) We shall identify two points of M belonging to the same manifold R. Then for every point $z \in M$ we can find a neighborhood U so that by this identification we obtain from U a symplectic manifold U' (the form ω' specifying the symplectic structure on U' can be determined by the relation $\pi^*\omega' = \omega$, where π denotes the natural projection of U onto U'). In global considerations we can encounter topological difficulties. (The global behavior of an integral manifold can be very complicated. Therefore, identifying points belonging to the same integral manifold, we can obtain from M a complicated topological space M'. The space is not necessarily a manifold; moreover one cannot assert that the points of M' are closed sets.) However, in physics we do not usually encounter these difficulties.

§3. Contact manifolds

We say that a 1-form α on the domain $U \subset \mathbb{R}^{2n+1}$ determines a *precontact structure* on U. If $\alpha' = F\alpha$, where F denotes a nonvanishing function on U, we say that the forms α and α' determine the same precontact structure. If the 1-form α is nondegenerate (i.e., the $(2n+1)$-form $\alpha \wedge (d\alpha)^n$ does not vanish anywhere), we say that α determines a *contact structure* on U. For instance, the form

$$\alpha = dz + \frac{1}{2}(p\, dq - q\, dp), \qquad z \in \mathbb{R},\ p \in \mathbb{R}^n,\ q \in \mathbb{R}^n, \tag{8}$$

determines a contact structure on $\mathbb{R}^{2n+1}$. This example is in some sense universal: any locally every nondegenerate 1-form is proportional to the form (8) in an appropriate coordinate system.

A transformation φ is said to be a *contactomorphism* if

$$\varphi^*\alpha = G\alpha, \tag{9}$$

where G is a nonvanishing function on U. In other words, a contactomorphism can be defined as a map preserving (pre)contact structure. If a precontact structure is defined on the domain $U \in \mathbb{R}^{2n+1}$ by means of the 1-form $\alpha = \alpha_i(z)\, dz^i$, we can determine a presymplectic structure on the domain $\mathbb{R}^* \times U \subset \mathbb{R}^{2n+2}$ by means of the closed 2-form $\omega = d\lambda$, where

$$\lambda = dt + t\alpha = dt + t\alpha_i(z)\, dz^i \tag{10}$$

(here $\mathbb{R}^*$ denotes the set of nonzero real numbers: $\mathbb{R}^* = \{t : t \in \mathbb{R}, t \neq 0\}$). The presymplectic structure on $\mathbb{R}^* \times U$ is said to be a *symplectization* of the precontact structure in U. It is easy to check that the 2-form

$$\omega = d\lambda = dt \wedge \alpha + t\,d\alpha \tag{11}$$

is nondegenerate if and only if the form α is nondegenerate. In other words, one can say that a contact structure in U can be defined as a precontact structure that gives a symplectic structure in $\mathbb{R}^* \times U$ after symplectization.

Let us consider a contactomorphism φ of the domain U. We shall construct a transformation $\tilde{\varphi}$ of $\mathbb{R}^* \times U$ by the formula

$$\tilde{\varphi}(t, u) = (G^{-1}t, \varphi(u)), \tag{12}$$

where G is defined by (9). It is easy to check that $\tilde{\varphi}$ is a symplectomorphism of $\mathbb{R}^* \times U$; one can say that $\tilde{\varphi}$ is obtained from φ by means of symplectization. The transformation $\tilde{\varphi}$ is homogeneous of degree 1 with respect to t: if $\tilde{\varphi}(t, u) = (t', u')$, then $\tilde{\varphi}(\lambda t, u) = (\lambda t', u')$. One can verify that every symplectomorphism $\tilde{\varphi}$ of $\mathbb{R}^* \times U$ satisfying this condition generates a contactomorphism φ of U by formula (12). This assertion allows us to give a description of infinitesimal contactomorphisms. Let us suppose for simplicity that $u = (z, p, q)$, $z \in \mathbb{R}$, $p \in \mathbb{R}^n$, $q \in \mathbb{R}^n$ and α is given by (8). It is evident that infinitesimal contactomorphisms correspond to infinitesimal symplectomorphisms (to Hamiltonian vector fields) in $\mathbb{R}^* \times U$ having a function of the form

$$H(t, u) = H(t, z, p, q) = tK(z, p, q) \tag{13}$$

as a Hamiltonian. We see that an infinitesimal contactomorphism can be written in the form

$$\delta z = K - \frac{p}{2}\frac{\partial K}{\partial p} - \frac{q}{2}\frac{\partial K}{\partial q}, \quad \delta p = -\frac{\partial K}{\partial q} + \frac{p}{2}\frac{\partial K}{\partial z}, \quad \delta q = \frac{\partial K}{\partial p} + \frac{q}{2}\frac{\partial K}{\partial z}, \tag{14}$$

where K is an arbitrary function on U.

Let us define a $(2n+1)$-dimensional contact manifold as a manifold glued together by contactomorphisms from domains in $\mathbb{R}^{2n+1}$ provided with contact structure. Without loss of generality one can assume that the contact structure on these domains is standard (determined by the form (8)). Given an arbitrary n-dimensional manifold M, one can construct a $(2n-1)$-dimensional contact manifold PT^*M in the following way. Let us consider the space T^*M of all covectors in M and the form $\omega = dp_i \wedge dq^i$ specifying a symplectic structure on T^*M (here $q^1, \dots, q^n$ are local coordinates on M and $p_1, \dots, p_n$ denote coordinates of the covector). Let us identify the points $(p_1, \dots, p_n, q^1, \dots, q^n)$ and $(\lambda p_1, \dots \lambda p_n, q^1, \dots, q^n)$ in $T^*M \setminus M$ (here $\lambda \neq 0$). The space obtained by means of this identification will be denoted by PT^*M. (One can say that PT^*M is the space of the *projectivized* cotangent bundle.) The space PT^*M can be provided with a contact structure in a natural way; the symplectization of this contact structure gives the natural symplectic structure on T^*M. The contact structure in PT^*M can be specified by means of a 1-form $Fp_j\,dq^j$. More precisely, let us consider the case when M is a domain in $\mathbb{R}^n$ and $q^1, \dots, q^n$ are coordinates on M. Then $T^*M \setminus M$ can be covered by sets $\widetilde{U}_1, \dots, \widetilde{U}_n$, where $\widetilde{U}_i$ is defined by the condition $p_i \neq 0$. By identification from $\widetilde{U}_1, \dots, \widetilde{U}_n$, we obtain the charts $U_1, \ldots, U_n$ covering PT^*M. We can introduce coordinates in U_i by the formula $\pi_k = p_k/p_i$. In other words, each point of U_i can be obtained from a point of $\widetilde{U}_i$ satisfying $p_i = 1$; the coordinates of this point of $\widetilde{U}_i$ are regarded as the coordinates of the corresponding point in U_i. The 1-form specifying the contact structure in U_i can be written as

$$\alpha^{(i)} = dq^i + \sum_{k \neq i} \pi_k \, dq^k = p_r \, dq^r . \tag{15}$$

It is easy to check that the forms $\alpha^{(i)}$ specify a contact structure in PT^*M (i.e., $\alpha^{(i)}$ and $\alpha^{(j)}$ are proportional in $U_i \cap U_j$). The definition of a precontact structure and the definition of a contactomorphism can be applied also in the case when the domain $U \subset \mathbb{R}^{2n+1}$ is replaced by a superdomain $U \subset \mathbb{R}^{2n+1|m}$. The symplectization of a precontact structure can be defined in the supercase and gives a presymplectic structure on the superdomain $\mathbb{R}^* \times U \subset \mathbb{R}^{2n+2|m}$. As usual, we say that the 1-form α determines a contact structure if α is nondegenerate, but the definition of nondegeneracy must be changed. One can say, for example, that α is nondegenerate if the 2-form (11) is nondegenerate. (In other words, a contact structure in U can be defined as a precontact structure giving a symplectic structure in $\mathbb{R}^* \times U$ after symplectization.)

§4. Symplectic and contact complex (super)manifolds

Let M be a complex (super)manifold. A presymplectic structure on M can be specified by a closed holomorphic $(2,0)$-form ω on M; if this form is nondegenerate, we say that M is a symplectic complex manifold. In local coordinates $(z^1, \ldots, z^n)$, the form ω can be represented by the formula (3), where $\omega_{ab}(z)$ are holomorphic functions.

In this way the definition of symplectic structure on a complex manifold repeats the definition in the real case, but ω must be regarded as a holomorphic $(2,0)$-form. The definition of a contact structure on a complex manifold also repeats the definition of a contact structure on a real manifold (the form α in this definition must be regarded as a nondegenerate holomorphic $(1,0)$-form).

If M is an arbitrary real or complex $(m|n)$-dimensional supermanifold, then the cotangent space T^*M can be regarded as a $(2m|2n)$-dimensional symplectic supermanifold and the projectivized cotangent space PT^*M as a $(2m-1|2n)$-dimensional contact supermanifold. (The construction of the symplectic structure on T^*M and of the contact structure in PT^*M is quite similar to the construction described above for the case where M is a real m-dimensional manifold).

Complex contact geometry is closely related to the superconformal geometry. Recall that an N-superconformal transformation of a $(1|N)$-dimensional complex superdomain U can be defined as a complex analytic transformation preserving up to a multiplicative factor the 1-form

$$\alpha = dz + \sum_i \theta_i \, d\theta_i . \tag{16}$$

Here $(z_1, \theta_1, \ldots, \theta_n)$ are complex coordinates in U (z is even, $\theta_1, \ldots, \theta_n$ are odd). The 1-form α is nondegenerate and therefore specifies a complex contact structure.

Superconformal transformations can be interpreted as complex contact transformations of the contact manifold U. An N-superconformal manifold can be defined as a manifold glued together from $(1|N)$-dimensional complex superdomains by means of N-superconformal transformations. We see that an N-superconformal manifold can be regarded as a $(1|N)$-dimensional complex contact manifold. Conversely, every $(1|N)$-dimensional complex contact manifold can be regarded as an N-superconformal manifold. The proof is based on the fact that locally every nondegenerate holomorphic form $\alpha_0(z,\theta)\,dz + \sum \alpha_i(z,\theta)\,d\theta_i$ on a $(1|N)$-dimensional superdomain in an appropriate complex coordinate system can be identified (up to holomorphic multiplier) with the form (16). We see that a $(1|N)$-dimensional complex contact manifold can be covered with local coordinates $(z,\theta_1,\dots,\theta_N)$ so that the transition functions are N-superconformal transformations. (These coordinates are called *superconformal coordinates*.) Using these coordinates, one can introduce covariant derivatives

$$D_i = \frac{\partial}{\partial\theta_i} + \theta_i\,\frac{\partial}{\partial z}\,.$$

It is easy to check that covariant derivatives in the superconformal coordinate systems $(z,\theta_1,\dots,\theta_N)$ and $(\tilde z,\tilde\theta_1,\dots,\tilde\theta_N)$ are connected by the relation

$$\widetilde{D}_i = \sum_j F_{ij}(z,\theta)\,D_j\,. \tag{17}$$

(This fact follows immediately from the remark that the orthogonal complement to the covector corresponding to the 1-form (16) is spanned by the vectors $D_1,\dots,D_N$.) In the case $N=2$, it is convenient to introduce the coordinates

$$\theta_+ = \frac{1}{\sqrt{2}}\,(\theta_1 + i\theta_2), \qquad \theta_- = \frac{1}{\sqrt{2}}\,(\theta_1 - i\theta_2)$$

and the covariant derivatives

$$\begin{aligned} D_+ &= \frac{1}{\sqrt{2}}\,(D_1 + iD_2) = \frac{\partial}{\partial\theta_-} + \frac{1}{2}\,\theta_+\,\frac{\partial}{\partial z}, \\ D_- &= \frac{1}{\sqrt{2}}\,(D_1 - iD_2) = \frac{\partial}{\partial\theta_+} + \frac{1}{2}\,\theta_-\,\frac{\partial}{\partial z}\,. \end{aligned} \tag{18}$$

One can verify that the behavior of D_+, D_- under an $N=2$ superconformal transformation is given by the formulas

$$\widetilde{D}_+ = F_+ D_+, \qquad \widetilde{D}_- = F_- D_- \tag{19}$$

in the untwisted case or by the formulas

$$\widetilde{D}_+ = G_+ D_-, \qquad \widetilde{D}_- = G_- D_+ \tag{20}$$

in the twisted case. An untwisted $N=2$ superconformal manifold M can be defined as a manifold obtained from domains in $\mathbb{C}^{1|2}$ glued together by untwisted superconformal transformations (covariant derivatives in different patches are related by (19)). Let us denote by $M^{k|l}$ the moduli space of $(k|l)$-dimensional complex supermanifolds, i.e., the space of classes of such manifolds with respect to holomorphic equivalence. This definition is not rigorous. More precisely one can define the moduli space by using as the starting point the superspace of all complex structures on a fixed real supermanifold and factorizing this superspace with respect to an appropriate equivalence relation. The moduli space defined in such a way is not necessarily a supermanifold. However, it can be regarded as a superspace.

The notion of the moduli space of complex manifolds is closely related to the notion of a family of complex manifolds. In some sense the moduli space can be regarded as the base of the universal family. Recall that a holomorphic family of compact complex $(k|l)$-dimensional supermanifolds can be defined as a holomorphic map of a complex supermanifold E onto a complex supermanifold S having $(k|l)$-dimensional compact complex manifolds as fibers. One says that S is the base of the family or that the family is parametrized by the points of S. The definition of a continuous family is analogous. (Speaking about families of manifolds, we have in mind continuous families.) For every continuous map $\rho\colon S' \to S$ of a space S' into the base S of a family F, one can construct in a natural way a family F' with the space S' as a base (the pullback of F).

In a similar way, we can define other moduli spaces, for example the moduli space M_N of N-superconformal complex manifolds. (This moduli space is related to families of N-superconformal manifolds.) The moduli space of untwisted $N = 2$ superconformal manifolds will be denoted by $\widetilde{M}_2$. Notice that there exists a natural involution in $\widetilde{M}_2$ (one can interchange the roles of D_+ and D_-). The quotient of $\widetilde{M}_2$ with respect to this involution is embedded into M_2.

For every $(1|n)$-dimensional complex supermanifold M, we can construct a $(1|2n)$-dimensional complex contact supermanifold PT^*M. Considering PT^*M as a $2n$-superconformal manifold, we obtain a map of the moduli space $M^{1|n}$ into the moduli space M_{2n}. Let us consider the case $n = 1$ more carefully. In this case we see that to every $(1|1)$-dimension complex supermanifold M we can assign the $N = 2$ superconformal manifold $\hat{M} = PT^*M$. If (z, θ) are complex coordinates in M, then the points of T^*M can be specified by coordinates (p, ρ, z, θ), where (p, ρ) are the coordinates of the covector on M. The behavior of these coordinates under the transformation $(z, \theta) \to (\tilde{z}, \tilde{\theta})$ is given by

$$\tilde{p} = \frac{\partial z}{\partial \tilde{z}} p + \frac{\partial z}{\partial \tilde{\theta}} \rho, \tag{21}$$

$$\tilde{\rho} = \frac{\partial \theta}{\partial \tilde{z}} p + \frac{\partial \theta}{\partial \tilde{\theta}} \rho\,. \tag{22}$$

A point of PT^*M can be specified by a triple (ρ, z, θ) and the behavior of this triple under the transformation $(z, \theta) \to (\tilde{z}, \tilde{\theta})$ is given by

$$\tilde{z} = \tilde{z}(z, \theta), \quad \tilde{\theta} = \tilde{\theta}(z, \theta), \quad \tilde{\rho} = \left(\frac{\partial \theta}{\partial z} + \frac{\partial \theta}{\partial \tilde{\theta}} \rho\right)\left(\frac{\partial z}{\partial \tilde{z}} + \frac{\partial z}{\partial \tilde{\theta}} \rho\right)^{-1}. \tag{23}$$

Using the general rule, one can write the 1-form specifying the contact ($N = 2$ superconformal) structure in PT^*M as

$$\alpha = dz + \rho\, d\theta \tag{24}$$

(we consider the form $p\, dz + \rho\, d\theta$ for $p = 1$).

In appropriate coordinates, this form coincides with (16). Using these coordinates, one can check that the transformation (23) can be considered as an untwisted

$N = 2$ superconformal transformation and therefore PT^*M is an untwisted $N = 2$ superconformal manifold. It is easy to check that every untwisted $N = 2$ superconformal manifold can be obtained by means of this construction (see [**2**]). The proof can be based on the remark that for every untwisted $N = 2$ superconformal manifold N one can construct a $(1|1)$-dimensional complex manifold $\alpha(N)$ factorizing N with respect to the vector field D_+. (In other words, we assume that there exists a natural map $\pi\colon N \to \alpha(N)$ and that the set of functions on N having the form $\Phi(\pi(x))$, where Φ is a function on $\alpha(N)$, consists of functions annihilated by the operator D_+.) One can check that $PT^*\alpha(N)$ is isomorphic to N. In this way we have proved that there is one-to-one correspondence between the classes of $(1|1)$-dimensional complex manifolds and the classes of $N = 2$ untwisted superconformal manifolds. In other words, the moduli space $M^{1|1}$ coincides with $\widetilde{M}_2$. Note that all moduli spaces at hand have natural complex structures; the isomorphism α is compatible with the complex structures in $\widetilde{M}_2$ and $M^{1|1}$. If we replace D_+ by D_- in the construction of the isomorphism $\alpha\colon \widetilde{M}_2 \to M^{1|1}$, we obtain another isomorphism $\beta\colon \widetilde{M}_2 \to M^{1|1}$. This means that there exists a natural involution $\iota\colon M^{1|1} \to M^{1|1}$ defined by the formula $\beta = \iota\alpha$. This involution can be described geometrically in the following way: if M is a $(1|1)$-dimensional complex manifold, we define $\iota(M)$ as the manifold consisting of all $(0|1)$-dimensional submanifolds of M.

The super Mumford form in superstring theory can be regarded as an object defined on $M^{1|1}$; using this fact and the connection between $M^{1|1}$ and M_2, one can discover the hidden $N = 2$ superconformal symmetry of the super Mumford form ([**2**]). A similar consideration leads to the explanation of the hidden $N = 2$ superconformal symmetry of the $B - C$ system discovered in [**8**].

§5. Axiomatics of $N = 2$ superconformal theory and topological quantum field theory

Let us briefly recall Segal's axiomatics of conformal field theory (CFT).

Let us consider a compact complex one-dimensional manifold M with $m + n$ embedded nonoverlapping parametrized disks $D_1, \ldots, D_m, D'_1, \ldots, D'_n$. (The set of disks is divided in two parts: "incoming" disks $D_1, \ldots, D_m$ and "outgoing" disks $D'_1, \ldots, D'_n$.) The moduli space of such objects will be denoted by $P_{m,n}$. There exist natural maps

$$P_{m_1,n_1} \times P_{m_2,n_2} \to P_{m_1+m_2,n_1+n_2}, \tag{25}$$

$$P_{m,n} \to P_{m-1,n-1}\,. \tag{26}$$

(To construct the first map, we take the disjoint union of manifolds. The second map corresponds to sewing D_m with D'_n. In other words, we identify the points $P \in D_m$ and $P' \in D'_m$ if the corresponding coordinates are related by the formula $z \cdot z' = 1/4$ and delete the points with $|z| < 1/2$, $|z'| < 1/2$. We assume that all disks are parametrized by points of the unit disk $D \subset \mathbb{C}$.) In Segal's axiomatics, we must fix a linear space H. The main object is a linear map $H^m \to H^n$ assigned to every point $M \in P_{m,n}$; this map must have a trace. In other words, we must have a map

$$\alpha_{m,n}\colon P_{m,n} \to B_{m,n} = \operatorname{Map}(H^m, H^n)$$

of $P_{m,n}$ into the space $B_{m,n}$ of trace class linear maps of H^m into H^n. (Here H^m stands for the tensor product of m copies of H.) Axioms are the conditions of compatibility of maps $\alpha_{m,n}$ and maps (25), (26). These axioms can be formulated in terms of the commutativity of diagrams containing the maps $\alpha_{m,n}$, the maps (25), (26), and the natural maps

$$B_{m_1,n_1} \otimes B_{m_2,n_2} \to B_{m_1+m_2,n_1+n_2}, \tag{27}$$

$$B_{m,n} \to B_{m-1,n-1}. \tag{28}$$

One must also require the compatibility of the maps $\alpha_{m,n}$ with the natural action of $S_m \times S_n$ on $P_{m,n}$ and $B_{m,n}$. (Here S_m denotes the symmetric group.) The maps $\alpha_{1|1}$ can be used to define the representation of the Virasoro algebra on H; more precisely, we obtain two copies of the Virasoro algebra with generators $L_n, \bar{L}_n$ obeying $[L_n, \bar{L}_m] = 0$. (The axioms above describe CFT with central charge $c = 0$. In this case the maps $\alpha_{1|1}$ yield representations of the Lie algebra Diff(S^1) of vector fields on the circle. To describe CFT with nonvanishing central charge, one should assume that the maps $\alpha_{m,n}$ are defined only up to a multiplier. Then the maps $\alpha_{1|1}$ give rise to projective representations of Diff(S^1), i.e., to representations of the Virasoro algebra, the central extension of Diff(S^1)).

It is important to emphasize that one can reformulate Segal's axiomatics by using families of one-dimensional complex manifolds with embedded disks instead of moduli spaces of such manifolds. We shall use the term (m, n)-*family* for a family of one-dimensional compact complex manifolds with m embedded "incoming" disks and n embedded "outgoing" disks. The base of an (m, n)-family F will be denoted by B_F. The map

$$\alpha^F_{m,n}\colon B_F \to B_{m,n} = \mathrm{Map}\,(H^m, H^n)$$

defined for every (m, n)-family F must be regarded as the basic object. Repeating the definitions of the maps (25), (26), we can obtain the corresponding maps for families. (For example, the construction of the map (26) allows us to assign an $(m-1, n-1)$-family to every (m, n)-family.) It is easy to translate the axioms above into the language of families. One must add the following axiom:

If the (m, n)-family F' can be obtained as a pullback of an (m, n)-family F by the map $\rho\colon B' \to B_F$, then $\alpha^{F'}_{m,n} = \alpha^F_{m,n}\rho$.

Similar axiomatics for superconformal theory can be obtained if we replace one-dimensional complex manifolds by N-superconformal manifolds. (The space H should be $\mathbb{Z}_2$-graded in this case.) It follows from the consideration above that $N = 2$ superconformal theory also has another description: one must replace one-dimensional complex manifolds by $(1|1)$-dimensional complex manifolds and disks by superdisks. The simplicity of this description leads to the conclusion that probably $N = 2$ superconformal symmetry is more fundamental than $N = 1$ superconformal symmetry.

To give an axiomatic description of topological conformal field theory, we introduce the notion of a $\mathcal{T}$-*manifold*. One can define a $\mathcal{T}$-manifold as a $(1|1)$-dimensional complex manifold equipped with a nondegenerate even exact 1-form ω (nondegeneracy means here that ω can be represented in the form $\omega = d\varphi$, where locally the odd function φ can be written as $\alpha(z) + a(z)\theta$ with invertible $a(z)$). The notion of $\mathcal{T}$-manifold is closely related to the notion of semirigid super Riemann surface

([**9**]). We define *topological conformal field theory* (TCFT) by replacing complex manifolds with $\mathcal{T}$-manifolds in Segal's axiomatics. Following the construction of the operators L_n, $\bar{L}_n$ in CFT we can construct operators in $\mathbb{Z}_2$-graded space H. We obtain even operators L_n, $\bar{L}_n$ generating two commuting copies of the Virasoro algebra, and odd operators b_n, $\bar{b}_n$, Q, $\bar{Q}$ obeying $L_n = [b_n, Q]_+$, $\bar{L}_n = [\bar{b}_n, \bar{Q}]_+$ (all other anticommutators of odd operators vanish). For every $N = 2$ superconformal theory, we can construct a TCFT in an obvious way, utilizing the fact that there is a natural map τ of the moduli space $M_{\mathcal{T}}$ of $\mathcal{T}$-manifolds into $M^{1|1}$ and that this map can be extended to a map of the moduli space of manifolds with embedded disks. (Every $\mathcal{T}$-manifold can be regarded as $(1|1)$-dimensional complex manifold.) Moreover, we can construct two different TCFT's corresponding to a given $N = 2$ superconformal theory by using the involution in $M^{1|1}$ described above (we use the map of $M_{\mathcal{T}}$ into $M^{1|1}$ defined as the composition of the map $\tau\colon M_{\mathcal{T}} \to M^{1|1}$ with this involution). The construction above can be regarded as the geometric counterpart of Witten's twisting. The two TCFT's corresponding to $N = 2$ superconformal theory are known as the *A-model* and *B-model.*

Note that in the definition of TCFT it is convenient to work with the version of Segal's axiomatics based on families of manifolds. (The moduli space of $\mathcal{T}$-manifolds is not a supermanifold.) If the map $p\colon E \to B$ determines a holomorphic family F of one-dimensional complex manifolds, then one can construct a family $\widetilde{F}$ of $\mathcal{T}$-manifolds in the following way. Let us consider the supermanifold $\widetilde{B} = \Pi TB$ (the space of the tangent bundle with reversed parity) and the natural map $\Phi\colon \Pi TB \times C^{0|1} \to B$. (In local coordinates Φ maps the point (m, μ, ν), where $m \in B$, μ is a tangent vector at the point m, and $\nu \in C^{0|1}$ to the point $m + \mu\nu$.) Then we can define $\widetilde{E}$ as the space consisting of points $(e, m, \mu, \nu) \in E \times \widetilde{B} \times C^{0|1}$ obeying $p(e) = \Phi(m, \mu, \nu)$. The maps $\tilde{p}\colon \widetilde{E} \to \widetilde{B}$ and $\pi\colon \widetilde{E} \to C^{0|1}$ are defined by means of projections of the direct product $E \times \widetilde{B} \times C^{0|1}$ onto the factors $\widetilde{B}$ and $C^{0|1}$. The map $\tilde{p}\colon \widetilde{E} \to \widetilde{B}$ determines a family of $(1|1)$-dimensional complex manifolds; the map π determines a structure of $\mathcal{T}$-manifold on the fibers of $\tilde{p}$. Therefore, we obtained holomorphic family $\widetilde{F}$ of $\mathcal{T}$-manifolds. Applying this construction to the family corresponding to the moduli space $P_{m,n}$, we obtain a family of $\mathcal{T}$-manifolds with embedded superdisks; the base of this family is $\widetilde{P}_{m,n} = \Pi TP_{m,n}$. The functions on $\widetilde{P}_{m,n}$ are forms on $P_{m,n}$; this fact permits us to relate the axiomatics based on $\mathcal{T}$-manifolds with axiomatics of TCFT formulated in [**10**].

The approach to TCFT in [**11**] was based on the consideration of two-dimensional manifolds equipped with the metric tensor $g_{\alpha\beta}$ and the odd field $\varphi_{\alpha\beta}$. The field $\varphi_{\alpha\beta}$ can be regarded as an infinitesimal odd variation of the metric tensor $g_{\alpha\beta}$; in other words, the space $\widetilde{Q}$ of pairs $(g_{\alpha\beta}, \varphi_{\alpha\beta})$ can be represented as ΠTQ, where Q is the space of metrics on a given two-dimensional manifold. After factorization of Q with respect to the group of diffeomorphisms and to the Weyl group, we obtain the moduli space M of complex structures. Similarly, after an appropriate factorization we obtain ΠTM from $\widetilde{Q} = \Pi TQ$. By using this remark, one can construct a $\mathcal{T}$-manifold for every two-dimensional manifold equipped with tensors $g_{\alpha\beta}$, $\varphi_{\alpha\beta}$.

§6. Odd symplectic geometry and the BV-formalism

We already mentioned that an odd symplectic structure on a supermanifold M is determined by a nondegenerate closed odd 2-form ω. Odd symplectic manifolds (P-manifolds) play an important role in the Batalin–Vilkovisky quantization procedure; here we formulate some results about the notion of P-manifold and related notions.

One can check that a P-manifold can be covered by local coordinate systems so that in every chart the form ω is represented as $dx^i d\xi_i$ (here $x^1, \ldots, x^n$ are even and $\xi_1, \ldots, \xi_n$ are odd coordinates). In other words, a P-manifold can be glued together from superdomains by transformations preserving the form $dx^i d\xi_i$. It is important to emphasize that the transformations preserving the form $dx^i d\xi_i$ are not necessarily volume-preserving. (By definition, a transformation of a superdomain is *volume-preserving* if the determinant of the Jacobian matrix is equal to 1. Of course we are talking here about the superdeterminant, or Berezinian). It is well known that there exists a natural volume element on an even symplectic manifold and symplectic transformations are volume-preserving; we see that in the case of P-manifolds the situation is different. We define an SP-manifold as a manifold M glued together from $(n|n)$-dimensional superdomains by means of transformations preserving the form $dx^i d\xi_i$ and the volume. An operator Δ acting on functions on an SP-manifold can be defined by the formula

$$\Delta = \frac{\partial^2}{\partial x^i \partial \xi_i}. \tag{29}$$

It is easy to check that this formula gives a globally defined operator with $\Delta^2 = 0$. Conversely, let

$$\Delta = \omega^{ij}(z) \frac{\partial^2}{\partial z^i \partial z^j} + \alpha^i(z) \frac{\partial}{\partial z^i} + \beta(z) \tag{30}$$

be an odd second order differential operator on supermanifold M satisfying the condition $\Delta^2 = 0$. One can prove that in the case when the matrix $\omega^{ij}(z)$ is nondegenerate, the operator Δ can be represented locally in the form (29). This means that such an operator determines an SP-structure in the manifold M. In particular, it determines a P-structure and a volume element on M.

We say that a submanifold L of a P-manifold M is Lagrangian if in a neighborhood of every point of L one can introduce a coordinate system $x^1, \ldots, x^n, \xi_1, \ldots, \xi_n$ such that ω has the standard form $dx^i d\xi_i$ and L is defined by the equations $\xi_1 = 0, \ldots, \xi_n = 0$. (Here x^i can be even or odd, the parity of ξ_i is opposite to the parity of x^i. Therefore $\dim L = (k|n-k)$, where $0 \leqslant k \leqslant n$.) In a more invariant way, one can characterize a Lagrangian manifold as a maximal submanifold of M on which the form ω vanishes. If M is an SP-manifold, one can introduce a volume element in a Lagrangian submanifold $L \subset M$. For example, one can use a local coordinate system $x^1, \ldots, x^n, \xi_1, \ldots, \xi_n$, where Δ is standard and L is singled out by the equations $\xi_1 = \cdots = \xi_n = 0$; the volume element $dx^1 \cdots dx^n$ in L does not depend on the choice of coordinates up to sign. (One can prove the existence of the coordinate system that we use.) In the Batalin–Vilkovisky formalism, physical quantities are represented by integrals of $\exp(S/\hbar)$ over Lagrangian submanifolds. Here S is the extended classical action that should satisfy the so-called quantum master equation $\Delta \exp(S/\hbar) = 0$. The following theorem constitutes the mathematical basis of the Batalin–Vilkovisky formalism.

Let H, H', K be functions on a compact SP-manifold M such that

$$\Delta H = \Delta H' = 0, \qquad H' - H = \Delta K .$$

Then for every two compact homologous Lagrangian submanifolds $L \subset M$, $L' \subset M$ we have

$$\int_L H d\lambda = \int_{L'} H' d\lambda'. \tag{31}$$

Here $d\lambda$ and $d\lambda'$ denote volume elements induced by the SP-structure in M. Compactness and homology are defined in terms of bodies of supermanifolds. Functions H and H' are even, K is odd. This theorem is proved in [**4**], and the proof is based on the classification of P-manifolds, SP-manifolds, and Lagrangian submanifolds given in this paper.

Let us also state a theorem that yields a description of symmetry transformations in the BV-formalism.

Let us consider two operators $\tilde{\Delta}$ and Δ on a supermanifold M that determine different SP-structures, but the same P-structure on M. Let us suppose that the corresponding volume elements $d\tilde{\mu}$ and $d\mu$ are related by the formula $d\tilde{\mu} = e^{\sigma} d\mu$. Then

a) $\Delta \exp(\sigma/2) = 0$;
b) if S satisfies the master equation $\Delta \exp S = 0$, then $\widetilde{S} = S - \sigma/2$ satisfies the master equation $\tilde{\Delta} \exp \widetilde{S} = 0$;
c) the action functional $\widetilde{S}$ in the SP-structure determined by $\tilde{\Delta}$ describes the same physics as the action functional S in the SP-structure determined by Δ. In particular,

$$\int_L e^{\widetilde{S}} d\tilde{\lambda} = \int_L e^S d\lambda \tag{32}$$

for every Lagrangian submanifold $L \subset M$.

It follows immediately from this statement that every theory is physically equivalent to the theory with the action functional $\widetilde{S} = 0$. In this representation a symmetry is simply an automorphism of the corresponding SP-structure. For an arbitrary SP-structure we have the following result.

Every function H satisfying $\Delta H + \{H, S\} = 0$ (every quantum observable) determines a symmetry in the following sense. Neither the action functional S nor the volume element $d\mu$ are invariant with respect to an infinitesimal transformation with the Hamiltonian H; however, the new action functional

$$\widetilde{S} = S + \epsilon\{H, S\}$$

and the new volume element

$$d\tilde{\mu} = d\mu(1 + 2\epsilon\Delta H)$$

describe the same physics as the old action functional S and the old volume element $d\mu$.

Here we shall not discuss the interesting questions arising in the analysis of the asymptotic behavior of $\int_L \exp(S/\hbar)\, d\lambda$ as $\hbar \to 0$ (semiclassical approximation in the BV formalism). Such an analysis was performed in [**5**]; the answer was expressed in terms a generalization of the Reidemeister torsion.

References

1. A. A. Rosly, A. S. Schwarz, and A. A. Voronov, *Geometry of superconformal manifolds*, Commun. Math. Phys. **119** (1988), 129–152; **120** (1989), 437–450.
2. S. N. Dolgikh, A. A. Rosly, and A. S. Schwarz, *Supermoduli spaces*, Comm. Math. Phys. **135** (1990), 91–100.
3. A. S. Schwarz, *Symplectic, contact and superconformal geometry, membranes and strings*, Supermembranes andd Physics in (2 + 1) Dimensions (M. Duff, C. Pope, and E. Sezgin, eds.), Proceedings of the Trieste Supermembrane Conference, World Scientific, Singapore, 1990, pp. 332.
4. ———, *Geometry of Batalin–Vilkovisky quantization*, Comm. Math. Phys. **155** (1993), 249–260.
5. ———, *Semiclassical approximation in Batalin–Vilkovisky formalism*, Comm. Math. Phys. **158** (1993), 373–396.
6. ———, *Symmetry transformations in Batalin–Vilkovisky formalism*, Lett. Math. Phys. **31** (1994), 299–301.
7. ———, *On the definition of superspace*, Teoret. Mat. Fiz. **60** (1984), 37–42; English transl. in Theoret. and Math. Phys. **60** (1984).
8. D. Friedan, E. Martinec, and S. Shenker, Nuclear Phys. B **271** (1986), 93.
9. J. Distler and P. Nelson, *Semirigid supergravity*, Phys. Rev. Lett. **66** (1991), 1955.
10. E. Getzler, *Batalin–Vilkovisky algebras and two-dimensional topological field theories*, Comm. Math. Phys. **159** (1994), 265.
11. R. Dijkgraaf, E. Verlinde, and H. Verlinde, *Topological strings in $d < 1$*, Nuclear Phys. B **352** (1991), 59.

Department of Mathematics, University of California, Davis, California 95616
E-mail address: `ASSCHWARZ@UCDAVIS.EDU`

Amer. Math. Soc. Transl.
(2) Vol. **177**, 1996

Remarks on the Delocalization Transition for Heteropolymers

Ya. G. Sinai and H. Spohn

ABSTRACT. We consider heteropolymers with either periodic or annealed random sequences of link types. The free energy is determined by exploiting first returns to the origin.

§1. Introduction

The statistical properties of a polymer chain in a solution are usually modelled as a random walk or as some other discrete version of Brownian motion (see [**1**]). Physically, the basic issue is to understand how self-avoidance of the monomers affects typical conformations of very long chains. In this note we consider a different sort of polymer chain, called heteropolymers ([**2**]) or copolymers ([**3**]). In an idealized fashion they consist of a sequence of monomers which can be either hydrophilic (prefer water) or hydrophobic (dislike water). The heteropolymer chain is immersed in a mixture of oil and water well separated by a sharp planar interface. Physically one would like to understand under what conditions the heteropolymer prefers to stick at the interface (localized phase), respectively, to be dissolved in one of the pure liquids (delocalized phase). More ambitiously, one would like to predict universal properties of the unbinding transition.

It is straightforward to set up a simplified statistical mechanics model of the physical situation ([**3**]). We ignore excluded volume effects. The oil/water interface is located at $\{z = 0\}$ with, say, $\{z > 0\}$ corresponding to oil and $\{z < 0\}$ corresponding to water. Clearly, it suffices to merely record the monomer location perpendicular to the interface. To be specific, we allow only discrete values. A chain of length N is then represented by $\omega(j) \in \mathbb{Z}$, $j = 0, \dots, N$, with the nearest neighbor constraint $|\omega(j+1) - \omega(j)| = 1$, i.e., by a one-dimensional random walk. Water and oil act as an external potential on the heteropolymer. Let $\sigma_j = \pm 1$,

1991 *Mathematics Subject Classification.* Primary 81V55.

$j = 0, \ldots, N$, be the link type with $+1$ standing for hydrophilic, -1 for hydrophobic, and let

$$V(z) = \begin{cases} u & \text{for } z > 0, \\ 0 & \text{for } z = 0, \\ -u & \text{for } z < 0, \end{cases}$$

$u \geqslant 0$. Then, for given link sequence $\sigma = \{\sigma_j \mid j = 0, \ldots, N\}$, the potential for the chain is given by

(1) $$V_\sigma(\omega) = \sum_{j=0}^{N} \sigma_j V(\omega(j)) .$$

The parameter u regulates the affinities between the monomers and the pure solvents. To model the energetics at the interface, we introduce in addition

(2) $$V_I(\omega) = u_0 \sum_{j=0}^{N} \delta_0(\omega(j))$$

with $\delta_0(0) = 1$ and $\delta_0(z) = 0$ otherwise. Clearly, other possibilities could also be envisaged.

According to the rules, for a given link sequence, the partition function equals

(3) $$Z_N = \sum_{\omega:\omega(0)=0=\omega(N)} (1/2)^N \exp[-V_\sigma(\omega) - V_I(\omega)]$$

with the inverse temperature absorbed into u and u_0. In fact, it will be more convenient to work with the generating function

(4) $$\widehat{Z}(\rho) = \sum_{N=0}^{\infty} \rho^N Z_N ,$$

$\rho > 0$. If ρ_+ denotes the smallest positive ρ such that $\widehat{Z}(\rho) \to \infty$ as $\rho \to \rho_+$ from the left, then the free energy of the heteropolymer equals

(5) $$f = \log \rho_+ .$$

Let f_+ and f_- be the free energies of the heteropolymer in the pure solvents. If $f < f_+, f < f_-$, then the chain is localized. Otherwise it is delocalized either above ($f = f_+$) or below ($f = f_-$) the interface. With some extra effort typical chain conformations can also be extracted from the Boltzmann weight in (3).

From the point of view of statistical mechanics there are three natural cases:

(1) The link sequence σ is *periodic*, e.g. $\sigma_j = (-1)^j$. This problem will be discussed in the following section.
(2) The link sequence is random, i.e., σ_j are independent random variables with $\text{Prob}(\sigma_j = 1) = p$, $\text{Prob}(\sigma_j = -1) = 1 - p$. The link type is frozen and V_σ is a *quenched random* potential. For the particular case $p = 1/2$, $u_0 = 0$, it is proved that, except for $u = 0$, the heteropolymer is *always* localized ([**4**]).
(3) Apparently there are heteropolymers where the link type may flip on a fast time scale ([**2**]). Physically, it is then more appropriate to regard V_σ as an *annealed random* potential. This problem will be discussed in §3.

The aim of our note is modest. We only want to convince the reader that the optimal strategy consists in using first returns to the origin. This insight is not new and has been exploited for (1 + 1)-dimensional wetting ([**5**]). In the present context such a strategy does not seem to be fully appreciated.

§2. Periodic potential

As the simplest case we consider $\sigma_j = (-1)^j$. For a simple random walk let $p(n)$ be the probability of a first return to the origin after n steps ($p(n) = 0$ for n odd). We have

$$\sum_{n=1}^{\infty} p(n) = 1 \quad \text{and} \quad p(n) \cong n^{-3/2}$$

for large n. We also introduce

$$\widehat{p}(\rho) = \sum_{n=1}^{\infty} \rho^n p(n), \qquad 0 < \rho \leqslant 1 .$$

Then

$$\widehat{Z}(\rho) = \sum_{r=1}^{\infty} e^{-u_0(r+1)} [\widehat{p}(\rho)(e^{-u}/2 + e^{u}/2)]^r . \tag{6}$$

Thus the free energy is determined implicitly by

$$\widehat{p}(\rho_+) = e^{u_0}/\cosh u , \tag{7}$$

provided $e^{u_0}/\cosh u < 1$. In the pure solvents, we have $f_+ = f_- = 0$. Therefore the phase boundary, u_c, is given by

$$e^{u_0} = \cosh u_c . \tag{8}$$

In particular, the interface potential has to be *repulsive*, $u_0 > 0$, in order to induce unbinding. Since $\widehat{p}(\rho) \cong 1 - \rho^{1/2}$ for $\rho \to 1$, if $u_0 = 0$, we have $f(u) \cong -u^4$ as $u \to 0$, a somewhat unusual behavior at a second order phase transition. For $u_0 > 0$, we have $f(u, u_0) \cong -(u - u_c)^2$ close to the phase boundary.

For a more general periodic potential, one arrives at matrix equations with rank depending on the period.

§3. Annealed random potential

Following the approach suggested in [**2**], we assume that the random potential is annealed and has the energy

$$H_N(\sigma) = -\beta \sum_{j=0}^{N-1} \sigma_j \sigma_{j+1} . \tag{9}$$

The annealed partition function then reads as follows:

$$\widehat{Z}(\rho) = \sum_{N=0}^{\infty} \rho^N \sum_{\sigma(N)} \sum_{\omega:\omega(0)=0=\omega(N)} (1/2)^N \exp[-H_N(\sigma) - V_\sigma(\omega) - V_I(\omega)] . \tag{10}$$

Let S, T be the transfer matrices of the one-dimensional Ising model with external field $\pm u$,

$$S = \begin{pmatrix} e^{\beta+u} & e^{-\beta} \\ e^{-\beta} & e^{\beta-u} \end{pmatrix}, \qquad T = \begin{pmatrix} e^{\beta-u} & e^{-\beta} \\ e^{-\beta} & e^{\beta+u} \end{pmatrix}.$$

Let $\lambda = \lambda(\beta, u)$ be their common maximal eigenvalue.

We now fix the link values at the first returns as $\eta_r = \pm 1$, $r = 0, 1, \ldots$. The statistical weight between one return with left endpoint $\eta_r = \eta$ and right endpoint $\eta_{r+1} = \eta'$ is given by

$$(11) \qquad A(\rho)_{\eta\eta'} = \sum_{n=1}^{\infty} \rho^n p(n) \Big\{ \frac{1}{2} e^{-u(\eta+\eta')/2} (S^n)_{\eta\eta'} + \frac{1}{2} e^{u(\eta+\eta')/2} (T^n)_{\eta\eta'} \Big\}.$$

The 2×2 matrix A is well defined for $0 < \rho \leqslant \lambda^{-1}$. Thus the partition function will be

$$(12) \quad \widehat{Z}(\rho) = \sum_{r=1}^{\infty} e^{-u_0(r+1)} \sum_{\eta_0, \ldots, \eta_r} \prod_{j=1}^{r} A(\rho)_{\eta_{j-1}\eta_j} = \sum_{r=1}^{\infty} e^{-u_0(r+1)} \sum_{\eta, \eta'} (A(\rho)^r)_{\eta\eta'}.$$

Let $\tau(\rho)$ be the maximal eigenvalue of $A(\rho)$. Since S and T are linked by flip symmetry, we have

$$(13) \qquad \tau(\rho) = \frac{1}{2} \sum_{n=1}^{\infty} \rho^n p(n) (\varphi, S^n \varphi)$$

with $\varphi_\eta = \exp[-u\eta/2]$. Clearly, $\tau(0) = 0$ and τ is strictly increasing in ρ. By using the spectral representation for S, one obtains the inequality $\tau(\lambda^{-1}) < 1$, provided $u \neq 0$. From (12) we conclude that the free energy is $f = \log \rho_+$, where ρ_+ is implicitly defined by

$$(14) \qquad \tau(\rho_+) = e^{u_0},$$

provided $\tau(\lambda^{-1}) > e^{u_0}$. In the pure solvents the free energy is given by

$$(15) \qquad f_\pm = -\log \lambda(\beta, u).$$

Therefore the boundary u_c between localized and delocalized phase is determined by

$$(16) \qquad \tau(\lambda(\beta, u_c)^{-1}) = e^{u_0}.$$

Since $\tau(\lambda^{-1}) < 1$, for zero interface potential the chain is always delocalized. One needs some *attractive* potential, $u_0 < 0$, to bind the heteropolymer.

Acknowledgements. We are grateful to S. Nechaev for teaching us many things about heteropolymers. H. S. thanks E. Bolthausen for explaining to him how to use the theory of large deviations in the case of a quenched random potential.

References

1. P. G. DeGennes, *Scaling concepts in polymer physics*, Cornell University Press Ithaca, 1979.
2. A. Grosberg, S. Izrailev, and S. Nechaev, *Phase transition in a heteropolymer chain at a selective interface*, Preprint (1994).
3. T. Garel, D. A. Huse, S. Leibler, and H. Orland, *Localization transition of random chains at interfaces*, Europhys. Lett. **6** (1988), 307.
4. Ya. G. Sinai, *Random walk with random environment*, Prob. Theor. Appl. **38** (1993), 457–459.
5. M. E. Fisher, *Walks, walls, wetting, and melting*, J. Statist. Phys. **34** (1984), 667.

Department of Mathematics, Princeton University, Princeton, N. J. 08544, USA and Landau Institute of Theoretical Physics, Russian Academy of Sciences

Theoretische Physik, Ludwig-Maximilians-Universität, Theresienstr. 37, D-80333 München, Germany

Selected Titles in This Series

(Continued from the front of this publication)

(See the AMS catalog for earlier titles)